LATIN AMERICA TRANSFORMED

LATIN AMERICA TRANSFORMED
Globalization and Modernity

Edited by

Robert N. Gwynne
School of Geography, University of Birmingham, UK

and

Cristóbal Kay
Institute of Social Studies, The Hague, The Netherlands

A member of the Hodder Headline Group
LONDON

Co-published in the United States of America by
Oxford University Press Inc., New York

First published in Great Britain in 1999 by
Arnold, a member of the Hodder Headline Group,
338 Euston Road, London NW1 3BH

http://www.arnoldpublishers.com

Co-published in the United States of America by
Oxford University Press Inc.,
198 Madison Avenue, New York, NY10016

© 1999 Arnold

All rights reserved. No part of this publication may be reproduced or transmitted in any form or by any means, electronically or mechanically, including photocopying, recording or any information storage or retrieval system, without either prior permission in writing from the publisher or a licence permitting restricted copying. In the United Kingdom such licences are issued by the Copyright Licensing Agency: 90 Tottenham Court Road, London W1P 9HE.

The advice and information in this book are believed to be true and accurate at the date of going to press, but neither the authors nor the publisher can accept any legal responsibility or liability for any errors or omissions.

British Library Cataloguing in Publication Data
A catalogue entry for this book is available from the British Library

Library of Congress Cataloging-in-Publication Data
A catalog entry for this book is available from the Library of Congress

ISBN 0 340 73191 5 (hb)
ISBN 0 340 69165 4 (pb)

2 3 4 5 6 7 8 9 10

Production Editor: Rada Radojicic
Production Controller: Iain McWilliams
Cover Design: Terry Griffiths

Typeset by Phoenix Photosetting, Chatham, Kent
Printed and bound in Great Britain by J W Arrowsmith Ltd, Bristol

What do you think about this book? Or any other Arnold title? Please send your comments to feedback.arnold@hodder.co.uk

To Catherine, Daniel and Sylvia

CONTENTS

About the authors	xi
List of tables	xiii
List of figures	xv
Preface	xvii

PART 1   **INTRODUCTION TO THE POLITICAL ECONOMY OF LATIN AMERICA**   1

Chapter 1   Latin America transformed: changing paradigms, debates and alternatives   2
Robert N. Gwynne, Cristóbal Kay

The contemporary relevance of structuralism and dependency theories	3
Globalization and a paradigmatic transformation	8
Why has neoliberalism become so popular?	13
Social bases of neoliberalism	17
Regional and local impacts of the neoliberal model	19
The neoliberal model evaluated	21
Concluding remarks: futures of the paradigm	25
Further reading	29

PART 2   **POLITICAL TRANSFORMATIONS**   31

Chapter 2   Authoritarianism, democracy and development   32
Eduardo Silva

Political economy and the state	33
Democracy, authoritarianism and development	35
Conclusions	48
Further reading	50

Chapter 3   The new political order in Latin America: towards technocratic democracies?   51
Patricio Silva

Neoliberalism, modernization and democracy	52
The depoliticization of society	54
Political legitimation, consumerism and macro-economic stability	57

	The technocratization of decision-making	58
	The disenchantment	60
	The future of democracy in Latin America	62
	Further reading	65

PART 3	**GLOBALIZATION AND ECONOMIC TRANSFORMATIONS**	**67**
Chapter 4	**Globalization, neoliberalism and economic change in South America and Mexico** Robert N. Gwynne	**68**
	Historical contexts	70
	Economic policy change: moves to a new consensus	77
	Impacts and problems of neoliberal reform	84
	The future of policy: hemispheric integration in the Americas?	92
	Further reading	96
Chapter 5	**Globalization, neoliberalism and economic change in Central America and the Caribbean** Thomas Klak	**98**
	Introduction	98
	A region of small states, economically dependent on the United States	100
	Development policies prior to neoliberalism	104
	The debt crisis and the neoliberal remedy	108
	The meaning and impacts of globalization in Middle America	109
	The current neoliberal development model in Middle America	112
	The special case of Cuba: island socialism amid global capitalism	117
	The role of migration in Middle America	120
	Regional trading blocs in Middle America	121
	A very capitalistic enterprise: Middle America's growing drug economy	124
	Conclusions	125
	Further reading	126

PART 4	**GLOBALIZATION AND ENVIRONMENTAL CHANGE**	**127**
Chapter 6	**Natural resources, the global economy and sustainability** Warwick E. Murray	**128**
	Latin America – the paradox of plenty?	128
	The history of resource exports and the role of theory	129
	Latin America as a contemporary global resource periphery	132
	Specialization in resource exports and the question of sustainability	138

	The devil's excrement? Case study of Venezuelan oil exports	143
	Fruitful exports? Case study of Chilean nontraditional fruit exports	146
	Conclusions – the relevance of old ideas?	150
	Further reading	152
Chapter 7	**The political economy of sustainable development** Robert N. Gwynne, Eduardo Silva	**153**
	The political economy of natural resource use	156
	The environmental question in the city	166
	Conclusions	177
	Further reading	179
PART 5	**CULTURAL CHANGE AND MODERNITY**	**181**
Chapter 8	**Modernity and identity: cultural change in Latin America** Jorge Larraín	**182**
	Introduction	182
	Historical trajectories to modernity	184
	The Latin American trajectory to modernity	186
	Some specific elements of Latin American modernity and culture	196
	Conclusions	201
	Further reading	202
Chapter 9	**Civil society, social difference and politics: issues of identity and representation** Sarah A. Radcliffe	**203**
	Defining social movements	204
	The emergence of social movements: structural context	207
	Between politics and the everyday	211
	Understanding social movements	213
	Understanding the (political) spaces of social movements	216
	Conclusions	222
	Further reading	222
PART 6	**THE CONTEXT OF SOCIAL CHANGE**	**225**
Chapter 10	**Population, migration, employment and gender** Sylvia Chant	**226**
	Population	227
	Migration	241
	Employment	251
	Gender and the urban labor market	261

	Conclusions	265
	Further reading	268

PART 7	**RURAL AND URBAN TRANSFORMATIONS**	**271**
Chapter 11	**Rural development: from agrarian reform to neoliberalism and beyond** Cristóbal Kay	**272**
	The lost promise of agrarian reform	273
	Globalization, neoliberalism and Latin American agriculture	285
	Peasant futures: a permanent semi-proletariat?	290
	Conclusions	302
	Further reading	304
Chapter 12	**Cities, capitalism and neoliberal regimes** Colin Clarke, David Howard	**305**
	Industrialization and the informal sector	306
	The state and the changing social stratification	311
	Poverty, the household and housing	317
	Social inequality, citizenship and identity	321
	Conclusions	324
	Further reading	324
	Bibliography	**325**
	Index	**355**

ABOUT THE AUTHORS

EDITORS

Robert N. Gwynne is Reader in Latin American Development at the School of Geography, University of Birmingham. His research interests focus on industrialization in the developing world and on the impacts of neoliberalism and globalization on regional and rural development in Latin America. He is the author of *Industrialisation and urbanisation in Latin America*, *New horizons? Third World industrialization in an international framework*, many articles in both geography and Latin American journals and chapters in a wide range of edited books.

Cristóbal Kay was a lecturer at the University of Chile in the early 1970s and thereafter in the University of Glasgow. He joined the Institute of Social Studies in The Hague in 1989. He was the editor of *The European Journal of Development Research* and is co-editor of the *European Review of Latin American and Caribbean Studies*. His books include *Latin American theories of development and underdevelopment*, *Labour and development in rural Cuba* (joint author), *Development and social change in the Chilean countryside* (co-editor) and *Disappearing Peasantries? Rural Labour in Africa, Asia and Latin America* (co-editor).

CONTRIBUTORS

Sylvia Chant is Reader in Geography at the London School of Economics and has specialist interests in gender and development in Mexico, Costa Rica and the Philippines. She is author of several books and articles including *Women and survival in Mexican cities: perspectives on gender, labour markets and low-income households*, *Women of a lesser cost: female labour, foreign exchange and Philippine development* (co-author), *Gender, urban development and housing* and *Women-headed households: diversity and dynamics in the developing world*. Aside from her continued interests in local and global aspects of female household headship, she has recently started research on masculinities in the household domain and their interrelations with changes in family structures, poverty and social policy.

Colin Clarke is Professor of Urban and Social Geography at the University of Oxford and an Official Fellow at Jesus College. His research interests focus on the urban and social geography of Latin America and the Caribbean. His books include *Kingston, Jamaica: urban development and social change, 1692–1962*, *East Indians in a West Indian town: San Fernando, Trinidad, 1930–1970* and *A geography of the Third World* (co-author). He has written many articles for both geography and Latin American journals and chapters in a wide range of edited books.

David Howard is a Postdoctoral Research Associate in Urban Social Geography at the School of Geography, University of Oxford. His doctoral research focused on ethnicity and race in the Dominican Republic. His research interests include the urban and social geography of Latin America and the Caribbean, international migration, ethnicity and racism.

Thomas Klak is Associate Professor of Geography at Miami University, Oxford, Ohio, and Adjunct Professor of Geography at Ohio State University, Columbus, Ohio. His research interests include the critical analysis of development discourses in government policy and in the news media and the assessment of the distributional impacts of structural adjustment and neoliberalism in countries of the South. He is the editor of *Globalization and Neoliberalism: The Caribbean Context*. His work has also appeared in such development- and planning-orientated journals as *World Development, Third World Quarterly, Geoforum, Antipode, Political Geography, Economic Geography, Tijdschrift voor Economische en Sociale Geografie*, and the *Journal of the American Planning Association*.

Jorge Larraín is Head of Sociology at University Alberto Hurtado, Santiago, Chile and Professor of Social Theory at the Department of Cultural Studies and Sociology, University of Birmingham. His present research interests focus on culture and identity in Latin America. His books include *The concept of ideology, Theories of development* and *Ideology and cultural identity*.

Warwick E. Murray is Lecturer in Human Geography at the University of the South Pacific, Fiji. He gained his PhD from the University of Birmingham in 1996, after spending a year in Chile carrying out research on the globalization of fruit, neoliberalism and small-scale farmers. His general areas of interest are economic geography, the global agro-food complex, rural change and Latin America.

Sarah A. Radcliffe is a Lecturer in Geography, University of Cambridge. Her interests include social difference and political cultures in the Andes, as well as feminist geography and social theory. Her recent research has focused on the social and spatial dimensions of Ecuadorian national identities and the intersections of race and gender. Recent publications include *Remaking the nation: place, politics and identity in Latin America* (with S. Westwood) and the edited collection *Viva: women and popular protest in Latin America*.

Eduardo Silva is Associate Professor of Political Science and a Fellow of the Center for International Studies at the University of Missouri-St. Louis, U.S.A. He is author of *The state and capital in Chile* and co-editor of *Organized business, economic change and democracy in Latin America* and *Elections and democratization in Latin America, 1980–1985*. His articles on Chilean political economy and environment and development have appeared in *World Politics, Comparative Politics, Development and Change, The Journal of Latin American Studies* and the *Journal of Interamerican Studies and World Affairs*.

Patricio Silva is Lecturer of Political Sociology at the Institute of Social and Cultural Studies, Leiden University. His recent research has focused on democratization processes in the Southern Cone and the technocratization of the political arena in several countries in the region. He has published many articles and contributed to several books. He is the author of *Estado, neoliberalismo y política agraria en Chile, 1973–1981* and co-editor of several books, including *Designers of development: intellectuals and technocrats in the Third World*.

LIST OF TABLES

1.1	Increasing asymmetries in the world economy, 1978–95	5
4.1	Population and economic indicators for South American Countries and Mexico	69
4.2	Percentage price and quantity changes for exports, net barter terms of trade and export purchasing power, 1932	71
4.3	Latin America's percentage shares of world and regional exports, 1946–75	73
4.4	South American countries: peak inflation years since 1970 and annual inflation average over the period 1984–93	74
4.5	Conditionality contents of International Monetary Fund (IMF) and World Bank programs (percentage of programs with particular conditions)	78
4.6	Latin America: prices of external debt paper on the secondary market (percentage of nominal value)	79
4.7	Chronology of trade liberalization in Latin America, 1985–91	81
4.8	Record of neoliberal economic reform in the 1990s	84
4.9	Employment (percentage of labor force) in Latin America: sectoral shifts, 1960–90	89
4.10	Employment and unemployment in Chile, 1973–96	91
5.1	Basic indicators for Middle American countries and territories	101
5.2	Trade dependency: selected Middle American countries in comparative perspective	103
5.3	Central American and Caribbean trading blocs: intraregional exports as a percentage of total exports, selected years	122
6.1	Proportional importance of primary product exports in total merchandise exports (%), 1970–94	133
6.2	Major export products in Latin America, 1994	135
6.3	The proportional role of agricultural exports in total merchandise exports (%), Latin America, 1970–94	136
6.4	Net value of agriculture, fish and forestry exports from Latin America, 1990 and 1994 (US$ billions)	137
6.5	Evolution of the terms of trade (percentage change) in Latin America, 1972–80 and 1980–92	138
10.1	Population growth rates in Latin America, 1970–2000	227
10.2	Fertility and contraception in Latin America	228
10.3	Life expectancy and mortality in Latin America	231
10.4	Women's education and economic activity rates in Latin America	239
10.5	Women's average age at marriage, and household headship, in Latin America	240
10.6	Latin America: population and urbanization	244
10.7	Urban sex ratios in selected Latin American countries	247

10.8	Latin America: labor force and earnings	254
10.9	Women's share in labor force and occupational categories	261
12.1	Urbanization in Latin America, 1965–95	306
12.2	The structure of nonagricultural employment in some Latin American countries, 1980–92	310
12.3	Employment in labor-market sectors: Kingston and St Andrew, Jamaica, 1977–89	314
12.4	Economic activity in Kingston and St Andrew, Jamaica	314
12.5	Female labor force participation in selected Latin American countries, 1960–90	316

LIST OF FIGURES

1.1	Hemisphere trade: percentage of total 1997 exports to the Americas from (a) Mercosur countries, (b) United States, (c) NAFTA countries, (d) CACM members and (e) the Andean Community	11
4.1	Loading logs onto a truck of the forestry company, Colcura, to supply the cellulose plant of the Sante Fe corporation	80
4.2	Average percentage tariff by type of good, 1995	82
4.3	Trade liberalization and export growth (in US$ billions) in Chile, by sector	86
4.4	Capital and consumer good imports (in US$ millions) in Chile, 1981–90	87
4.5	Exports (in US$ billions) from Chile to East Asia, 1975–96	87
4.6	Schemes of economic integration in the Americas	94
5.1	The small countries of Central America and the Caribbean	99
5.2	The weekly banana shipment from Dominica to Britain	107
5.3	Electronics assembly in Costa Rica	111
5.4	Dominica's nontraditional industrial exports, on display at the Dominica Export Import Agency (DEXIA)	114
5.5	Cuban economic growth rates, 1985–97	118
6.1	Value of manufactured and primary product exports (in US$ billions), 1970–94	133
6.2	Value of Chilean nontraditional fruit exports (in US$ millions), 1977–95	147
6.3	Seasonal female labor in one of Chile's fruit-packing plants	149
7.1	Components of sustainable development	155
7.2	The water supply system in San Juan de Lurigancho, a poor squatter settlement of 100 000 inhabitants outside Lima, early 1980s; water fuelling point and tanker supplying (costly) water to poor households	167
7.3	Overcrowded slum housing by the edge of the heavily polluted Rimac River, Lima, early 1980s	172
7.4	The urban environmental planning process	176
10.1	Young mother, Puerto Vallarta, Mexico	229
10.2	Home-based shoe workshop, León, Mexico	253
10.3	Home-based production of Christmas lanterns out of Coca-Cola© cans, Querétaro, Mexico	257
10.4	Home-based tortilla making for door-to-door sale, Querétaro, Mexico	258
11.1	A group of Cuban manual cane cutters having their lunch break, 1985	279
11.2	Field of lilies, Chile, 1994. A rural extension officer provides technical assistance to Mapuche peasant smallholders in the south of Chile	294
11.3	Inside a smallholder's greenhouse, Peumo, Chile, 1996	295

PREFACE

Since the debt crisis of the early 1980s there has been a radical series of transformations in the economic, political, social and cultural spheres of Latin America and the Caribbean. This book aims to provide an holistic approach for students wishing to understand these transformations and relate them to the wider processes of modernization and globalization. It does this by integrating authors from a range of disciplines – cultural studies, economic geography, political science, sociology and social geography. This enables the book to explain the nature of recent transformations from a range of perspectives – political, economic, environmental, cultural, social, rural and urban.

Part 1, Chapter 1, by Bob Gwynne and Cristóbal Kay, attempts a broad overview of the political economy of transformation in Latin America. It explores the nature of the new neoliberal paradigm as well as the contemporary relevance of the previous paradigm based on structuralism and dependency theory. It attempts to relate the paradigm shift to processes of globalization and investigates the social bases of neoliberalism. It is argued that neoliberalism is driven primarily by macro-economic arguments despite broadly negative impacts on labor markets, income distribution and poverty. The paradigm is perhaps being modified through the emergence of center-left governments that, whilst adhering to the macro-economic reforms of neoliberalism, have nevertheless boosted social spending and poverty-alleviation programmes.

Another major transformation since the early 1980s in Latin America has been the transformation from authoritarian governments to democratic systems. Part 2 of this book therefore explores the relationship between authoritarianism, democracy and development. Successive waves of authoritarianism and democracy in Latin America raised enduring questions about the relationship between economic development and political change in Latin America. Does economic development inevitably lead to either democracy or authoritarianism or is the connection more tenuous? Will the current trend toward democracy in Latin America persist? What are the chances for the consolidation of emerging polyarchies? Chapter 2, by Eduardo Silva, traces the intellectual history of responses to this problem by examining the evolution of two competing schools of thought: modernization theory and political economy. Buoyed by a wave of democratization in the 1950s and 1960s, modernization theorists initially argued that economic development inevitably led to democracy. After a wave of democratic breakdowns and military dictatorships in the more economically advanced nations of the Southern Cone in the 1960s and 1970s, political economists concluded the opposite held true. Middle stages of economic modernization required authoritarianism. The outbreak of democracy in the 1980s and 1990s has led to more cautious hypothesizing. The post World War Two period shows that both authoritarianism and democracy are compatible with economic change. But the modernization and political economy schools of thought differ on the reasons why. Those differences have a profound impact on policy prescriptions to help the consolidation of democracy.

This political transformation has created a variety of new democratic orders in Latin America. In Chapter 3, Patricio Silva explores the process of democratic transition taking place in the region since the early 1980s, highlighting the main characteristics of the new democracies. He stresses the continuities from the old authoritarian regimes in the new democratic order such as the application of neoliberal economic policies, the attempts by governments to depoliticize civil society and to legitimize their rule by means of increasing consumerism. Technocratic decision-making, which was inaugurated by the former military regimes, has become even more accentuated under the new political scenarios. Economists and financial experts have assumed key positions within governments and have set the political agenda with issues related to economic and financial reforms and the modernization of the state. At the same time, strong presidentialism seems to have become a distinct feature of the new democracies in the region, symbolized in the figures of Menem, Fujimori and Cardoso, among others. The scarce possibilities for political participation as well as the inability of many governments to satisfy the basic needs of the largest part of the population have generated in many countries an increasing disenchantment with politics and politicians. He concludes that the consolidation of democratic rule in Latin America is still far from being a reality, as the persistence of extreme poverty, political corruption and increasing alienation between technocratic leaders and masses remain serious obstacles for this purpose.

In Part 3 of the book the extent of the economic transformation is explored. It is argued that as countries become more closely inserted into the global economy as a result of neoliberal policies they experience varying impacts depending upon the characteristics of the national economies in question – such as resource endowment, location and social structure. Another important distinguishing factor is that of country size. In Chapter 4, Bob Gwynne focuses on economic change in the larger economies of South America and Mexico, whilst in Chapter 5 Tom Klak explores economic transformations and their impacts in the smaller economies of Central America and the Caribbean.

More specifically, Chapter 4 starts by outlining the historical background behind the previous paradigm of inward orientation and asks why it lasted so long and why it is now so criticized. The impact of the debt crisis in provoking the paradigm shift is explored and how policy was subsequently changed (including the shift to outward orientation) in order to find a way out of the crisis. The theoretical and ideological elements of neoliberal reform are then discussed before an analysis of its impacts and problems – particularly with regard to the record of economic growth, inflation, trade liberalization, employment change, income distribution and poverty alleviation. Outward orientation has become increasingly associated with open regionalism during the 1990s and the chapter concludes by assessing the future of hemispheric integration in the Americas.

Central America and the Caribbean, collectively called Middle America, is a region of small, underindustrialized, economically vulnerable and trade-dependent countries. In an era of globalization and neoliberalism, Middle America finds itself locked in a core–periphery relationship with the United States. Over recent decades Middle Americans have learned that the willingness of the United States to intervene in their region severely restricts their political and economic options. The region's hurried experiments with various democratic socialist paths to development are now largely defunct. Today Middle America pursues dependent capitalist development under the tutelage of the United States. Only Cuba remains outside of the fold, but even its development strategy focuses on courting foreign investors and tourists. Chapter 5 argues that the current global political economy presents Middle America with many risks of further vulnerability and marginality, but also some narrow options and opportunities. The greater economic openness associated with neoliberalism has meant that many domestic firms are being outcompeted by larger and more experienced firms from abroad. However, international economic integration can potentially have positive impacts in Middle America if it helps to expose and dislodge internal obstacles to social and economic development.

The increased exploitation of resources, both renewable and nonrenewable, under the neoliberal model has important environmental dimensions, which have not been fully recognized by many international organisations and national governments. So far the emphasis has been on economic growth without much regard to the environmental consequences. Part 4 examines these issues of resources and the environment. Latin America has served as a resource periphery for the global economy since colonial times. Given the abundance of their natural resources, Latin American countries might be expected to be among the most economically advanced in the world. However, a number of theories have indicated that resource abundance might actually operate as a 'curse' which, under certain conditions, can prejudice sustainable development. Chapter 6, by Warwick Murray, attempts to describe and account for Latin America's historical and contemporary role as a resource periphery in the global economy, analyses some of its impacts and assesses its relationship to sustainable development.

Latin American governments have almost bridled at demands from advanced economy governments and environmental movements for the preservation of natural resources, such as forests and watersheds. Chapter 7 attempts to set the resolution of environmental problems within a social context in which decisions and outcomes are the result of trade-offs between different institutional agents and actors. Bob Gwynne and Eduardo Silva examine these trade-offs not only in relation to use of natural resources but also in terms of environmental conditions in the rapidly expanding cities of Latin America. In the neoliberal transformation of the state, expenditure on social and infrastructural programmes have been curtailed (often severely), putting further stress on managing severe environmental problems, particularly in poor urban areas. The greater reliance on market forces far from ameliorating these problems has further intensified them.

The neoliberal model can also be interpreted as part of an ideological and cultural transformation with the aim of modernizing and more closely identifying with the social traditions and values of Europe and North America. Such an ideological vision is highly problematic and is contested in Part 5 of this book. In Chapter 8, Jorge Larraín tries to show that from the beginning of the nineteenth century, modernity has been presented in Latin America as an alternative to identity, as much by those who are suspicious of Enlightened modernity as by those who badly wanted it at all costs. This means that modernity is supposed to be totally alien to Latin America and can only exist in the region in conflict with its true identity. The chapter seeks to show that contrary to these absolutist extremes which present modernity and identity in Latin America as mutually excluding phenomena there is continuity and interconnection between them.

Meanwhile, Chapter 9 complements this analysis by discussing the social, economic and cultural dimensions of 'informal' politics at the end of the twentieth century. In the context of formal democratization and increased globalization of political activity, Sarah Radcliffe considers the enduring legacies of modern social hierarchies for the practices and identities of nondominant groups. The case studies discussed range from human rights groups to peasant patrols, from indigenous social movements to feminism. Theoretical frameworks for understanding these political forms are also introduced.

Economic, political and cultural transformations are taking place at the same time as the continued rapid expansion of population, putting further pressure on the need to create further employment opportunities. In Part 6, Chapter 10, Sylvia Chant provides an overview of recent trends in population, migration and employment in Latin America and refers to their interrelations with changing patterns of gender in the continent. Particular attention is given to low-income groups during the period of economic crisis and neoliberal restructuring in the 1980s and 1990s. Current demographic and labour-market trends, together with diminishing state resources for social expenditure, are likely to lead to growing economic and social polarization in the twenty-first century. Multisectoral policy initiatives sensitive to the needs of disadvantaged groups are essential to redress these tendencies and to improve existing processes of self-exploitation among low-income households, particularly in respect of the labour burdens carried by women.

Latin America can still be meaningfully divided into rural and urban dimensions even though there are increasing linkages between them. Part 7 investigates separately transformations in the rural and urban spheres. In Chapter 11, Cristóbal Kay discusses the causes, objectives and outcomes of Latin America's agrarian reforms in terms of their impact on production, income distribution, employment, poverty, gender relations and socio-political integration. He then examines the changes in the technical and social relations of production in the countryside ushered in by government policies supportive of the modernization of large-scale farms and their increasing production for global markets. The impact of these transformations on the peasantry is then analysed. Finally, the emergent social movements, which are challenging the imposition of neoliberal policies in the countryside, are examined.

Most Latin American countries were already highly urbanised by the time the shift to more neoliberal policies occurred. However, this shift has changed the patterns of social stratification and social mobility. In Chapter 12, Colin Clarke and David Howard examine the implications of neoliberal policies for urban poverty, household organization, housing tenure, social inequality, citizenship and identity. The poorest members of urban society have experienced a double disadvantage: loss of secure employment, including various social benefits, and worsening home circumstances resulting from the withdrawal of government from a concern for housing, health and education. Women and children have been seriously disadvantaged by these changes, which have turned urban economics inside out. Social polarization has ensued and has undermined many of the gains in citizenship achieved as a result of the democratization of Latin America in recent years. Urban poverty has given rise to protest movements involving ethnic outcasts and the socio-economically disadvantaged, though the latter's aims are more material than ideological.

Part 1

Introduction to the political economy of Latin America

1

LATIN AMERICA TRANSFORMED: CHANGING PARADIGMS, DEBATES AND ALTERNATIVES

ROBERT N. GWYNNE, CRISTOBAL KAY

The proposition that an economic, political, social and cultural metamorphosis is occurring at the end of the twentieth century in Latin America is widely accepted amongst academics, yet the conceptualization of this change is less so. This book aims to explore the various components of this metamorphosis. However, it can be argued that studies on societal transformations in contemporary Latin America have been characterized by their strong fragmentation along disciplinary lines. Since the early 1980s, political scientists have almost exclusively directed their attention to the processes of democratic transition and consolidation taking place in a large number of countries (O'Donnell *et al.*, 1986; Malloy and Seligson, 1987; Mainwaring *et al.*, 1992). During the same period, economists have focused on analysing the policies of macro-economic adjustment and trade liberalization implemented in order to regenerate economic growth across the continent (Devlin, 1989; Williamson, 1990; Edwards, 1995). Meanwhile, several sociologists and social anthropologists have begun to examine the nature of the social and cultural transformations generated by modernization and the increasing globalization of Latin American societies (Rowe and Schelling, 1991; García Canclini, 1992; Brunner, 1994; Larraín, 1996). Geographers have tried to integrate ideas around certain themes such as industrialization or urbanization but without an attempt at providing an holistic vision of the overall transformations in Latin America (Storper, 1991; Gilbert, 1994; Preston, D., 1996).

This book aims to provide a more holistic approach to contemporary transformations in Latin America. It does this first by integrating authors from a range of disciplines – cultural studies, economic geography, political science, sociology, social anthropology and social geography. Second, all authors attempt to contextualize their different disciplinary foci within a broad political economy approach that consciously tries to integrate political, economic, social and geographical phenomena. Third, the focus is on attempting to explain the nature of recent transformations from a range of perspectives

– political, economic, environmental, cultural, social, rural and urban. Fourth, this will be done by making reference to the prevailing theories that have been used to interpret not only these changes but also those of past paradigms. Finally, the aim has been to contextualize such transformations (as far as it is possible to do so) within the wider processes of modernization and globalization.

As a result, it is argued that a new political economy is being constructed in Latin America, as national economies become radically restructured and transformed and new social arrangements are being created within national societies. It would seem that the dynamic nature of the world capitalist market is being seen in a new and more positive light in much of Latin America, at least by the new governing classes. Latin American economies and societies are reacting to these changes and relinking to the demands of an increasingly competitive and interdependent world. Such changes are taking place within a continent of democratic governance, providing channels to challenge the new paradigm.

This introductory chapter will examine the broad nature of this new paradigm and the relevance of the previous paradigm based on structuralism and dependency theory. Latin America's changing relationships to global processes, both economic and political, will be explored and the question posed as to why this new paradigm (sometimes called neoliberal) has become so popular. Brief analyses of the social bases and regional impacts of neoliberalism will precede an evaluation of the new paradigm. Finally, the other chapters in the book will be briefly introduced.

THE CONTEMPORARY RELEVANCE OF STRUCTURALISM AND DEPENDENCY THEORIES

Since the debt crisis of the early 1980s there has been a radical series of transformations in the economic, political, social and cultural spheres of Latin America. This can be termed a paradigmatic shift, such has been the scale of the ideological transformation, particularly of governments and their advisers. The previous paradigm can be said to have lasted from the early 1930s to the mid-1980s and developed in a similar way as a response to an economic crisis. It was characterized by greater state involvement in the management of the economy, an attempt to reduce the linkages with the wider world economy and promote industrialization. This paradigm spawned structuralist and dependency theories (Kay, 1989) which tried to interpret events that had already occurred.

The neoliberal policies introduced throughout most of Latin America over the past one or two decades opened a new development era which could be referred to as the globalization phase succeeding the earlier import-substitution phase. There is nothing inevitable about this phase as it is the outcome of powerful struggles between different social forces in the world system, in general, and within Latin America, in particular. This globalization reveals the defeat of the socialist project and the triumph of capitalism in Latin America which had been challenged in a variety of ways by the Cuban revolution of 1959, Allende's Chilean road to socialism in the early 1970s and by the Sandinista revolution in the 1980s. Chile's and Nicaragua's attempts at socialist transformation failed, and Cuba's is barely surviving and no longer inspires those forces seeking a progressive alternative to the neoliberal project. While neoliberalism can point to some successes, especially in its ability to become the dominant ideological force among policy-makers, it has so far been unable to resolve Latin America's endemic problems of vulnerability to external forces, social exclusion and poverty and has even aggravated some of them, as discussed in the chapters of this book.

In view of the crisis of socialism and neoliberalism's failure to address the social question it is imperative to develop an alternative development paradigm which is able to tackle the problems mentioned. Although it is beyond the scope of this book to develop this alternative paradigm it is our belief that such an alternative has to build upon Latin America's contribution to development theory while considering other contributions as well. These are principally the structuralist and dependency theories. Latin America's structuralist theory, sometimes referred to as the center–periphery paradigm, was mainly developed by staff working in the United Nations Economic Commission for Latin America (UNECLA) during the 1950s and 1960s under the inspired leadership of Raúl Prebisch. Meanwhile Latin America's dependency theorists were more widely dispersed throughout a variety of institutions all over the region, although some of the key neomarxist thinkers were at one time working in the Centre for Socio-Economic Studies (CESO) of the University of Chile. Both theories grew out of a critique of existing development paradigms which these authors saw as being unable to uncover let alone deal with Latin America's problems of underdevelopment and development. While structuralism argued in favor of an inward-directed development policy largely through import-substituting industrialization, dependency theory proposed a new international economic order and, in one of its strands, a transition to socialism as a way out of underdevelopment. It is not the purpose here to review these theories as many writings have already done so[1] but to explore briefly their contemporary relevance for developing an alternative to the existing neoliberal paradigm. While structuralist and dependency theories have many shortcomings (which are not explored here; see Kay 1989, among others) their contemporary relevance has been clouded by the inadequate knowledge of them and by the often misplaced criticism, particularly in the Anglo-Saxon world.[2]

Structuralism might provide more relevant ideas for thinking about alternative development strategies for those with a more pragmatic bent while those with a more radical mind and long-term (and possibly utopian) vision might find the ideas of dependency theorists more appealing. Structuralism and the structuralist strand within dependency sought to reform capitalism both internationally and nationally while the neomarxist version of dependency strived to overthrow capitalism as socialism was seen as the only system able to resolve the problems of underdevelopment. Given the collapse of the East European socialist system and China's transition from a planned to a market economy the dependency's theorists' socialist alternative is unable to command much support in the less developed world while the structuralist view of reforming the capitalist system is seen as a more feasible option for those searching for an alternative to the existing neoliberal model. How far a neostructuralist alternative development process within capitalism is able to deal with the problems of underdevelopment remains to be seen, but judging from previous structuralist attempts the outlook does not look too hopeful either. It appears that, at most, Latin American countries can aspire to achieve rates of growth similar to those of the post-war import-substitution period but driven this time mainly by a shift towards nontraditional exports rather than by the domestic market as under import substitution industrialization (ISI). The conclusion seems that although exports and economic growth have been enhanced this has not been sufficient for significantly reducing income inequalities nor the levels of extreme poverty, although absolute poverty has been reduced from its high level of the lost decade of the 1980s.

Increasing asymmetry in the world economy

These days of increasing globalization, which appears as an unstoppable and relentless process, should point to the continuing relevance of structuralist and dependency theories as they view the problems of underdevelopment and development within a global context. A central vision of structuralism is its conceptualization of the international system as being constituted by asymmetric center–periphery relations. Similarly, dependency theory took as its starting point the world system, rooting underdevelopment within the unequal relationships within that system. The economic divide and income gap between the center or developed countries and the periphery or underdeveloped countries has widened continually, especially during the debt and adjustment decade of the 1980s, thereby vindicating the predictions of structuralist and dependency theories as opposed to the neoclassical and neoliberal theories which foresee convergence.

Table 1.1 gives some indication of the increasing asymmetries between the center economies and the periphery of Latin America, as represented by the six countries with the highest and lowest per capita incomes in 1995. The evidence for increasing divergence between Latin American countries, on the one hand, and the center or developed economies, on the other hand, is indisputable. Already in 1978, the per capita income enjoyed by inhabitants of the center countries of the world economy was virtually 5 times that of the highest income economies and 12 times that of the lowest income economies of Latin America; however, by 1995 the ratio had increased to virtually 7 and 30, respectively.

Nevertheless, within the peripheral or dependent countries a few have succeeded in achieving remarkable and consistent high rates of economic growth over the past three or four decades, as well as improvements in equity. This is the case of the Southeast Asian newly industrializing countries (NICs): South Korea, Taiwan, Hong Kong and

TABLE 1.1 Increasing asymmetries in the world economy, 1978–95

Year	Per capita income of the six center economies (A, in US$)	Per capita income of the six Latin American countries with the highest GNP[b] (B, in US$)	Per capita income of the six poorest economies of Latin America[c] (C, in US$)	Ratio A:B	Ratio B:C	Ratio A:C
1978	7899	1602	676	4.9:1	2.4:1	11.7:1
1995	27 870	4015	917	6.9:1	4.4:1	30.4:1

Source: World Bank (1980, 1997).
[a] United States, Japan, Germany, United Kingdom, France, Italy.
[b] In 1995: Argentina, Uruguay, Chile, Brazil, Mexico, Venezuela.
[c] In 1995: Haiti, Nicaragua, Honduras, Bolivia, Guatemala, Ecuador.

Singapore. In particular the larger countries of South Korea and Taiwan, through their spectacular export-orientated-industrialization success, have acquired semi-peripheral status and may soon be considered as developed countries. In this sense the structuralist and 'associated-dependent development' view of Cardoso (1973) is more relevant as compared with Frank's (1967) 'development of underdevelopment' version of dependency, which is at odds with the development achieved by these countries.

However, there are at least two points to stress here. First, such a dramatic transformation in East Asia was possible as a result of the central role played by a national developmentalist state with a forceful industrial policy (imposed after sweeping land reform) in the pursuit of international competitiveness and growth. This has confirmed the position of structuralists and *dependentistas* who pointed to the importance of the state in promoting development. The East Asian model has shown that this state intervention has to be selective and temporary, ensuring that firms acquire international competitiveness within a specified period. Contrary to initial claims by neoliberals (World Bank, 1987), the success of the NICs of East Asia was state-induced rather than market-driven, as expressed so well by Wade's (1990) phrase of 'governing the market'.[3] The second point from the late 1990s is, however, that the rapid growth of the 'peripheral' economies of East Asia has brought its own vulnerabilities, particularly in the financial sector. World Bank advisers often accused Latin American governments of providing a state bail out or further protectionist measures for industrial firms in Latin America, and that this inability to contemplate bankrupting inefficient firms distinguished Latin American from East Asian governments. The role of the state in peripheral economies is not only crucial but also must be constantly changing and aware of the increasing vulnerability of each country in a competitive world economy.

The Latin American debt crisis of the 1980s, which also affected Africa and many Asian countries, can be seen as an illustration of the contemporary relevance of dependency theory. With the vast increase in capital mobility and its availability in the world economy since the 1970s (see Gwynne, 1990) the economies of developing countries have become more and more dependent on foreign capital. This greatly increased their exposure and vulnerability to changes in world capital markets and substantially reduced their room for policy maneuver. In the aftermath of the debt crisis the international financial institutions were by and large able to dictate economic and social policies to the indebted countries, especially the weaker and smaller economies, through structural adjustment programs (SAPs). While Brazil and Mexico were able to negotiate better terms with the World Bank and foreign creditors, Bolivia and other countries were unable to do so. Peru, during the government of Alan García, tried to defy the international financial institutions but was severely punished for it and, after a change of government, the country had to accept the harsh reality of the new power of global capital and implement a SAP. SAPs were used as vehicles for introducing neoliberal policies (see Chapter 4); they had particularly negative consequences for the poor of Latin American economies as unemployment soared and wages and social welfare expenditures were drastically reduced (see Chapters 4, 5 and 10).

The center–periphery view of the world economy

Dependency writers put particular emphasis on technological dependence. Structuralists had pointed to the weakness of the Latin American ISI process in the 1960s and 1970s (Jenkins, 1977; Gwynne, 1985) caused by the difficulties it was experiencing in moving from consumer goods industries to capital goods industries. However, the larger countries such as Brazil had managed to develop a substantial intermediate goods industrial sector, for example the steel and chemical industries (Baer, 1969). Despite the increasing presence of transnational corporations (TNCs) in Latin America there has been little technological diffusion, which has confirmed dependency theory's critique of TNCs. Government policy has failed to develop an indigenous technological capacity in Latin America and could have acted more decisively to ensure that TNCs made a contribution to this process. Nevertheless, Brazil and to some extent Mexico have acquired some competitive technological capacity largely as a consequence of a purposeful industrial policy (Gereffi, 1994). With the new electronics and communications technological revolution the more advanced economies have gained a further competitive advantage over the less developed countries (LDCs), increasing the dependence of those LCDs (Castells and Laserna, 1995). Through the remittances of royalties, profits and interest payments the LDCs continue to transfer a significant net economic surplus to the advanced economies. Such surplus transfers arising from foreign investment and the unequal exchange in foreign trade means a significant reduction in funds which could have been used for domestic investment in the LDCs themselves. However, this does not mean that underdevelopment is due to these factors, although they make the task of achieving development that much harder. It is in the internal class configuration and the role of the state within the peripheral country that the main reason for the continuance of underdevelopment is found. This much has to be learnt by structuralist and dependency writers from the experience of the NICs. Although geopolitical factors played an important role in the success of the NICs it was the developmental role played by the state and its ability to achieve certain autonomy or dominance over class forces which were the key element.

Neither structuralism nor dependency theory foresaw the rapid growth of world trade in the post-war period. This has acquired a new dimension in the present phase of globalization, with its time and space compression, and the more recent impetus to liberalization of the world economy, with the reduction of barriers to the mobility of goods, services and capital across frontiers thereby creating new opportunities for international trade and foreign investment. These globalization forces have certainly reduced even further the room for maneuver for national development policies as compared with the ISI period, thereby confirming one of the key tenets of dependency theory. International market forces rule with even greater strength than in the past, and national states have to take even greater consideration of these global market forces than before as otherwise they can be faced with large withdrawals of foreign capital (as in the case of Chile and Mexico during the 1982/83 and 1994/95 financial crises, respectively), the wrath of the international financial institutions and difficulties with international firms and investors.

Meanwhile, the reinforcing processes of globalization and liberalization have opened up new export opportunities for Latin American economies and have attracted increasing amounts of foreign investment to the region. In some Latin American countries the export sector has been able to give a new dynamism to the national economy. This dynamic capacity of the world trade system has been underestimated by structuralists and seen as having negative consequences by some dependency writers. While some of these misgivings are justified it has detracted from focusing more firmly on the key issue of the domestic policies pursued by the state and the class and other social forces which shape those policies as well as the internal market forces in the periphery.

Recent studies have continued to confirm the deterioration of the periphery's terms of trade in relation to those of the center countries[4], a fact first highlighted by structuralism and incorporated into dependency theory's unequal exchange theory. This does not necessarily mean that foreign exchange earnings have declined – often the case has been the contrary because of the continued rise in the volume of commodity exports from the periphery. But it does mean that a substantial part of the periphery's economic surplus is transferred to the center countries further strengthening the power of the center's capitalist class. The lesson continues to be that LDCs should shift their export structure to higher value added commodities and services rather than continuing to export basic primary commodities which can lead to resource depletion and negative environmental consequences while furthering the periphery's dependency (see Chapter 6). It should not be forgotten that structuralist theorists were among the first to argue that Latin American governments should encourage industrial exports, which they saw as the next phase of the region's industrialization process (Kay, 1989: 40). However, governments (apart from Brazil) failed to act or did so too timidly.

GLOBALIZATION AND A PARADIGMATIC TRANSFORMATION

Capitalism is the dominant socioeconomic system in the global economy, and capitalism has always been an international system. However, at the close of the twentieth century, the international integration of the world market economy is progressing at a very rapid pace. Perhaps it is because of the speed of this integration that the process has become termed 'globalization'. This process encompasses economic transformations in production, consumption, technology and ideas. It is also intimately linked with transformations in political systems as well as socio-cultural and environmental change (Johnston *et al.*, 1995).

What is globalization? O'Brien (1992: 5) argues that it is a tendency that refers to 'operations within an integral whole', since 'a truly global service knows no internal boundaries, can be offered throughout the globe, and pays scant attention to national aspects'. This gives globalization a clear definition but it is also a definition that means that globalization is not yet operating outside those global corporations that have a presence in most countries of the world – as in international banking. This is essentially an approach that sees geographical variations in the world as becoming less and less significant. Indeed, O'Brien (1992: 5) argues that 'the closer we get to a global, integral whole, the closer we get to the end of geography'. Globalization will become synonymous with an homogenized world of global corporations according to this view.

Other views on globalization stress the notion of the 'shrinking world'. This indicates a growing integration and compression of the world's peoples, places and nation-states and a blurring of their territorial boundaries (Klak, 1998). Territory and geography are still important but there is an increasing connection between local events and global phenomena. Thus Giddens (1990: 64) defines globalization as an 'intensification of world-wide social relations which link distant localities in such a way that local happenings are shaped by events occurring many miles away and vice versa'. Complex networks now link the locality with global spaces. In the economic sphere, this can lead to increasing competition between localities and to an intensification of uneven development. In this way, globalization can be seen to be transforming geographical space at a variety of scales – regional blocs, nations, regions, localities.

It could be argued that nation-states in Latin America must increasingly pursue national goals and objectives within globally-defined parameters and structures (Watson, 1996). For developing countries in particular, the impact of being more fully inserted into the global economy increasingly reduces the room for policy maneuver. In part this is because the governments of developing countries are more dependent on the policy approval of the global institutions that 'supervise' the world economy [such as the International Monetary Fund (IMF), World Trade Organization (WTO) and the World Bank] and on the investment decisions of multinational companies that can be strongly swayed by the verdicts of international institutions.

The fall of the Berlin Wall and the crisis of the Soviet world in the late 1980s has reasserted the dominance of the world capitalist system and emphasized the importance of economic success in establishing nodes of power in the world. The demise of the bipolar world, which had been based around Cold War political ideologies, shifted the emphasis to the variations of political economy within the capitalist world system. Some have argued that the world is now tripolar (Preston, P.W., 1996), centered on:

- North America, with the United States, in particular, re-emphasizing its global hegemonic power both in political and in economic matters;
- Japan and the NICs of East Asia, which had formerly been closely linked politically to the United States in the bipolar world but which are in a region which has now emerged as a global economic pole, deriving its power from its success in manufacturing in general and knowledge-intensive industries, in particular (as in Japan);
- the European Union, a regional bloc in the process of both enlargement and deepening.

There are at least two important characteristics of this tripolar world relevant to Latin America. First, firms in these poles have identified hinterlands or spheres of influence, where they can subcontract labor-intensive operations and extend the regional markets for their products: European firms have identified Eastern Europe; firms from the key East Asian NICs have targeted the second-generation NICs of East Asia (such as Malaysia and Thailand); and, to a certain extent, North American firms have identified Latin American countries in this respect.

Second, the great majority of direct foreign investment during the 1980s and 1990s of the large multinational firms originating in the three poles has been in countries of the other two poles (Hirst and Thompson, 1996). This cross-investment has been particularly noticeable for firms from the East Asian pole (Japanese and South Korean firms in North America and Europe); North American firms have tended to concentrate on direct investment in Europe; and European firms in North America. However, the important issue for Latin America has been that major corporations from the three poles have cross-invested rather than invested in regions of the periphery, such as Latin America. There seems now to be one exception in Latin America, and that is Mexico. Since Mexico negotiated to become a signatory of the North American Free Trade Agreement (NAFTA) in the early 1990s, European and East Asian firms (as well as North American ones) have dramatically increased their investment in the country – primarily in order to gain favored access to the key US market from a location with relatively low labor costs. Thus, Mexico seems to have tied itself intimately to one of the three economic poles and is now receiving considerable inward investment.

In this globalizing tripolar world, how then can we interpret the geography of Latin America? How are its component parts being integrated within the global economy and the global political system? The key political and economic relationship is that with the

United States, the dominant player in the global economic and political system of the latter half of the twentieth century. There are, then, important political and ideological issues at stake. However, it would seem that Latin American countries see themselves, after the demise of both the Second World and the military dictatorships of Latin America, as more influenced by US policy. Countries as distant as Mexico and Argentina have become much closer politically to the United States. Meanwhile, in economic terms, Latin America is shifting to the economic reforms of the Washington consensus (see below) and seems to be more closely following the US model of capitalism rather than other models of capitalism, such as the state-driven models of East Asia or even the welfare-state models of continental Europe. Meanwhile, US corporations are becoming an even more important economic force in Latin America, particularly in terms of market access and investment opportunities.

It is thus increasingly popular to claim that the traditional division between the center and periphery of the world economy is no more (Klak, 1998) and to justify this claim through reference to the process of globalization. According to Kearney (1995: 548), 'globalization implies a decay in the distinction' between center and periphery. The East Asian model of rapid economic growth through increased trade, manufacturing production and technological capability has been significant in this respect. In Latin America, the rise of substantial manufacturing sectors and technological capability in Brazil and, to a lesser extent, Mexico and Argentina has also served to obscure the center–periphery model, at least in terms of its original formulation based around the location of manufacturing capacity (Prebisch, 1962). What is true is that the global periphery is becoming increasingly differentiated. Those spaces in the periphery (whether at the scale of the nation-state, region or city) that are becoming more fully inserted into the global economy and able to achieve a sustained improvement in international competitiveness seem to be operating like new growth centers within the periphery, attracting both capital and labor (if the latter's mobility is allowed). However, to what extent are these new centers linked to the growth in manufacturing activity? The conceptualization of a global economy integrated by commodity chains is among the ways in which dependency theory has evolved (Gereffi and Korzeniewicz, 1994). The analysis of commodity chains in relation to Latin America as a whole demonstrates that:

- the export profile of virtually all of the smaller countries of Latin America is dominated by primary products, much as it was in the 1950s;
- the export profile of the larger more-industrialized countries of Latin America is characterized by labor-intensive consumer products or components. Certainly the case of Mexico and particularly the type of industrialization experienced in its northern cities has been well documented in this respect (Sklair, 1988).

It should be pointed out that economic relationships between North America and Latin America are asymmetrical. Exports from Latin American countries to the United States (outside those of Mexico and Brazil) are mainly in the form of primary products, with manufactured products dominating in imports from the United States. Exports to the United States are also lower than US exports to Latin America. The US trade surplus with Latin America is in contrast to its long-standing trade deficit with Japan and East Asia. Nevertheless, the importance of Latin American trade to the US economy is low – only 8 per cent of its exports go to Latin America and the Caribbean (excluding Mexico) (Fig. 1.1). Meanwhile, the three countries of NAFTA (the United States, Canada, Mexico) are the dominant destinations for exports from the Central American and

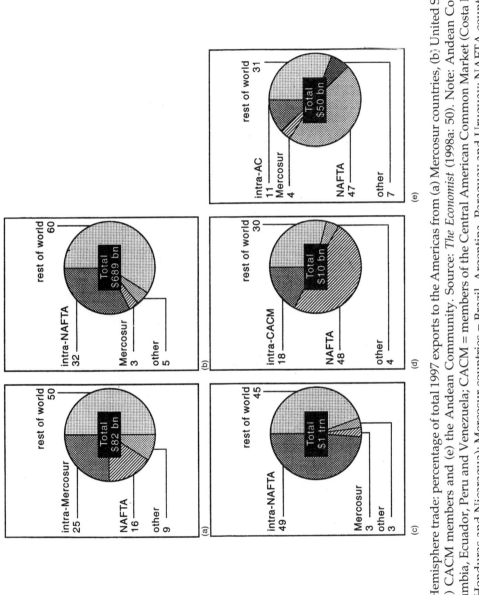

Figure 1.1 Hemisphere trade: percentage of total 1997 exports to the Americas from (a) Mercosur countries, (b) United States, (c) NAFTA countries, (d) CACM members and (e) the Andean Community. Source: *The Economist* (1998a: 50). Note: Andean Community (AC) = Bolivia, Columbia, Ecuador, Peru and Venezuela; CACM = members of the Central American Common Market (Costa Rica, El Salvador, Guatemala, Honduras and Nicaragua); Mercosur countries = Brazil, Argentina, Paraguay and Uruguay; NAFTA countries = signatories to the North American Free Trade Agreement (United States, Canada, Mexico).

Andean countries – nearly 50 per cent of their exports go to NAFTA countries. Only the export trade of Mercosur countries (Brazil, Argentina, Paraguay, Uruguay) is not dominated by NAFTA countries.

There is considerable capital mobility between the United States and Latin America but very limited labor mobility. In a truly globalized and market-orientated world, labor should also be free to move. There is some limited labour mobility to the United States (from Mexico, Central America and the Dominican Republic), but this is often illegal, with major restrictions on any legal flows of unskilled labour. There is more flexibility surrounding labour flows of professionals and those with capital.

The reality of contemporary politics is that internal political forces in the United States (most notably as represented in Congress) wish to restrict not only labor flows from Latin America but also free trade agreements as they are seen to prejudice local US interests. In contrast, the Clinton administration has set its sights on promoting economic integration in the Western Hemisphere (with a hemispheric free trade agreement pencilled in for 2005). The NAFTA experience has not been particularly encouraging for advancing this goal. Perhaps unfairly, US citizens see NAFTA as prejudicing their interests – with employment being reduced as US firms locate plants in northern Mexico and as flows of illegal and legal immigrants have not abated.

Paradigm shift

There are, therefore, at least three interesting points to develop here concerning comparisons between recent paradigms of development in Latin America. First, the inward-orientated paradigm replaced one that had lasted since colonial times and was based on the construction of open economies relying on the export of resources to industrialized economies; in some ways the new neoliberal paradigm is reverting back to this previous paradigm. Second, there is the comparison between the theoretical sources for the two recent paradigms; whereas important elements of structuralism and dependency originated from within Latin America, the new neoliberal paradigm has been driven more by external sources. A third point would be that the inward-orientated paradigm has been the dominant one of the twentieth century (stretching from the 1930s to the 1980s); this may lead one to see the new neoliberal paradigm as the one that will be more representative of the twenty-first century.

It can thus be seen as opening a new chapter in Latin America's evolution, particularly in terms of forming new relations with the world economy. It can be termed a paradigmatic change and be related historically to Latin America's insertion into the global economy of the nineteenth century. Whereas Latin America's economies at that time could rely on the comparative advantages of its natural resources, the important issue today is how competitive advantages can be generated and created – at the level of both the nation-state and the firm. This requires new conceptualizations. Structuralism underestimated the key importance of competitiveness of the world market in transforming economies and societies. Structuralism thought that Latin American economies could shield themselves from global forces and that they could continue to rely on comparative advantages in minerals and basic primary products whilst promoting inward-orientated industrialization.

In contrast, the 'pure' form of the neoliberal model believes in completely opening up national economies to global markets without state mediation. It therefore seems willing to sacrifice uncompetitive sectors (most notably in industry) to foreign competition.

The corollary for this has been a return to relying on natural resource advantages and what has come to be known as nontraditional exports. Some key leaders in Latin America at present (such as Cardoso in Brazil) see a need for the state to bring about the necessary institutional changes for the Latin American economies to build up competitive advantages. The need to be part of the world market is now fully accepted but it is also identified that there is a crucial role for the state in (for example) developing human resources. This can be seen as an interpretation of the East Asian model of economic success based on industrial competitiveness and its application to Latin America (Gereffi and Wyman, 1990; Gwynne, 1990).

Such social reconstruction can be very painful, affecting many layers of society – the industrial working classes (as industrial plants are closed or modernized), the state-employed middle classes (as governments privatize and reduce employment in the public services) and uncompetitive (often inward-orientated) sectors of the capitalist class. In the main, this process has been driven by highly centralized national governments and has often operated in the form of a state-driven social restructuring. This has occurred in authoritarian governments, most notably that of the Pinochet dictatorship in Chile (1973–90). However, as Chapters 2 and 3 argue, democratically elected governments have also initiated these market-orientated reforms and have even managed to be re-elected on such a platform (Menem in Argentina and Fujimori in Peru). It could be argued that such governments have required strong presidential systems in order to succeed.

This model of state-driven social restructuring has responded to the exigencies of the global market and the pulling down of the economic barriers between the national economy and the world market. In a way, it has represented a repressive approach to the demands of the social losers of the new economic model. This social restructuring has had varying impacts on different social groups and has varied from country to country. On the whole, less protection has been given to certain sectors (such as the industrial working class and peasantry) than to others (such as the entrepreneurial middle class and the new financial groups that have emerged). The capitalist class has been more able to readjust to the changing circumstances and realities of the international market and, as a result, has not only expanded in size and influence but has become the key national winner of the paradigmatic shift. These represent some of the new social forces, particularly significant in the finance and export sectors.

WHY HAS NEOLIBERALISM BECOME SO POPULAR?

Definitions, economic characteristics and variations

The political economy of Latin American countries seems increasingly characterized by neoliberal approaches. The use of the term neoliberal has numerous problems because of its ideological connotations. For example in international policy circles the term Washington consensus (Williamson, 1990) tends to be used, indicating virtually the same package of reforms. In their original formulations, neoliberal reforms have normally emphasized economic reforms as opposed to social policies or political reform (Kay, 1993). Hence perhaps some writers have talked about 'the new economic model' (Bulmer-Thomas, 1996a). The economic package of reforms has focused on at least five main areas: fiscal management, privatization of state firms, labour markets, trade and financial markets.

As governments have become committed to neoliberal policies, they have tended to stress the political and economic advantages of creating a more technical, strict and transparent approach to macro-economic management in order to improve the running of the national economy. Thus, fiscal reform has emphasized the need for the reduction of budget deficits, the creation of strong budget and tax offices and, even, an independent central bank (as in Chile in 1989). In countries such as Argentina, Chile and Peru treasury ministers have used this policy in order to justify the slashing of public expenditure, particularly in economic sectors but also in social areas. However, as neoliberal policies have evolved, the need to increase public spending in social areas, such as education, health and welfare, is often identified – as in Chile during the 1990s (though not so evident in Argentina and Peru).

The reduction in the powers of the state in the neoliberal model is further justified through privatization. Indeed, in some countries, such as Argentina, policies of privatization have been intimately linked to those of fiscal reform. This is because of two reasons. First of all, privatization has had the objective of eliminating inefficient and insolvent state enterprises, thereby reducing government expenditure. Second, the sale of these firms to the private sector has boosted income for government during restructuring, when government finances are at their most vulnerable; however, this can only be a short-term palliative, lasting as long as there are state enterprises to privatize. Furthermore, strong regulatory bodies are needed so that in areas of potential monopoly (such as electricity production and distribution) it is ensured that the private sector companies will actually work more efficiently than those of the former public sector.

Another key, though less publicized, neoliberal reform is that of restructuring labor markets. New wage and employment bargaining systems have been introduced, giving more power to employers and less to trade unions. New employment laws have been passed in order to make labor markets more flexible and to reduce the social security contributions and responsibilities of employers. Overall, these reforms have restructured labor markets in favor of employers, as they have gained a more flexible system of hiring and firing and lower wage and nonwage costs.

Private sector employers are seen as the key targets of trade reform. In essence, trade reforms are concerned with making Latin American economies more outward looking and private sector firms keener on becoming more competitive in the international market place. Trade liberalization has emphasized the need to promote exports (through policies that create more effective exchange rates) and to reduce tariffs on imports. Such reform is deemed to create more international competition for firms so that they change from producing for just the home market and raise their horizons to global markets. At the same time, governments are supposed to avoid industrial policy and encourage the inward flow of direct foreign investment from multinational companies.

Financial market reform has also had the objective of reducing government intervention and aiming for the operation of free markets, in which national markets become increasingly influenced by global investors and speculators. However, working towards market-determined interest rates can have consequences that are both favorable (increased inflow of capital) and unfavorable (increased volatility of capital inflow from global financial institutions). The Mexican crisis of end-1994 showed up the unfavorable impacts of high volatility very clearly and, to a certain extent, the policy emphasis changed in its aftermath. It is now understood that financial deregulation needs to be combined with stronger oversight of banks through an effective and efficient superintendency of banks as well as curbs on short-term speculative capital.

These are the core of the neoliberal reforms that are being put in place to varying degrees in Latin American countries (Edwards, 1995). It is worth emphasizing that the paradigmatic transformations in political economy have not been similar in all countries. The commitment to and the extent of neoliberal reform in Latin America varies substantially – for example, from Chile (with over two decades of reform and a shift from authoritarian to democratic governance), to Argentina and Peru (late but committed converts) and on to Venezuela (where conversion to neoliberal reform between 1989 and 1992 was short-lived and became closely involved with corruption). There are considerable variations then in the practice of neoliberal reform in Latin America. However, the Brazilian conversion to a form of neoliberalism under President Cardoso in the mid-1990s (Cammack, 1997) seems to indicate that a new continent-wide paradigm is being adopted. Thus, as a continent, Latin America is shifting towards a closer integration with world markets but its constituent parts are doing so at different rates. Mexico and Chile are integrating fastest, Venezuela perhaps the slowest. Nevertheless, overall the shift in policy package is distinctive. Why have these reforms become so popular now?

Global factors

Why has neoliberalism become the dominant paradigm of the 1990s in Latin America? There are perhaps two relevant geographical scales upon which to attempt an answer: the global and Latin American. At the global scale, the package of economic reforms is strongly supported by international institutions such as the World Bank and IMF – hence the relevance of the label that the consensus was forged in Washington (Edwards, 1995). It is worth pointing out that over the past 50 years the IMF and World Bank have made many such recommendations for the liberalization and management of Latin American economies, recommendations which have often been ignored. Nevertheless, these international institutions gave strong external support for the adoption of a neoliberal framework. The technocracies of these institutions combined with networks of economic and political advisers throughout Latin America actively to push for reform, particularly in the wake of the debt crisis.

The neoliberal model has had surprising converts in other parts of the world. The late 1980s and early 1990s saw the collapse of the Soviet system and the 'alternative' economic model of state-directed, centrally-planned economies. The introduction of market reforms in Eastern Europe and the countries of the former Soviet Union and the apparent vigor with which governments and the populace shifted from planned to market economies gave neoliberal reform considerable impetus in Latin America.

At the global level, Latin America could look to the economic success of certain East Asian countries, which had embarked on outward-orientated policies since the 1960s (though with strong state involvement), thus justifying more export-orientated strategies for Latin American countries. It was often argued that open economies and market orientation had led to the economic success of the NICs of East Asia and their own rapid recovery from the debt crisis of the 1980s (Jenkins, 1991; Gereffi and Wyman, 1990; Edwards, 1995). Nevertheless, the East Asian model was 'consumed' in Latin American policy circles because East Asian countries were seen as being more successful in developing dynamic and technologically innovative economies and because this was related to their insertion into the world economy through trade, investment and technology flows.

Continental factors

Meanwhile, at the Latin American scale, there seem to be a number of historical and comparative factors to point out. First of all, in the 1980s, neoliberal policies provided a framework to extricate Latin American economies from the severe debt crisis of that decade, in which access to external finance was suddenly curtailed (see Chapter 4). Neoliberal economic policies that favored rapid export growth, high interest rates, privatization and reductions in government spending were seen to ameliorate the severe constraints of the sudden drop in external finance and increased government indebtedness.

The adoption of neoliberal policies could thus be seen as a specific response to the impact of the 1980s debt crisis. In many countries, the adoption of a new paradigm also constituted a wider response to the perceived economic failure of the previous political economy paradigm of inward orientation (Kay, 1989; Dietz, 1995). The intellectual justification of inward orientation came from structuralism and dependency theories. It was seen as necessary for governments to give protection in the home markets for industrial enterprises and hence to mediate between the national and global economies. However, in the wake of the debt crisis, this policy package demonstrated two key economic problems. The first problem was that of stagnant export trade, partly linked to the historical pattern of overvalued exchange rates and partly linked to governments ignoring the importance of export growth during inward orientation. Second, the inward orientated model had left a legacy of very high inflation in many countries, particularly in the 1980s, and increased the problems of economic instability in Latin America (see Chapter 4).

The decade of the 1990s has witnessed big advances in the globalization of the Latin American economy, with capital flows, trade and investment increasing significantly (Edwards, 1995). The inward orientated model was effectively cutting Latin American economies off from the advantages (and problems) of being more fully inserted into a globalizing world economy. Neoliberal policies provided the framework for Latin American economies to increase trade with other world regions and increase inward investment and capital inflows from firms and banks in those regions.

Furthermore, there is the question of the link between neoliberal reform and governance. During the late 1980s and 1990s the link between neoliberal policies and democratic governance has become particularly strong in Latin America (Haggard and Kaufman, 1995) – especially through transitions to democracy in former authoritarian governments. There have been significant shifts from authoritarian to democratic governance in all Southern Cone countries and Brazil during the 1980s and 1990s. In all cases there has been either a shift towards or a maintenance of neoliberal economic policies in the aftermath of the democratic transition. Shifts to neoliberal reform did not always come immediately. In the mid-to-late-1980s, heterodox stabilization plans were attempted in Argentina (the Plan Austral of Alfonsín) and in Brazil (the Plan Cruzado). However, these plans met with failure and thus allowed the neoliberal paradigm to gain further influence. It could be argued that the failure of these stabilization plans helped to persuade the population that the bitter pill had to be swallowed. There was no soft option to the shock treatment in order to stop the trend of rampant inflation. As a result many governments in Latin America have justified their shift to neoliberal policies through the 'lack of an alternative' argument.

Political parties that have come to power after the demise of the authoritarian governments that instigated neoliberal policies and have subsequently maintained them (as in the case of the *Concertación* in Chile and the governments of Aylwin and Frei). These parties have stressed that democratic governance allows for and encourages greater public participation and representation in the policy process. Democratic transitions have been important because they have allowed responses from the citizenry to policies that have often created widespread hardship – for example, increased unemployment and poverty and a more unequal distribution of economic benefits (Bulmer-Thomas, 1996a).

SOCIAL BASES OF NEOLIBERALISM

In order to consider the present nature of neoliberalism and its future sustainability, it is important to assess how firm are the bases of consensus for this paradigm and to determine what are the challenges to this consensus. Has any social consensus been achieved in order to support the neoliberal order or is it just a technocratic consensus of government circles and their advisers (see Chapter 3)? As a result, it is useful to envisage the potential consensus for neoliberalism in terms of not only the technocratic but also the wider social and political base.

The technocratic base

It has been argued that the growth of technocratic support for the neoliberal model emerged as a reaction to the deficiencies of the previous inward-orientated paradigm based on protected markets and industrialization. Economic growth based on ISI had encountered both economic and political difficulties. The technocratic argument was that, owing to the power of the state in the ISI model, opportunities for private investment were crowded out, fiscal budgets became characterized by permanent and large deficits, inflation tended to be high or very high and the firms engaged in ISI production (whether public or private) had become inefficient and uncompetitive internationally. These economic difficulties were compounded by the political. Industrial firms continued to demand higher rates of protection in order to survive, which discriminated against exporters and agricultural producers. In some countries, the emergence of a substantial industrial base had given rise to an industrial working class that was gaining in political significance. Meanwhile industrial firms emphasized the high social contributions that they were burdened with and complained about the expensive nature of the rudimentary welfare states that were being created.

The technocratic forces that came to favor neoliberal strategies were not only defined by what they were against but also by what they were in favor of. The theoretical attraction of free market models, a smaller state and the importance of achieving macro-economic stability were some of the main themes. The great majority of technocrats had been research students in the economics and business schools of US universities (Centeno and Silva, 1998). Before the debt crisis, such technocrats had presented neoliberal policy alternatives but had been unable to command sufficient political support for their implementation. After the debt crisis, this changed dramatically and they became the main agents of economic change not only through direct political appointments (such as treasury ministers) but also through the range of advisers and civil servants required by government. They became part of an international network of advisers all broadly sympathetic to market-orientated solutions, macro-economic reform and outward orientation as a way out of the debt crisis.

In spite of delays, the evolution of this new government technocracy has occurred in most Latin American countries. In Argentina and Peru the delay was until the early 1990s, in Brazil until the mid-1990s. In these countries, the technocratic elements supporting more market-orientated policies had to struggle for the policies to come into place – against the continuation of both populist and inward-orientated policies. Indeed in Peru it should be remembered that Fujimori actually came to power on the back of a populist agenda in the 1990 election; it was only after extensive consultations with influential international institutions and networks of Latin American (and Peruvian) technocrats that he became converted (and forcefully) to the neoliberal agenda. Thus, technocrats became influential agents in the installation of the new paradigm.

There was an important distinction as to whether the technocratic force pushing the neoliberal agenda was linked to democratic or authoritarian governments. Within democratic structures, government ministers and technocrats needed to explain and justify the concepts behind radical policy changes to a wide public. Within authoritarian governments, such changes were imposed from above, often with little justification or consultation. Technocrats within authoritarian governments tended to become more ideological as a result, able to impose theoretically consistent policies but unwilling to listen and react to the many who suffered from the fundamental restructuring of the economy. Meanwhile, technocrats within democratic governments have often been both less rigidly ideological in their policy formulation and more willing to adapt policy to political realities.

The social and political base

It is worth emphasizing that the neoliberal model had little social and political base in the early stages of its evolution – apart perhaps from a limited number of entrepreneurs associated with export industries. In general, entrepreneurs in the protected sectors of agriculture, finance and industry were not supportive of more outward orientated policies as this would bring increased competition and would change their political influence in relatively closed markets. How did the social and political base develop?

In many countries it developed as a response to the impacts of the debt crisis and the need to shift power to export-producing sectors and, subsequently, to foreign investors – both in terms of finance and in terms of productive capital. In the 1990s, there has been a surprising extension of the political base as center–left coalitions and governments (as in Chile and Brazil, for example) have been converted to neoliberal economic reform. A range of social democratic parties have come to adopt the Washington consensus (Bresser Pereira, 1996), although emphasizing the need for social policies and welfare programmes to smooth over the hardships of the transition and the restructuring process. The widening of the political base supporting neoliberal reform has given it a wider sense of social support and legitimacy.

Thus, although there was little original social support for the neoliberal model, this has steadily gained ground through the past decade. The expansion of social support through the adoption of a revised and modified model (more socially concerned but still market orientated) has been notable. In conclusion, one can point to three elements:

- social support for the neoliberal model, although limited originally, has steadily gained ground, both in terms of the number of countries adopting the policies (the late arrival of Brazil to reform has been important) and in terms of internal social support;
- the achievement of macro-economic stability has helped the neoliberal program achieve wider legitimacy (as in the governments of Fujimori, Menem and Cardoso, where they achieved stability after previous more heterodox and populist models had been tried and failed);
- after the initial substantial shock in which poverty and unemployment increase dramatically, gradual improvement *can* occur in these variables (as in the case of Chile but not yet Argentina). As employment increases, the neoliberal framework can perhaps gain social support within those poorer groups that are gradually finding waged employment.

REGIONAL AND LOCAL IMPACTS OF THE NEOLIBERAL MODEL

With the consolidation of the neoliberal model in Latin America, localities and regions within countries are now becoming more and more integrated into global (rather than just national) markets. Thus, there is the need to study the relationship between global markets, on the one hand, and regional and local responses in Latin America, on the other. The study of global–local relations in contemporary Latin America poses a number of interesting questions. To what extent are patterns of uneven development – most notably the core–periphery model so characteristic of Latin America's previous phase of inward-orientated industrialization (Gwynne, 1985; Morris, 1981) – being changed? Has the shift to market-led and outward-looking development expanded the manufacturing or primary resource sectors in Latin America? What has been the nature of the nontraditional exports that have emerged in Latin America and what have been their locational characteristics? Have peripheral regions gained from the increased role of the private sector in export-orientated activity? Furthermore, at what levels and geographical scale do we need to discuss globalization?

The nature of export-orientated growth at the regional level

The shift to outward orientation in Latin America has normally been associated with the growth of what have been termed nontraditional exports. These are distinguished from traditional exports that were able to be traded internationally under the inward-orientated model. These traditional exports were normally raw materials that were traded on world markets despite suffering from overvalued exchange rates in their country of origin. They tended to be nonrenewable resources (oil and minerals in particular) whose international price reflected the global balance of supply and demand in the commodity rather than the costs of production. In contrast, the growth of nontraditional exports (as opposed to traditional) is very much influenced by the costs of production of potential exports. The growth of nontraditional exports in both manufacturing and primary product sectors has benefited from trade liberalization and the shift to more effectively-valued exchange rates. Quality and reliability of the export product are important considerations as well but the relative cost of production has been the crucial factor behind the early growth of a nontraditional export.

Within this context, non-traditional exports can best be divided into two groups – those based on renewable resources and those based on manufacturing products. Renewable resources include such sectors as agriculture (Gwynne, 1993b), fishing (if extraction rates are controlled), aquaculture and plantation forestry (Clapp, 1995; Gwynne, 1996a). Exports within these sectors had tended to be discriminated against during the inward-orientated phase. With the shift to neoliberal policies, these exports have become particularly important for the smaller countries of Latin America (Barham et al., 1992; Gwynne, 1993a). Carter et al. (1996) point out that from the middle to late 1980s, nontraditional agricultural exports grew at rates of 222 per cent in Chile, 78 per cent in Guatemala and 348 per cent in Costa Rica. However, because of the renewable nature of these primary exports, the question of environmental sustainability of resources becomes an important political and geographical issue.

The importance of primary resource exports in smaller countries is partly because these countries have generally been unsuccessful in promoting manufacturing exports (Gwynne, 1985). For smaller countries the previous inward-orientated phase based on manufacturing had been characterized by high-cost and small-scale industrial plants that were weak in terms of industrial competitiveness on a world scale. Firms found it difficult to lift their horizons from domestic to international markets.

The shift to neoliberal policies has also boosted nontraditional manufacturing exports from the larger countries of Latin America, particularly Mexico and Brazil. In these two countries, the inward-orientated phase of development was much more successful in creating manufacturing sectors that came close to international levels of competitiveness. In the shift to outward orientation, many firms have been unable to compete in international markets and have closed down plants. Other firms, however, have been able to restructure successfully and achieve international levels of competitiveness. Such firms have required access to capital, new technology, best practice in management and a range of labor skills. However, the key factor in the international competitiveness of these firms has been low labor costs. Wage levels in Mexican industrial plants in the late 1980s were approximately one tenth of those in equivalent plants north of the border in the United States (Shaiken, 1994). Such labor-cost differentials have attracted much investment from foreign firms, particularly since the signing of NAFTA in 1993 and the privileged access of Mexico to the US market; labor markets, the gender division of labor and social relations have dramatically changed in Mexico's northern border towns as a result (Sklair, 1988).

Economic change at the regional and local level

In broad terms, the outward-orientated nature of economic growth has had an impact on peripheral regions according to the ability of producers in those regions to export successfully to international markets. In those regions where producers have made the shift from supplying domestic markets to supplying international markets (normally with the assistance of international intermediaries) significant increases in regional investment and labor productivity have often followed. However, in regions where producers remain geared to the domestic market no such transformation occurred. Thus, in countries that have experienced outward-orientated g development has been the consequence – although without unduly affe ited core–periphery relationship (Scott, 1996; Uribe-Echevarría, 1996).

Neoliberal reform has tended to accentuate the economic importance of the region or main city of each country. According to de Mattos (1996), foreign investment in Chile (outside mining) has been heavily concentrated in the metropolitan region of Santiago during the period of neoliberal reform, and particularly during the late 1980s and 1990s. Between 1974 and 1993, nearly two-thirds (62.7 per cent) of foreign investment in manufacturing and services was concentrated in the Santiago region, causing that region to receive the largest benefits in terms of regional economic growth, employment, construction and labor productivity.

Away from the core region, a complex patchwork of regions and subregions has evolved in terms of the comparative advantage and factor endowments of regions in world markets. Prosperity has been linked to an area's ability to attract investment and produce for export markets. In regions that did well under the inward-orientated model but found it difficult to attract export-led capital and restructure production for global markets, economic stagnation and decline relative to other regions has occurred, particularly in terms of labor productivity. Furthermore, the old reliance on supplying domestic markets has become more difficult as regional producers have had to face competitive import goods and products. Thus, it is often at the regional and local scales of analysis that the impacts of globalization can best be seen in terms of changing social relations – for example through changing labor markets (both paid and unpaid labor) and land markets. However, one then has the methodological problem of generalizing from the particular impacts of globalization at one locality to those at other localities.

THE NEOLIBERAL MODEL EVALUATED

Despite the increasing social consensus surrounding it, the neoliberal model is nevertheless being contested, particularly in the social area. Peasant movements in Southern Mexico, Brazil and Ecuador, movements representing the urban poor in squatter settlements and ecological movements taking up environmental issues at the local scale provide examples of this. This section will first assess any economic shortcomings of the model before assessing the social and environmental problems associated with it.

Economic shortcomings of the model

The neoliberal model has extended its influence largely as a result of the economic gains that it has provided. However, are there any economic shortcomings to the model? Apart from the Chilean case, the model still suffers from low savings rates; this problem was a fundamental factor in explaining the Mexican crisis of the mid-1990s (Otero, 1996a). In terms of the public sector, the tax base is relatively low and tax evasion high. With regard to the private sector, sweeping reform to private pension funds have been associated with a significant rise in the savings rate in Chile (Barrientos, 1996). However, other countries have not followed the example and the Chilean pension funds suffered heavily from declining stock values during the late 1990s. High interest ates on their own do not seem to have changed savings habits in most of Latin America. Indeed, greater access to credit in the wake of financial liberalization has appeared to fuel consumerism and high indebtedness (Sklair, 1994). As a result, Latin American economies still depend heavily on external finance, either in the form of private capital flows or foreign investment.

By making Latin American economies more closely integrated into the global economy, the neoliberal model has also made them more dependent on, and hence vulnerable to, global economic shifts. As with the structuralist arguments of the 1950s, Latin American economies are still concerned about the wide fluctuations in world prices for primary commodities. With exports of these products growing rapidly at the end of the twentieth century, particularly in small countries, their potential vulnerability is being accentuated – as in 1998 when most primary product prices were at low historic levels. In addition, economies are now more vulnerable to changes in strategy (and profitability) of international finance.

Furthermore, although neoliberal economic policies have provided reasonable rates of economic growth once they have become firmly established, this has been associated with increasing inequality. Within the framework of market-orientated economics, benefits have been concentrated within the more successful entrepreneurs and executives of the private sector. Entrepreneurs specializing in exports and finance and large national companies that have been able to restructure successfully have been some of the main beneficiaries of the reforms.

Labor markets transformed

In contrast, labor has suffered much more heavily than have holders of capital during economic restructuring. The adoption of an outward-orientated economic policy has normally been associated with large increases in unemployment in key industrial sectors, at the same time as the privatization of state firms has been characterized by a significant loss of labor. Growth in export-orientated sectors has taken much longer to generate adequate employment opportunities. This has created the need to restructure labor markets radically in order to lower wage costs, to have a more flexible hiring and firing system for employers and to lower employers' nonwage costs (as in employers' insurance contributions). Employers have further been able to reduce costs by adopting short-term contracts and more subcontracting for the supply of parts and services (Thomas, 1996). This has increased the importance of informal arrangements in productive activities (see Chapters 10, 11 and 12).

The state has also tried to reduce the power of trade unions in order to reduce worker protection and to lower labor costs (as in Chile and Peru). Increased employment of female labor (particularly in areas of agricultural exports and assembly industries) has been another feature (see Chapter 10). Labor has increasingly suffered reduced bargaining power with the acquiescence or indeed active support of the state. These processes have often been perceived as the necessary prerequisites to produce a more flexible labor market and to create more competitive labor conditions for employers in the international market place. Overall, labor has become more vulnerable and insecure as a result of the growth of short-term contracts, the shift to more competitive labor markets and the decline of social security. Unless workers are skilled and/or possess a marketable knowledge, they become destined for either low wages or, even worse, underemployment and periods of unemployment.

Social impacts of reform

The transformations of labor markets introduces the wider theme that neoliberal reform has been associated with negative effects in such social areas as income distribution and poverty. These negative effects can be seen in the impact of neoliberal reforms in at least five areas of the labor market (Bulmer-Thomas, 1996a).

- Unemployment rate: trade liberalization, fiscal and labor-market reform have combined to increase unemployment substantially during the economic crisis and the process of economic restructuring. Those companies unable to compete with foreign firms in the domestic market lay off workers, governments drastically reduce the numbers of civil servants and short-term contracts make temporary unemployment more common (see Chapter 4).
- Real minimum wage: labor-market and fiscal reforms have normally operated to reduce the minimum wage in real terms – both to save government spending on social provision and to maximize employment during economic restructuring. Although the real minimum wage declines during the economic crisis, it can subsequently increase once economic growth becomes more sustained (as in Chile since the late 1980s).
- Real wages: trade liberalization, fiscal and labor-market reform have all tended to exert downward pressure on real wages – as companies face more competition from overseas firms, as governments increase wages and salaries at lower rates than inflation and as greater flexibility enters the labor market. Again a distinct sequencing can be found, with real wages declining during the first phase of economic restructuring but with slight increases occurring once the labor market subsequently tightens.
- Wealth effects: the impact of fiscal reform, the liberalization of trade and domestic capital markets and increased inflow of foreign capital has been to increase substantially the wealth of the top two deciles of income earners – the capitalist class in general and entrepreneurs in particular.
- The urban informal sector: this corresponds to that part of the urban economy that is small-scale, avoids regulation and covers a wide variety of activities (see Chapter 12). During the phase of economic restructuring the informal sector tends to expand as more enterprises wish to enter the unregulated sector. However, subsequently, it can decline as it becomes easier for small-scale enterprises to comply with the more limited regulations required of a deregulated formal sector. It has been argued (de Soto, 1989) that the urban market does offer opportunities for many (as in petty commerce). However, as Thomas (1996) and Roberts (1995) point out, these are basically survival strategies, and enterprises will normally remain with low levels of capital accumulation and therefore income. It would be interesting to know the level of support for the economic model from these sectors. Again support would probably emerge when economic growth resumes. Increased subcontracting from larger firms to small-scale informal enterprises would be one example of such trickle-down mechanisms operating.

Thus, the social impacts of neoliberal reform are both considerable and substantial, although it is important to indicate a certain sequencing – normally a period of drastic change (increased unemployment, declining wages) followed, once economic growth picks up, by a period of gradual improvement. Does this period of gradual improvement reduce inequalities as well? It is difficult to judge at present. In the longest historical surveys of the relationship between neoliberal reform and inequality (Altimir, 1994; Scott, 1996), there is a tendency for improvement *after* the crisis of economic restructuring – during which income distribution becomes considerably worse. Even so, it is the upper two deciles which have performed consistently well during economic reform, because of the great advantages enjoyed by those owning capital and earning high salaries as a result of business skills. The middle four deciles tend to be relatively static or even declining, whilst the lower four deciles remain with low and declining percentage proportions of national income.

Within states that have shifted from authoritarian to democratic governance, there is greater evidence of integrating social policies into neoliberal reform packages with the objective of achieving greater social equity – or 'neoliberalism with a human face' as it has been called. The democratic transition in Chile after 1990 saw a significant shift in social priorities, as tax increases were directed to pay for greater spending on social welfare, education and health. However, there seems to be less commitment to social policies in other countries experiencing neoliberal reform.

Poverty and the changing social provision of the state

One of the strongest criticisms against the neoliberal model has been its inability to tackle poverty. Indeed there has normally been a substantial increase in poverty as structural adjustment policies (the shock treatment) have been enforced. After the debt crisis of the 1980s, there was discussion of a social debt, society's debt to the poor and underemployed (between 30 per cent and 50 per cent of the population of most countries). There was the idea that this debt had to be paid alongside that of the foreign debt. However, whereas many countries have arranged the latter debt, the social debt continues. The poor are characterized by poor health and high infant mortality rates (see Chapter 10). When epidemics break out, as with cholera in Peru in the 1980s, it is the poor with their low levels of sanitary infrastructure that most suffer. The social debt remains high, and even the Inter-American Development Bank (IADB, 1996) recently emphasized the need to rebuild the continent's social infrastructure and social services.

However, in general, the state has tried to reduce its long-term commitment to social provision and to create more market-driven forms of social support. Notable here is the case of pension reform in which the private sector takes control of workers' contributions, the investment of those contributions and the delivery of social and pension benefits (Barrientos, 1996). This reduces the fiscal burden and shifts resources from the state to the private sector, giving greater opportunities for the private sector to invest. The private sector has also been encouraged to invest in the health and education sectors. However, this has normally been associated with two-tier systems of social welfare with only the upper and wealthier middle classes able to afford the high costs of private schools and health provision. The poorer majority is left to fend for itself within an underfunded and low-quality public service. With the reduction of social welfare provision from the state, there is an increased role for nongovernmental organizations (NGOs) in helping with skills and livelihoods for the poor both in rural and in urban areas. However, overall, inequality of access to social welfare has become a characteristic of the new economic model.

Are there contradictions to the neoliberal model?

The operation of neoliberal reforms can be seen as a contradictory process. It is achieving rapid economic growth but with increasing income inequality, more exclusion and less social protection. But this is taking place within a Latin American continent that has shifted to democratic frameworks. However, as Gills and Rocamora (1992) argue, in the transition from authoritarian to democratic regimes in Latin American institutions have failed to broaden popular political participation in a meaningful way. In these elite democracies, social reform agendas that could have established the basis for broader popular participation and greater social equity have been abandoned. Indeed, Green (1995: 164) argues that the application of the new economic model 'has ripped the heart out of democratization, turning what could have been a flowering of political and social participation into a brand of "low-intensity democracy"'.

Many question how the economic model can be orchestrated within democracies in which large numbers of the electorate are not enjoying the benefits of economic growth. Is democracy sustainable under such conditions? Or does the continuation of the economic model rely on technocratic governments? Does the model rely on the necessity for economic growth and the increasing integration within global consumerism? Has neoliberalism become responsive to local needs?

Some neoliberal policies have tried to become closer to local needs by decentralizing powers and functions from the central state. Reforms to local government (Nickson, 1995) have tried to provide services more responsive to local needs and with clearer local targeting. These reforms have attempted to increase efficiency whilst legitimating the reduced state at the local level. NGOs have in the past helped in this process. However, NGOs have tended to be captured by the state during the democratic transition. Although there are some positive factors in this (qualified personnel moving into democratic local government), there are also substantial negative factors. These include: much reduced budgets and staff; lack of alternative perspectives; loss of a certain degree of autonomy as a result of reliance on public funding (as opposed to international organizations); not being that closely linked to local needs.

However, the realities of decentralization in many countries that have attempted it (such as Chile and Bolivia) are more akin to deconcentration – the shifting of functions down to the local scale without also giving significant decision-making powers. In this way, deconcentration has been more closely linked to the idea of maintaining a small state but being able to provide cheaper and more efficient social services at the local scale without any increased resources.

CONCLUDING REMARKS: FUTURES OF THE PARADIGM

During the 1990s in Latin America, globalization has been intimately linked with the shift to neoliberal policies. During this decade, the governments of most of the countries of mainland Latin America have integrated their national economies more closely with the global economy. In particular this has been achieved through trade liberalization and the deregulation of financial markets; increased trade, capital flows, investment and technology transfer have normally resulted. The more global framework for Latin American economies has coincided with a shift from authoritarian governments (that were still significant in the 1980s) to democratic governance so that at present all

16 mainland Latin American countries have governments elected through the ballot box. Thus, the Latin American state in the 1990s has transformed itself into a democratic system at the same time as reducing its direct influence over the economy (through privatization and deregulation) and cutting the size of the public sector through fiscal reform.

Globalization, or Latin America's closer integration with global markets, has thus been associated with a shift to a more representative and participatory political system (Haggard and Kaufman, 1995). To a certain extent this may have obscured the negative social impacts of neoliberal reform. Increased unemployment and poverty, an even more unequal distribution of income and a further rise in the informal sector have resulted. However, democratic governments have attempted to explain or justify this in two ways. First, there is the argument that the negative social impacts reflect a short-term adjustment to new conditions and will be soon turned around. Unemployment and poverty will increase as the economy adjusts to new external realities and as the country forges a more competitive economy. The second justification concerns the 'lack of alternatives' argument. Latin American governments point to the political economy of neoliberalism becoming the basis for policy in other areas of the world that are identified as 'competitor' regions in the world economy – Eastern Europe and East Asia, in particular. It becomes paramount, according to Latin American treasury ministers, to 'modernize' their economies in order to make them more competitive in world markets and so that they can better take advantage of global forces (Foxley, 1996). Such modernization is necessary in order successfully to attract foreign investment from global corporations that have a wide range of options of where to invest.

To what extent will the shortcomings of the neoliberal model be recognized and social movements be created outlining alternative development strategies and socio-political scenarios? It could be argued that, in order to make Latin American countries more competitive in a globalizing world, neoliberal reform cannot simply be about making economies more market-orientated. The Chilean case shows that substantial and critical institutional reforms have to take place over a long period of time in order for a Latin American country to become more competitive and less prone to international crises. Institutional reform in Chile has stretched over a period dating from 1964 and has emerged from a wide variety of political ideologies. Reforms to landholding, to the ownership of national mineral wealth (notably copper), to health and personal pensions, to financial institutions and to taxation have been notable examples that have occurred under governments of widely different ideologies. Martínez and Díaz (1996) argue that it is the combination of these profound institutional reforms with market-orientated neoliberal policies that lie behind Chile's sustained economic success during the 1990s. This has a great significance for the theme of sustaining economic growth within an increasingly competitive world.

The future relationship of the state with the process of economic change is thus a key issue. The ideological shift to limited government involvement in the economy may not produce the modernized, competitive economy that is anticipated from neoliberal reform. If this is the case, sustained economic growth will not occur – which is seen as the prerequisite for governments to address the social debt and begin to rectify the highly unequal patterns of income distribution.

There is also the question of the relationship between economic integration, neoliberalism and globalization. For year 2005, it is planned that the Americas will be one large free trade zone. This will involve integrating the dominant economy of [the 21st] century with 16 much smaller but highly diverse countries in La[tin America]. Geopolitical reasons have become important additional factors in this [process (see Chapter 4). Neoliberal reform and the opening up of formerly inwa[rd-looking] economies have produced a more successful record of economic integratio[n in the 1990s] than in the 1960s, the previous decade in which economic integration was seen as a key international policy in Latin America (Gibb and Michalak, 1994). In geopolitical terms, it will still be necessary to resolve the problems inherent in a strong center–periphery pattern that economic integration of the Americas (in contrast to other schemes) will be characterized by.

It is important to emphasize that the neoliberal model has evolved – from an often narrow and economistic interpretation to the Washington consensus and on to a more social democratic interpretation in Chile and Brazil. Indeed some form of convergence between neoliberalism and structuralism seems to have occurred in some parts of Latin America. There is a reappraisal of the theories of the 1950s and 1960s and the evolution of a neostructuralist position since the late 1980s.[5] It could be argued that neostructuralism has gained some influence on government policy in Latin America, such as with the Concertación regimes of Chile and the administration of F.H. Cardoso in Brazil in the 1990s.

Neostructuralism has taken on board some elements of neoliberalism while retaining some of the core structuralist ideas. While some authors have dismissed neostructuralism as being merely the human face of neoliberalism and its second phase (Green, 1995: 189) it is certainly true that there has been a shift of structuralism towards neoliberalism. However, there are differences, some of which have already been mentioned when discussing the contemporary relevance of structuralism and dependency theories. These differences concern mainly their respective views on the relationship between developed and developing countries as well as between state, civil society and the market.

The neoliberal view is that further liberalization of the world economy is required and that this will benefit the developing countries considerably. In contrast, neostructuralists, as well as dependency writers, view the world economy as a hierarchical and asymmetric power system which favors the center countries and the TNCs in particular. They are thus more sceptical about further liberalization, believing that it will act to enhance the inequalities between and within countries; powerful global groups located in developed countries would ensure that the benefits of global liberalization would be channelled in their favor.

As for the relationship between state, civil society and the market, neostructuralists give a more important role to the state in the process of social transformation and are eager to involve the disadvantaged groups of society in this process, particularly as it has tended to exclude them. Meanwhile, neoliberals desire a minimalist state, putting the market at center stage as they believe it to be the most effective transformative force; the fewer constraints that are put on the free operation of the market, the better for the national economy, society and polity.

Neostructuralism should not be interpreted as caving in to neoliberalism nor as an indication that structuralism was wrong but rather as an attempt to come to terms with a new reality. In this sense structuralism is showing an ability to adapt to changing historical circumstances rather than remain frozen in the past. Certainly, structuralism made mistakes such as its trade pessimism and its technocratic conception of the state.

Despite the shortcomings of neostructuralism it is the only feasible and credible alternative to neoliberalism in present historical circumstances. The main lesson neostructuralists take from the East Asian NICs is the need selectively to integrate into the world economy and create competitive advantages through a well-designed industrial policy. Such an industrial and export policy tries continually to exploit niches in the world market and shift upstream to more skilled, technologically advanced and higher value added industrial ventures. Policies to improve the knowledge base of the economy and national technological capability are seen as crucial for long-term growth. Thus, the importance of education continues to be stressed, although less mention is made of the need for land reform as this has become a politically sensitive topic in many Latin American countries (see Chapter 11).

Neostructuralism gives more importance to market forces, private enterprise and foreign direct investment as compared with structuralism but argues that the state should govern the market. However, in neostructuralist thinking the state no longer plays the pivotal developmental role that it did under structuralist ISI as state enterprises are largely limited to provide essential services such as health care and education but no longer undertake direct productive activities through ownership of industrial enterprises or otherwise. Also the ability of the state to steer the economy is limited as protectionism and subsidies are used only in a restricted and sporadic fashion in stark contrast to the ISI period. The imperative of achieving and maintaining macro-economic balances is recognized as now price and fiscal stability are a condition for growth which was not necessarily the case in the past.[6] Another key element of neostructuralism is its greater concern for equity and poverty reduction requiring special action by the state and involving also NGOs.

The position with regard to the world market is much changed as export orientation rather than import substitution is now the strategic direction which the economy has to take. Neostructuralists propose a strategy of 'development from within' as argued by Sunkel (1993a). This means that it is the country through the guidance of the state and its intermediary organizations which decides in which particular direction it wishes to develop its links with the world economy. Choices are, however, restricted because of the forces of globalization as mentioned earlier. Another key element in neostructuralism is the achievement of competitive advantages in certain key productive areas in the world market by selective liberalization, integration into the world economy and an export-orientated industrial and growth policy. Neostructuralists are keen advocates of 'open regionalism' which they hope will enhance Latin America's position in the world economy while at the same time reducing its vulnerability and dependence (see ECLAC, 1994a, 1995).

The neoliberal model has restructured the political and economic system but has created new interest groups, particularly in finance capital and exporting companies. In addition, it has become apparent that a closer relationship with the global restricts the internal room for maneuver of virtually all Latin American governments. Opening up to the global economy has been a disciplining force for both capital and labor in Latin America. Mistaken policies, or those policies *perceived* as mistaken by international capital, are penalized, such as through a rapid withdrawal of finance capital. However, if the neoliberal model is to continue, it must also continue to evolve in terms of providing improved social conditions.

ENDNOTES

1 For useful analyses of these Latin American theories, see Blomström and Hettne (1984), Kay (1989), Larraín (1989), Lehmann (1990) and Love (1994), among others. For insightful review essays of some of these books, see Slater (1990) and Frank (1991, 1992).

2 It is surprising to find that even today many of the books and articles referring to structuralist and dependency theories continue to display a limited and often mistaken knowledge of these theories as they fail to consider key Latin American contributors. Not necessarily among the worst offenders are Arndt (1987), Harrison (1988), So (1990), Peet (1991), Packenham (1992), Spybey (1992), Hout (1993), Kiely (1995), Leys (1996), and P. Preston (1996). While this may have been excusable when few of the original texts were available in English this is no longer the case; since 1979 some of these were translated, and comprehensive and thoughtful analyses of these theories have now become available in English (see footnote 1).

3 The World Bank has tried to accommodate some of the many critics of their initial interpretations of the NICs with their 'East Asian miracle' study (World Bank, 1993) but this has in turn spawned further criticisms. Among the most perceptive of these new critiques are those done by Wade (1996) and Gore (1996).

4 See, for example, Singer (1991), Diakosavvas and Scandizzo (1991), Cuddington (1992) and Ocampo (1993).

5 For some key writings on neostructuralism, see Rosales (1988), Ffrench-Davis (1988), Sunkel and Zuleta (1990), Fajnzylber (1990), ECLAC (1990, 1992a), Lustig (1991) and Ramos and Sunkel (1993). For a comparison between neoliberalism and neostructuralism see Bitar (1998) and Sunkel (1994). For a critical assessment of neostructuralism see van der Borgh (1995).

6 For an analysis of the debate between structuralists and monetarists regarding inflation, see Kay (1989: 47–57).

FURTHER READING

Cubitt, T. 1995 *Latin America society.* 2nd edn. Longman, Harlow, Essex. Aimed at undergraduate students this book provides a helpful overview of the major social issues facing Latin America.

Dietz, J.L. (ed.) 1995 *Latin America's economic development: confronting the crisis.* Lynne Rienner Publishers, Boulder, CO. A valuable textbook which has been updated and revised since it was first published in 1987. Contributions deal with various aspects of economic development such as growth, equity, development strategy, inflation and employment.

Green, D. 1995 *Silent revolution: the rise of market economics in Latin America.* Cassell, London; Monthly Review Press, New York. A popular and well-written critical analysis of Latin America's neoliberal economic revolution sweeping through the region since the debt crisis of 1982. A most useful introductory text which has gained a wide readership, particularly with students.

Halebsky, S., Harris, R.L. (eds) 1995 *Capital, power, and inequality in Latin America.* Westview Press, Boulder, CO. Discusses the critical issues facing Latin America in the 1990s from a largely Marxist perspective focusing on the impact of globalization on Latin America's political economy. The book highlights the struggles for survival and empowerment by rural and urban workers, women and indigenous peoples.

Kay, C. 1989 *Latin American theories of development and underdevelopment.* Routledge, London. A comprehensive study of structuralist and dependency theories as well as of the theories on marginality and internal colonialism. These theories are contextualized within Latin America's transformation since 1945 and the wider theoretical debates of the period.

Preston, D. (ed.) 1996 *Latin American development: geographical perspectives.* 2nd edn. Longman, Harlow, Essex. Written by geographers this student-friendly textbook analyses a variety of topics including industrialization, rural development, environment, race and gender, migration, health, education, urbanization and geopolitics.

Sunkel, O. (ed.) 1993 *Development from within: toward a neostructuralist approach for Latin America*. Lynne Rienner Publishers, Boulder CO. An important effort at developing an alternative development paradigm to neoliberalism written by key Latin American social scientists and edited by one of the key contributors to structuralist, dependency and neostructuralist thinking.

Smith, W.C., Acuña, C.H, Gamarra, E.A. (eds) 1994 *Latin American political economy in the age of neoliberal reform: theoretical and comparative perspectives for the 1990s*. Transaction Publishers, New Brunswick, NJ. Written by prominent social scientists from Latin America and the United States this book discusses the successes and failures of neoliberal reform from a political economy perspective.

Part 2

Political transformations

2

AUTHORITARIANISM, DEMOCRACY AND DEVELOPMENT

EDUARDO SILVA

The challenge of economic modernization often places the states of developing countries under great pressures, to which democracies seem particularly vulnerable. Democracy implies broad societal representation in policy-making, accountability of executive branches to legislatures, deliberation, compromise and tolerance. Yet economic development frequently requires harsh trade-offs between savings for investment on the one hand and redistribution for social needs on the other; economic crises demand swift, decisive, comprehensive responses; and economic restructuring imposes steep costs on losers. Between 1964 and 1976, such trials generated political tensions that caused many Latin American democracies to give way to authoritarianism. Ironically, with few exceptions, the military governments that followed proved equally incapable of managing the political economy of their nations. A wave of democratization – which crested in the 1980s – ensued. Today, most of the states in the region are democratic. Yet despite much congratulatory rhetoric, uncertainty over the deepening, consolidation and permanence of these new democracies persists as Latin America confronts the demands of globalization. These cyclical bouts of authoritarianism and fragile democracy raise an enduring question about the relationship between economic modernization-cum-globalization and the state. Is there a fundamental incompatibility between economic development and democracy?

The response to this question has varied widely. One school of thought believed that economic modernization inevitably generated its political correlate: democracy. Others disagreed, arguing that the middle stages of economic development required authoritarian states: the economic and social crises associated with these levels of development snuffed out democratic impulses. A third contingent was less deterministic. It held that beyond a certain point the level of economic development did not determine whether states were democratic or authoritarian. Which system prevailed depended on the domestic impact of external crises, the severity of social tensions and the structure of political institutions.

What prompted such divergent responses? Will the current trend toward democracy in Latin America persist? What are the chances for the consolidation of emerging polyarchies? There are no hard and fast answers. But we can begin to understand the diversity of opinion by closely examining competing theories about the relationship between economic development, dictatorship and development on three dimensions. Those dimensions include the theorists' assumptions, what they mean by democracy and authoritarianism and differences in what they believe to be the causes for the outcomes. Understanding those theories, however, first requires briefly defining what we mean by the state, democracy and authoritarianism and to explain the relationship between economics and politics.

POLITICAL ECONOMY AND THE STATE

States are long-term expressions of a society's political power through which some social groups and individuals dominate others within a given territory (Poggi, 1990). Max Weber (in Gerth and Mills, 1958) stressed that states exercise a monopoly over the legitimate sources of coercion and that social domination is based on unequal access to them. The modern state is organized in a coherent set of bureaucratic institutions based on impartial, rational rule-making which differentiates it from the individual, charismatic style of earlier states (Nettl, 1968). Other theorists add that state managers have their own interests, over and above those of the social groups they control. Those interests include economic policy, the state's organization, taxation and territorial expansion through war (Skocpol, 1979; Tilly, 1975). Their capacity to impose those interests depends on the porosity of state institutions to social groups – the less porous a state is to social forces the more it is autonomous from them (Evans *et al.*, 1984).

These studies focused on defining the state, established degrees of stateness and differentiated the state from society while recognizing that it is embedded in society. They also set criteria for distinguishing between modern states and their predecessors. Poggi (1990) further refined categories of modern states by classifying them into liberal democracies and communist one-party states. His effort invites further refinement in the categorization of Third World states, whether they be liberal democracies, dictatorships or communist states.

These approaches to the state, however, have their limitations. They are not suited for explaining change from one form of state to another in concrete cases, say from democracy to dictatorship in the Southern Cone of Latin America between 1964 and 1976. Instead, these studies focus on how ideal–typical premodern states have given way to modern ones, whatever their form, over several ideal–typical historical eras such as the Age of Absolutism and the Modern era. Explanation focuses on cultural change (from charismatic traditionalism to rationalism), the rise of the idea of constitutionalism (liberalism and the limitation of absolute power), the concept of citizenship and the ideology of Communism (Weber, in Gerth and Mills, 1958; Poggi, 1978, 1990). Democracy in these analyses is usually seen as an improvement over the absolutist state or the one-party state.

Political economy approaches, of both Marxist and non-Marxist inspiration, are better equipped to explain concrete examples of change in the form of the state. They focus more directly on the effects of economic and social tensions on the state by examining relationships between economic structure, class-based social groups and politics. Frieden (1991: 16) argues that modern political economists share a general analytic framework. They all study how rational, self-interested actors combine to affect economic and social policy and to influence the form of the state. Those actors pursue their political aims both within and outside of established institutional settings. For political economists, the economic system itself, the structure of production, is a fundamental starting point for the analysis of economic and social policy and political change. It is the principal building block for defining actors and their interests; it shapes relations of domination and subordination among social groups; and it determines some key functions of the state (Anglade and Fortín, 1985).

In market economies, capitalists (the owners or controllers of production and money) and labor are usually considered to be the main social actors. Capitalists, in the first instance, dominate labor by virtue of their economic power. To simplify drastically, capitalists seek to maximize profits, which often means paying less to labor in the form of wages and benefits. It also means obstructing labor's attempts to organize to increase their share of the wealth generated by their work. For Karl Marx, this essential tension, or conflict, between the two fundamental social classes of capitalism constituted the basis for economic, social and political change.

Many political economists since then have argued that classes must be broken down into smaller units (or disaggregated) to understand properly processes of economic and political change (Gourevitch, 1986; Frieden, 1991). Thus, capitalists are most commonly subdivided by economic sector: financiers, those with commercial interests, industrialists, landowners, those in the construction sector, mineowners and so forth. Economic sectors can be subdivided further depending on the capacity of subsectors to compete in international markets. Labor is frequently separated into urban industrial workers, service sector workers, government employees and peasants and/or rural labor. Middle classes – usually professionals – are also treated as a distinct category; and they are especially important for political change in Latin America (Rueschemeyer et al., 1992). Disaggregation along these lines permits analysis of how divisions within classes or of alliances between specific sectors of capital, labor and middle classes affect the direction of economic and political change.

Politics and economics are intertwined in another way as well. A country's economic fortunes have an impact on political stability. Prosperity – economic good times – usually reduces tension between class-based social groups. If economic policies are generating economic growth, and if most social groups perceive a benefit from that growth, social pressure for change will be slight. By the same token, deep or recurrent economic crises exacerbate tensions between class-based social groups and generate conflict as those groups seek the establishment of alternative policies conducive to a resumption of economic growth (Gourevitch, 1986; Haggard and Kaufman, 1995). If the crisis is severe enough, social groups may perceive that a change in the form of state is the only solution.

What are the implications of these features of capitalist societies for theories of the state and the problem of democracy, authoritarianism and development? One very important ramification is that they point to two essential functions that the state – whether democratic or authoritarian – must perform if it is to remain stable. First, it is compelled to maintain the overall conditions for capitalist economic development (Lindblom, 1977; Przeworski and Wallerstein, 1988). Second, the state must maintain the social order in which capitalists dominate and labor and middle classes are subordinate to them (Mandel, 1978; Poulantzas, 1973).

How state managers approach the task of fulfilling those functions varies. In a synthesis of extensive academic debates, Christián Anglade and Carlos Fortín (1985, 19–23) have argued that state managers intervene in several ways to mediate class tensions that could interfere in the accomplishment of those functions. First, state managers have to maintain conditions to ensure that business people generate sufficient profits to maintain adequate investment levels. Naked repression, however, cannot accomplish this over the long term. Thus, state managers also have to protect labor from excessive exploitation in order to avoid rebellion and to keep up productivity. Second, state managers also intervene on behalf of specific economic sectors of capital to ensure the economic health of the nation through economic policies such as taxation, tariffs, subsidies and monetary and fiscal policy. In this manner, the state may promote industry over agriculture, or export sectors over domestic market-orientated sectors. Third, state managers may involve the state directly in production through public enterprise if the private sector is unwilling or incapable of significant investment of its own.

How state managers navigate these interventions also has an effect on economic growth and political stability. Too much protection for labor may pit capitalists against them. Excessive favoritism for some economic sectors over others can generate alliances between disgruntled business people and other social sectors (middle classes and some labor groups). This too could have destabilizing consequences, especially for conservative dictatorships. By the same token, substantial public investment can turn a united private sector against state managers because it feels threatened by unfair competition from the state for markets and credit.

DEMOCRACY, AUTHORITARIANISM AND DEVELOPMENT

The functions of the capitalist state apply equally regardless of whether the state is democratic or authoritarian in form. But how state managers fulfil those functions varies depending on state form. In democratic capitalist states governments are elected by contestation between two or more political parties with discernable differences on major policy issues. Electoral participation – voting – should be by secret and universal ballot. Sufficient civil liberties must be guaranteed to ensure full and fair contestation and electoral participation (Dahl, 1971). Participation can also extend to the policy-making process through negotiation between political parties in the congress with the executive or directly through access to the executive by the organizations of labor, capital and other groups. In some instances, democratic capitalist states may also exhibit a high degree of concern over the economic rights of subordinate class-based groups. This effectively extends the egalitarian principle of democracy to the economic sphere (Held, 1996).

Given these characteristics, in principle, democratic states protect subordinate social groups (middle classes and urban and rural labor) from excessive exploitation by business people and landowners. Political parties may represent some of their interests and try to advance them from either the executive (if the party has won the presidency) or the legislature. Greater freedom to organize may allow subordinate social groups to negotiate more directly with capital and state managers. Less able to rely on repression, and in the interest of maintaining political order, state managers have more of an incentive to give some concessions to subordinate groups. All of these allowances must be allocated in ways that do not upset the maintenance of a good business climate for investment and economic growth; nor should they threaten the basic social order.

Authoritarian capitalist states, to some degree or another, are lacking on one or more of the dimensions of contestation and participation. The most closed authoritarianism would have no contestation or participation, either electoral or in policy-making. Policy is formulated by state managers themselves. Capitalist authoritarianism in developing countries is also closely associated with repression of labor and other subordinate class-based and ethnic groups. State managers may ignore most of their concerns because of the lack of contestation for political power and the relative absence of societal participation in policy-making. Instead, state managers concentrate on manipulating tensions between different capitalist and landowning sectors. At times, they also seek to co-opt middle classes and minority sectors of labor. Given these characteristics, it was widely assumed that the managers of authoritarian states can concentrate more exclusively on maintaining social order and a good business climate.

Development places great strain on the states of developing countries (Chilcote, 1981; Handelman, 1996). Development occurs when societies less economically advanced than the great industrial powers of Europe, North America and Asia are drawn into political and economic relationships with them. In other words, development is a process in which societies that had been peripheral to the global market economy are pulled into it. The economic changes that result from the experience generate social change. Political tensions flare as new class-based groups compete with established ones; for example new industrialists, middle classes and urban labor may clash with traditional landed and commercial elites. A state that expressed the domination of the latter two classes may find itself under pressure to include the new social groups. Depending on how the conflict is resolved, the form of the state may suffer a more or less violent change.

As societies are drawn into the world economy, state managers may have great difficulty devising policies that promote economic growth. Shifts in the international economy have a strong effect on the economies of underdeveloped nations. Those countries are dependent on economically advanced nations for capital and technological know-how. Thus, booms, depressions and technological breakthroughs upset established patterns of production more deeply than in the advanced countries. Developing countries must scramble to adjust their economies through drastic means. This too places inordinate strains on states; subsequent conflicts over appropriate policy responses can lead to a change in state form.

In short, the process of development can generate situations in which state managers have difficulty fulfilling the functions of the state in market economies. They may not be able to generate policies conducive to capitalist economic development. Moreover, they may be overwhelmed by social tensions that make it hard to maintain a social order in which the private sector clearly dominates over other class-based social groups, such as labor, peasants and middle classes.

Because development is such a wrenching process, it poses difficult questions for political order. Is democracy compatible with economic development? Democracy permits contestation for power and participation in policy-making by political forces with differences in their policy agendas. What if those agendas are perceived to be disadvantageous for capitalist development and the maintenance of its social order? In other words, what happens to democracy when the state cannot fulfill its primary functions? Are some stages of capitalist development more compatible with authoritarianism because the lack of contestation ensures that economic and social policy will not scare off scarce investment? Or do development and democracy go hand in hand after a certain plateau of economic growth and diversification has been achieved? How have competing theoretical perspectives responded to these questions?

Underdevelopment, dependency theory and state form

In the 1960s, dependency theory took a nondeterminist view to this conundrum: there was no mechanistic relationship between levels of development and state form. Fernando Henrique Cardoso and Enzo Faletto (1979: 199–212) expressed the view most forcefully. Although heavily critiqued by Marxists (Chilcote, 1982), their analysis was rooted in a Marxist political economy. Thus, for them, states expressed the domination of the propertied classes over the rest of society. Their primary functions were to maintain social order and the conditions for capitalist development; democratic states were safe as long as they fulfilled those functions. Whether or not they managed to accomplish those functions depended on a number of factors. These included the composition of political alliances within and across class-based groups, their bases of power within state institutions, the organizational capacity of subordinate classes and the nature of their demands and the presence or absence of internationally and domestically induced economic crisis.

In some countries the mix of those factors – the specifics of class conflict and its politics – made it difficult for the state to fulfill its primary functions. Those nations succumbed to military dictatorship, as occurred in Brazil, Argentina, Uruguay, Chile and Peru. Meanwhile, in Colombia and Venezuela class struggles had worked themselves out in a manner compatible with democracy, where democracy did not threaten the state's principal functions.

Cardoso and Faletto argued that economic change lay at the root of the crises of the Latin American state that began in the 1960s. They concluded that the globalization of the world economy forced states and domestic capitalists to accommodate transnational corporations. This required a shift in alliances between state actors, international capital, domestic capital and the middle and working classes. The states of Brazil, Argentina and Chile, for example, collapsed under these tensions, ushering in military government. In these cases, labor had its own political party (or parties). When they won the presidencies of their nations, they used the state's power and authority to expand the inclusion of labor and peasants into the political system. Moreover, their governments heavily attacked foreign capital, which detracted from the business climate necessary to attract transnational corporations and to link domestic capitalists to them.

By contrast, the Colombian and Venezuelan states weathered the storm. Colombia was aided by its party system, dominated by two multiclass, catch-all political parties. Neither felt beholden to a lower-class constituency and, thus, were in a better position to mediate between the external sector, domestic entrepreneurs and lower classes. The country's system of power sharing between the two political parties (the *Frente Nacional*) also helped. Meanwhile, the Venezuelan state was aided by oil money, which lubricated class tensions through redistribution of resources for all classes, upper and lower. The larger integration of business into the major populist party, *Acción Democrática*, also helped.

Dependency theorists, however, were not primarily interested in explaining the relationship between development and the prospects for democracy, authoritarianism or totalitarianism. They wanted to understand the socio-economic and political roots of underdevelopment and the difficulty of overcoming it (Frank, 1966; Dos Santos, 1970).

As a result, there was little analysis of which intra-class and inter-class tensions might destabilize a particular state form. Nor was there much specification of how political institutions could be structured to moderate class tensions. The same applied to the strategies of organized social groups. Thus, the assertion that development was compatible with either democracy or dictatorship rested mainly on the theorists' assumptions about the historical specificity of the class struggle in different nations, and empirical observation. They did not, by and large, uncover the conditions that moderated or exacerbated the class struggle and that, therefore, might explain why one developing country had a democratic state form while another yielded to dictatorship.

Nevertheless, dependency theorists made important contributions to political economy studies of development and its politics. A fundamental insight was the argument that a developing nation's domestic politics and the prospects for development had to be analysed in the context of that country's insertion in the international economy. The economically developed nations of the world had pulled them into the world economy. They dominated developing countries because of superior investment capability, technological advancement and military capability (in the case of the United States). This condition defined a situation of dependency. It was also the cause of persistent underdevelopment because of the assumption that the asymmetrical power relations between developed and developing nations perpetuated the condition. Development might occur, but it would always be lagging and different in its social, political and cultural consequences. This was called 'associated dependent development' (Cardoso and Faletto, 1979; Evans, 1979).

Critics argued that dependency theory was not theory at all. Jackson et al. (1979) maintained that the object of study, dependency, lacked specification. If its opposite could not be identified, then how could anybody know whether a country was dependent or not? Moreover, they held that the explanatory factors – economic and social variables – were not used systematically enough to pass the test of scientific rigor. In other words, the same variables did not lead to similar outcomes, or significant differences failed to produce different outcomes. Other critics simply claimed that dependency theory was ideology masquerading as theory (Packenham, 1992).

As a result of these critiques, a lively controversy erupted over the possibilities of overcoming underdevelopment; a debate that intensified with the economic success of some East Asian countries. Stephan Haggard (1990) and others (Yoffie, 1983) argued that developing countries were not condemned to their situation of dependency. Under the right circumstances they could develop economically. For Haggard, whether they did or did not depended on the timing of industrialization, class structure, resource endowments, relative autonomy of the state and their relationship to great powers. Haggard maintained that Taiwan and Korea industrialized for two reasons. First, their states had greater autonomy from social groups. This allowed them to formulate and implement policies for economic change. It also helped that the United States had a strategic interest in them and conferred trade and financial advantages not available to Latin American nations. The lack of state autonomy from upper classes in Mexico and Brazil has hindered economic adjustment, and, therefore, development, in those countries. Yoffie (1983) and Haggard (1986) argued that the strategies of state actors and the timing and sequencing of policy were also crucial elements.

These critiques were not without merit. Nevertheless, dependency theory made useful, lasting contributions. The concept of the situation of dependency helped to refine our understanding of class structure, class relations and alliances and economic tendencies in Latin America. Analysis of these factors, in turn, promoted a greater understanding of the functions the state had to perform to meet the challenge of underdevelopment. In other words, dependency theory brought attention to what was different about the class structure and state functions of developing countries in relation to more developed nations, whether capitalist or socialist. The emphasis was on examining how different class alliances helped states to fulfil the functions of social dominance and economic growth in specific historical periods. This was an important contribution given the proliferation of competing theories that held that all countries went through similar stages of economic, social and political development: what was adequate for explaining European or North American history was also applicable to Latin America.

According to dependency theorists, a country's class structure and the economic and social problems facing the state had to be understood in terms of the nation's connection to economically advanced countries. Dependency theorists divided the major economic sectors of national economies (industry, commerce, finance and agriculture) into subsectors. These subsectors were then tightly linked to either the international economy or the subsectors that were more exclusively reliant on the domestic economy. Accordingly, dominant classes were categorized into the internationalized and the national bourgeoisie. Urban and rural labor and peasants were also classified in relation to this distinction. Dominant trends in the world economy at different historical periods tended to favor either externally or domestically orientated forces. Equally important, the choices and actions of national states and classes were conditioned by the preferences of the foreign sector, meaning international capital and the governments of developed nations that supported them. These foreign entities had to be accommodated.

In conclusion, dependency theory drew attention to the functions the states of developing nations needed to perform in specific historical periods for the maintenance of capitalist development and social order. Those functions, and the class alliances required for them to succeed, had to be understood in terms of a developing country's insertion in the world economy. Moreover, the foreign sector was an integral component of class alliances and a key shaper of the options open to states. The problems that states face, the difficulties they encounter in fulfilling their functions in market economies, the role of the foreign sector – all of these points are relevant for the political economy of political change. They help to refine key variables in the study of the relationship between development, authoritarianism and democracy. But they tell us little about that relationship. This would be the task of future studies in political economy.

Stages approaches: modernization theory

Dependency theorists had argued that dependent capitalist states could be either democratic or authoritarian. Which prevailed hinged on the resolution of the class struggle in a given country. More or less contemporaneously – in the 1950s and 1960s – modernization theorists took a much more optimistic, and deterministic, view. In their analysis, the causal arrows pointed in one direction only. As countries modernized socio-economically, authoritarian and other traditional forms of government would give way to democracy.

The intellectual underpinnings of this conclusion were diametrically opposed to those of political economy. They were based on the sociology of Max Weber and Talcott Parsons, drew heavily on the systems theorizing of David Easton (1965) in political science and borrowed from the stages approach to economic development of Simon Kuznets (1976) and Walt Rostow (1960). From these sources emerged a stages approach to political development. It emphasized cultural change, diversification of social groups and institutional development through role differentiation and specialization.

Modernization theorists assumed that developing countries would follow the same evolutionary patterns as the economically advanced nations. In the developed nations, they observed that economic modernization had brought about a transition from traditional culture, social organization and political authority to more modern versions. In advanced countries, economic progress produced more rational, goal-orientated thinking, it diversified social structure (giving rise to middle classes, urban industrial and service labor and the transformation of agrarian social relations), and generated institutions to process increased demands and political democracy (Handelman, 1996, 11–14).

Modernization theorists rejected the state both as an idea and as a real construct. Instead, they embraced David Easton's (1965) concept of the political system, which drew heavily on the sociology of Talcott Parsons (1951). Embedded in a socio-economic and cultural environment, political systems consisted of the authorities and institutions of government that handled inputs and delivered outputs. Governments processed inputs in the form of political demands and produced outputs in the form of decisions and actions. Governments also needed legitimate sources of support to function properly. This was a thoroughly US-centric, pluralist approach to the study of government. Government institutions (executive, legislative and judiciary) processed the demands of social groups that cut across class, and they mediated group conflict.

Based on this schema, modernization theorists reasoned that changes in the socio-economic and cultural environment produced changes in political systems as a result of shifts in demands and the sources of support for government. As economically and politically advanced countries integrated developing nations into the world economy traditional agricultural societies would become modern societies. In the process, traditional societies would go through a sequence of changes that mirrored those of Western Europe (Nisbet, 1969).

Rostow (1960), a US economic historian, argued that in their encounter with advanced industrial societies traditional economies were destined to go through five stages in their journey toward development: traditional society, preconditions for take-off, take-off, drive toward maturity and the age of high consumption. Taking a page from this schema, Gabriel Almond and Bingham Powell (1966) suggested that the advanced democratic countries of Western Europe had experienced three political stages. Stage I (akin to traditional society and preconditions for take-off) was the age of absolutism and early nation-building based on traditional, authoritarian forms of government. In stage II (akin to take-off), the successes of liberal capitalism in the nineteenth century bred a process of democratization that established limited monarchies. In the twentieth century the further development of capitalism led to stage III (akin to the age of maturity): full political democracy with greater attention to economic equality through welfare.

Almond and Powell also distinguished between economic, social and political modernization. Economic development referred to productivity increases and more equitable distribution of income. Social modernization involved an increasing division of labor, differentiation of the social structure, urbanization, industrialization, the development of mass education and communication and better health care and welfare. Political development consisted of two dimensions: first, the evolution of 'specialized political executive and bureaucratic agencies capable of setting collective goals and implementing them' (Almond and Powell, 1978: 358); second, 'the rise of broadly articulating and aggregating agencies [such as] political parties, interest groups, and communication media that serve the purpose of relating groups in the population to these goal-setting and goal-implementing structures' (358).

How are economic, social and political development related? According to Almond and Powell (1978: 362–3), in stage I traditional societies are integrated into the world economy. They experience a rise in productivity and some new social groups appear. In order to conduct relations with advanced countries the authorities must consolidate a nation-state and competent bureaucratic institutions of government capable of processing new domestic and international policy demands. For most Latin American countries outside of Central America and the Caribbean, this would be equivalent to the period between independence and the 1860s. In stage II, market economic development intensifies social modernization. These new social groups organize political parties and interest group associations, and they use the communications media and other means to press their demands for greater welfare and inclusion into the political system's decision-making processes. This leads to an opening of political contestation and participation where traditional authoritarian elites share power with new elites. South America and Mexico between the 1870s and the 1960s were thought to be in this stage. After economic 'take-off' and during the drive for economic maturity begins in stage III, full political democracy appears. With poverty vanquished, the question of liberty and distributive issues no longer threaten elites. Some Latin American countries (Argentina, Brazil, Chile, Colombia, Venezuela, Uruguay and Mexico) were thought to be poised for this stage.

Stages approaches: bureaucratic authoritarianism

Modernization theorists took a benign view of the effects that economic, social and political integration into the world economy had on developing countries. Perhaps its greatest utility lay in its optimistic message. First, Third World nations could overcome economic underdevelopment. Second, political stability depended on highly manipulable factors such as the proper structuring of political demands by new social groups, the use of communications media and education. In short, leadership, strategic decisions – action and choices – sufficed to overcome the limitations of both economic and class structure.

The greatest limitation of modernization theory lay in its assumption that developing nations would follow the idealized sequence of economic, social and political stages of advanced, developed nations. Disturbing events in the 1960s severely challenged that assumption. Two of the most economically advanced countries of South America, Argentina and Brazil, succumbed to a new type of military dictatorship. Other economically advanced cases in the Southern Cone, Chile and Uruguay, followed suit in the early 1970s. By the middle of the 1970s, most South American countries were ruled by military governments.

These events caused Guillermo O'Donnell (1973) to stand modernization theory on its head, proposing a reverse determinism. Developing nations were not destined to replicate the developmental path of advanced industrial countries. Greater levels of economic development did not lead to democracy; instead, they were equated with a new form of military dictatorship, a new authoritarianism. O'Donnell dubbed them bureaucratic authoritarian regimes.

O'Donnell built his hypothesis on three different sources: stages theory, Barrington Moore's (1966) class analysis and dependency theory. Of the stage theorists, O'Donnell borrowed the most from A.F.K. Organski (1965). Of particular interest was Organski's idea that political development involved increasing the state's efficiency in mobilizing human and material resources toward national ends. O'Donnell also drew on Moore's notion that there were different pathways to economic and political development. Moore, drawing on the political economy of Alexander Gerschenkron (1962), maintained that the timing of industrialization heavily affected outcomes. Most importantly for Moore, it conditioned the structure of class alliances. This was the key determinant of whether nations took a democratic, authoritarian or communist route to modernity. The later the period of economic modernization, the less likelihood nations would take the democratic path. Dependency theory provided the basis for O'Donnell's thinking about the nature of the Latin American state, class structure, the economic problems facing the state and the functions it had to perform in light of the region's insertion in the world economy.

For O'Donnell, the state stood at the center of the dilemma. In the 1960s, the old economic development model, import substitution industrialization (ISI), had reached a crisis. Substantially more investment was required to go beyond the light assembly stage to the more capital and technologically intensive stages of industrialization, which he called the deepening of industrialization. The problem for the state lay in how to mobilize human and material resources to accomplish this end, an end that fulfilled one of the key functions of the capitalist state: providing for the growth and development of the economy.

At the same time, Latin American states were faced with another problem. They were hard-pressed to fulfil the function of maintaining social order. The evolution of class conflicts under democratic regimes threatened both the dominance of propertied classes and the ability of states to maintain a good business climate. In democracy, populist and marxist social movements, unions and political parties strongly pressured governments and even ran them on occasion. Their redistributionist welfare policies, and sometimes revolutionary stances, threatened the established order.

This threat from below in newly urbanized societies frightened the upper classes. Dependency had generated weak industrial, financial and commercial classes that could not establish economic, social, political and ideological dominance over lower classes. Thus, they allied with landowning elites in defence against populists and would-be revolutionists. This coalition, reminiscent of the authoritarian alliance of industrialists and grain growers in nineteenth century Germany, needed a dictatorial state to impose economic and social order.

Enter the armed forces. The military in Latin America had developed the doctrine of national security. This anti-communist doctrine held that without economic growth there could be no political order. Thus, the military as an institution took it upon itself to provide political order through repression and to undertake conservative economic modernization strategies. They alone, went the argument, possessed the power and authority to mobilize material and human resources for national development. The eventual success of economic strategies would ultimately recast the society and polity and make democracy safe for capitalism. The military also believed that recreating a good business climate would entice foreign investors back to their countries. And, given a situation of dependency, foreign investment was key for economic success.

In short, the military entered into an alliance with upper classes, the foreign sector and middle classes that also felt their lifestyles were threatened by populism. The military would rule as an institution in authoritarian fashion and through modern bureaucratic methods. This set bureaucratic authoritarianism apart from previous styles of military rule, where a charismatic 'strong man on horseback' ruled in a personalized and clientelistic manner.

O'Donnell's theorizing was useful for a number of reasons. Most importantly, it refocused attention on *explanations* for the strong correlations observed between levels of economic development and their impact on political development. Thus, his work not only challenged modernization theory but also refocused attention on how the structural conditions of dependency affected class conflict and political change. Moreover, O'Donnell firmly anchored class tensions in the state. Whether those tensions destroyed a state or not depended on whether the upper classes, the military and the foreign sector believed the state capable of fulfilling its functions with respect to capital accumulation and social control. Given the acuteness of the crisis of economic development and the political strength of populist political forces, O'Donnell pessimistically concluded that repressive military governments, not democracy, were Latin America's political future.

O'Donnell's critics also made telling observations. The most important one was the unilineal direction of change: that higher levels of economic development produced political authoritarianism. If that was the case, why were Colombia and Venezuela still democracies? Moreover, how could one explain the fact that bureaucratic authoritarian military governments, such as the Argentine government, were also subject to collapse? Apparently military governments could fail in their modernizing missions.

Another important critique focused on the generalizability of the theory (Hirschman, 1979). Exhaustion of import substitution industrialization as a factor applied best to Brazil and Argentina in the early 1960s, two cases on which O'Donnell drew heavily for his hypotheses. There were problems in extending the thesis to Chile and Uruguay, the closest cases. Chile had had sluggish economic growth for a number of years. But hyperinflation and economic crises were politically induced during the socialist government of Salvador Allende in the early 1970s. By the same token, Uruguay's economy was stagnant, but not in crisis (Kaufman, 1979). More tellingly, the thesis does not help to explain the political travails of civil war-torn Central America. Nor is it useful for explaining the persistence of the one-party-dominant Mexican state. Critics also noted that the economic and social objectives of military governments were far from uniform. This was especially the case in Chile (1975–89) and Argentina (1976–82) where free-market economics, not the deepening of industrialization, became the primary economic objective (Serra, 1979; Whitehead, 1985; Schamis, 1991). In short, critics uniformly noted a lack of fit between the presence of necessary explanatory factors, their sequencing and their effects.

Modernization theory revisited

The wave of democratization that swept Latin America underscored some of the bureaucratic authoritarianism literature's weaknesses. With very few exceptions, between the end of the 1970s and the middle of the 1990s authoritarian regimes gave way to democracy. This trend invalidated the argument that higher levels of socio-economic development had an elective affinity for military government. Democratization also revealed that the literature on bureaucratic authoritarianism had difficulty explaining those transitions, shortcomings which O'Donnell himself recognized (O'Donnell and Schmitter, 1986), in the wake of attempts to do so by Philip O'Brien and Paul Cammack (1985), Cardoso (1979) and O'Donnell himself (1979).

These new directions in political change spawned a literature that drew heavily on a

ed version of modernization theory to explain them. These works built heavily
ier attempts to explain the breakdown of democracy, when they had argued that
s from the left and center of the political spectrum had made ill-advised choices. Too many demands had overloaded government's ability to process them. This so reduced the effectiveness and efficacy of government that the military stepped in (Linz, 1978; Crozier *et al.*, 1975).

Samuel Huntington (1991) and Larry Diamond, *et al.* (1989) provide two of the most representative works of modernization-theory-based efforts to explain recent transitions to democracy. Both studies identified numerous preconditions for democracy drawn from modernization theory. With some differences in accent, they emphasized the significance of levels of socio-economic development and political culture. Huntington argued that the midrange of economic development, measured as per capita income of $1000 to $3000, correlated very highly with countries engaged in transitions to democracy. Diamond *et al.* added that steady growth and broad distribution of wealth were more important than the fact of high levels of socio-economic development alone.

In any case, in these works, socio-economic development – especially in the 1950s and 1960s – is key for the promotion of civic culture, raising education, broadening access to communications media and providing resources for distribution. Having resources to distribute nurtures democratic values such as tolerance and compromise. Economic development, according to Huntington, is also key for social change, especially the emergence of middle classes, the harbingers of democracy.

However, Huntington and Diamond *et al.* took earlier critiques of modernization to heart. Socio-economic development by itself does not produce democracy. Other factors mediate, such as the policies of external actors – whether they are supportive of dictatorship or democracy. Especially significant for Latin America was the shift in support for military government to democracy by the United States with the demise of the Cold War. Huntington further argued that authoritarian regimes had to go through a legitimacy crisis before democratization became possible. He also noted a Christian religious culture was more propitious for democracy than other religious cultures.

Another significant difference between these works and the original modernization theorists involved two issues: the specification of political culture as a variable and concern over the problem of democratic consolidation. In the revised theory, transitions to democracy and the prospects for democratic consolidation were inextricably linked to changes in the political culture of elites. This was an important departure from the view that socio-economic modernization inevitably fostered mass civic culture, which was necessary for democracy to flourish. Now only elites (whether socio-economic, political or of middle class and labor organizations) needed to value the democratic virtues of tolerance, moderation in political demands and of compromise.

In addition to these sources of support for democratic political systems, Diamond *et al.* added an institutional dimension for the consolidation of democracy. Building on previous work by Linz (1978), they maintained that institutional factors must also be considered when assessing the prospects for the consolidation of democracy. These referred to the institutions that processed society's political demands in the political system. They argued that prevailing presidential systems exacerbated confrontation; thus, they should be replaced by parliamentary systems because building governing coalitions requires moderation and compromise (Linz and Valenzuela, 1994). The same rationale applied to prescriptions for changes in electoral systems.

In sum, the revival of modernization theory addressed three of the major critiques leveled against it in the past. In contrast to previous writings, they now argued that socio-economic development, while necessary, was not sufficient for the establishment of democracy. International factors needed to be considered. The conditions conducive to the consolidation of democracy had to be understood and changes in the specification of political culture along with the institutional conditions were key to such understanding.

From this perspective democracy in Latin America is, in the first instance, the result of increased economic development – measured in per capita income – between the 1950s and the 1990s. Moreover the moderation of political elites, especially of populist and leftist parties in Brazil, Chile, Argentina and many Central American countries has been key for the wave of democratization that has swept the continent. US support for polyarchy has also been crucial, especially in Chile, the Andean countries, and the Spanish-speaking Caribbean.

Nevertheless, the approach still suffered from some of modernization theory's original shortcomings in addition to some new ones. One of the lingering problems was that the preconditions – such as socio-economic development and religious doctrine – do not explain why a country has a democratic or authoritarian form of government. They may be correlated with democracy under the right circumstances, but the causal linkages are still missing and the right combination of factors is not understood. A second problem was the number of factors used to understand transitions to democracy and democratic consolidation. Recognition of the complexity is welcome, but there was little theoretical integration of the variables; the relationships between variables are not clarified. In part, this was because they did not systematically compare cases to establish whether the presence or absence of the same variables had the same effect.

By the same token, the emphasis on the political culture of elites offers hope that leadership can forge democratic outcomes where structural conditions may not be very favorable. But these studies offer no explanation for why political elites embrace democratic values. Are they voluntary decisions or the product of historical pressures (Espinal, 1991)?

Recent political economy theorizing

Recent theorizing in political economy concurs that economic development does not necessarily produce either democratic or authoritarian state forms. Unlike dependency theory, however, these studies frequently focus on the dual problem of economic and political change. What conditions allow Latin American and other developing countries to encourage free-market economic reforms and democratic states simultaneously? In their analysis of these issues, political economists address a number of the questions left unanswered by revived studies in modernization theory.

In one or another form, all of these studies examine the interaction of three clusters of variables: class-based social groups, degree of state autonomy and international factors. Beyond this convergence, two strands exist, depending on the weight assigned to these variables. One strand argues that class alliances essentially determine whether a country has a democratic or authoritarian state form. The other maintains that political institutions are especially important for the consolidation of democracy.

Jeffry Frieden (1991) offers an elegant explanation for the connection between the Latin American debt crisis, economic change and the recent wave of democratization. Frieden argues that international pressure for free-market economic adjustment was relatively constant. Therefore, domestic class alliances, the extent of class conflict and the degree of state autonomy were more important than external factors in both economic and political change.

For Frieden, capitalists are the key class-based actors. Because of their control over the public functions of investment and employment (also called structural power) their support for any particular state form is crucial for political stability. Moreover, in countries with moderate tension between labor and business (low class conflict), state managers should heed the policy preferences of dominant economic groups, or else economic elites may turn against the government. Conversely, in situations of high class conflict, state managers may have more autonomy in policy decisions. The dominant sectors of the upper classes are more willing to submit to economic policy changes because they need the state to keep social order.

For the post World War Two period, Frieden divided Latin American capitalists into two categories. One category consisted of business elites with assets concentrated in fixed-asset sectors. This meant their investments were sunk in plant and equipment, usually producing for domestic markets given their rise during the heyday of ISI. Because of the ubiquity of this development model in Latin America, Frieden assumes they are the dominant faction of capital in virtually all cases. The other category held most of their investments in liquid assets, meaning the financial sector. Frieden argues that the Brazilian and Argentine economic elites, concentrated in fixed assets, supported democratization because the military governments refused to adopt the sectoral economic policies demanded by dominant business and landowning groups. The implication is that their endorsement of democracy strengthened other political and class forces that were struggling for democracy and took away the military government's reason for being. Conversely, Chile had high degrees of class conflict. Thus, the government's turn to free-market policies did not alienate the dominant faction of capitalists and landowners (fixed-asset owners) from the regime. They still sought its protection from labor and leftists, which gave the military government a higher degree of autonomy. Again the implication is that the absence of a multiclass alliance for democracy retarded the end of authoritarianism in Chile. And, when the transition to democracy took place, it did in accordance with the military's agenda.

Frieden concentrated more on the political economy of economic change, whether countries shifted to free-market policies or not. Although he extended the logic to political change, Frieden confined his analysis to the regime loyalties of capitalists, assuming that their regime preferences weighed heavily in how transitions from authoritarianism unfolded. The limitation here is that by themselves the regime loyalties of upper classes did not determine the outcome. Mexico remained largely authoritarian despite the pro-democracy posture of business people. Chile eventually became democratic despite business support for the military. In other cases, such as Brazil, Argentina, Peru, Bolivia, Ecuador and the Dominican Republic, transitions to democracy required broader class alliances.

Dietrich Rueschemeyer *et al.* (1992) address those issues. Following Barrington Moore (1966) they argued that democracy or dictatorship are explicitly the result of class alliances. The power of those alliances is mediated by state autonomy and international factors. The authors stress that capitalist development weakens labor-repressive authoritarian classes, especially traditional landowners. It strengthens classes that have an interest in democracy: the bourgeoisie (capitalists), middle classes and labor.

But the mere emergence of those classes is not enough to explain whether democracy or dictatorship will prevail. In Europe, the bourgeoisie had an interest in creating limited contestation for power in systems that included them in policy-making but excluded other social classes. It was the organization of labor that expanded capitalist struggles for protected democracy into fully-fledged democracy. However, the bouts with fascism suggest that high degrees of state autonomy are not propitious for democracy. When the state is highly independent of society and cohesive in its institutions and ideology it may overpower civil society, especially if the latter is ideologically and organizationally weak and fragmented.

The situation was slightly different for Latin America. The bourgeoisie, like its European counterpart, also had an interest in limited democracy. But it was the struggles and organizational capacity of the middle class, not labor, that turned the tide in favor of full democracy. Labor in Latin America was simply organizationally too weak to perform the role. Thus, a middle class that did not feel threatened by labor could ally with it, and some sectors of the bourgeoisie, to build democracy. Conversely, when middle classes make common cause with upper classes, authoritarianism might result. Highly organized and institutionally and ideologically cohesive autonomous states may still threaten bouts of authoritarianism, especially if civil society is weakly organized.

The impact of transnational forces on the prospects for democracy or authoritarianism depends, in the first instance, on which classes the colonial metropole supported, that is the colonial power's impact on class formation. Democracy is more brittle in countries formerly ruled by Spain because that power supported landed elites. This was the class least interested in democracy because, following Moore (1966), extensive landholdings and extractive activity required greater amounts of labor repression. Conversely, a metropole's support for democracy usually helped it to flourish in the colony or former dependency. Thus, former British colonies have more stable democratic states than have Latin American ones. By extension, US support for democracy as of the middle of the 1980s also strengthened the movement to democracy in the region.

Rueschemeyer *et al.* developed a framework for analysis based on the relationship between class structure and organization, state institutions and transnational forces. Their aim was to tease out how long-term processes of capitalist development affected those variables and their impact on state form. This approach, however, has difficulty with shorter-term analysis of the dynamics of democratic consolidation. Moreover, the relationships between such complex clusters of variables over long periods of time are hard to keep track of in individual cases.

Stephen Haggard and Robert Kaufman (1995) developed a framework for analysis more suited to short-term studies of transitions from authoritarianism to democracy and the prospects for democratic consolidation. They also squarely address the question of the socio-economic and institutional conditions under which elites embrace democratic values. These were the very issues the new modernization theorists reviewed (see above) and that the increasingly popular strategic choice analysts (O'Donnell and Schmitter, 1986) had ignored. Haggard and Kaufman grounded their model of authoritarian breakdown in the dictatorship's economic performance and elite cohesion. Transitions to democracy were most likely in cases with poor economic performance. Economic crises tended to undermine the state managers' ability to purchase the compliance of social groups and elite fragmentation weakened the state's capacity to manage the crisis. Under these conditions of regime decomposition, the political organizations of middle-class groups and labor could press for full political democracy and even ally with some disgruntled upper-class groups; but only if they moderated their policy platforms. In cases without economic crisis and where elites remained cohesive in support of military government, authoritarians were able to hold out longer and impose more restrictions on the democratic regime that was to follow them.

The prospects for the consolidation of democracy depended on two institutional factors according to Haggard and Kaufman. The first one hinged on the structure of executive authority. Centralized executives were better because they facilitated the initiation of economic reforms necessary to resume economic growth. Renewed economic growth is crucial because long-term political stability is closely linked to healthy economic development. Economic crises breed political instability. Having initiated economic reforms, state managers must find or build support coalitions for the implementation of those reforms in order to sustain them. Without such support coalitions economic reform will most likely fail and a downward spiral of political instability might ensue. The second institutional factor deals with the problem of how to mobilize such support coalitions. Here Haggard and Kaufman suggest that the structure of the party system is key. Specifically, they advocate building two-party systems. These tend to reinforce compromise and moderation.

Haggard and Kaufman provided a crucial insight. The consolidation of democracy requires a shift in both strategic thinking (moderation and compromise to achieve democracy) and an institutional change that reinforces it. Additional work, however, is required on a few fronts. Most importantly, institutional arrangements can be overwhelmed by social mobilization and conflict. Even two-party systems have been overwhelmed in Latin America, as occurred in Uruguay in 1973. Economic good times by themselves cannot generate social compliance; the distribution of wealth is also crucial. Hence political economists also need to address the conditions under which social justice can be advanced without threatening elites to the point where they withdraw their support for democracy. The same applies to efforts to increase social participation in policy-making, which has been severely curtailed in many Latin American democracies. The various efforts to protect democracy from the 'masses' has produced a burgeoning literature on Latin America's penchant for democracy with adjectives (Smith *et al.*, 1994; O'Donnell, 1992; Loveman, 1994; Collier and Levitsky, 1997).

Conclusions

Since the 1980s, a consensus across methodological perspectives emerged that there is no unilineal relationship between economic development and the form of the state. Economic modernization does not necessarily lead to democracy or dictatorship. However, the analysis of which prevails varies according to modernization or political economy perspectives. Perhaps the greatest value of modernization theory, especially in its current incarnation, lies in its emphasis on elite strategies and the structure of political institutions. We have the sense that we can control them; we believe they are a matter of choice. Whether political elites value democracy, their tolerance and their willingness to compromise are clearly important. Whether political institutions mediate or exacerbate social and political conflicts is also a crucial consideration.

But the modernization approach does not address why political elites make the choices they do, whether for good or ill. They cannot answer how international economic or political factors and power relations among social groups help to shape the strategies of elites. Nor does the modernization approach help to distinguish how the unique features of regions affect political outcomes. This is because it aspires to be a generalizing theoretical approach. Thus, analysts use cases only to illustrate the influence of variables rather than as systematic units of comparison capable of validating theoretical propositions.

Political economy approaches are better suited for answering such questions. They recognize the existence of the state and its role in society. Understanding the functions of the state in capitalist economies provides a sound foundation for analysing the effects of economic and social tensions on political stability over time. New directions in the disaggregation of class-based social groups adds finesse to analysis. It allows for more fluid and varied coalition-building. The addition of factors such as international variables and the structure of political institutions has also advanced our understanding of the prospects for democracy and the danger of backsliding to authoritarianism.

Although most analysts focus on either modernization-inspired or political-economy-orientated approaches, this has not been the case of one of the most influential Latin American scholars, Guillermo O'Donnell, who has spanned the full methodological spectrum. In the end he has tried to blend the two. O'Donnell first used political economy to turn modernization theory on its head. Yet, by the early 1980s he felt that this approach could not explain the process of transitions from authoritarianism that had begun to sweep over Latin America. Somehow democratic transitions were occurring despite the fact that economic hard times had beset nearly every country as a result of the Latin American debt crisis. As a result, he embraced a strategic choice approach. Drawing mainly on the cases of Argentina, Brazil, Chile and Uruguay he focused on the revaluation of democracy by political elites. For analytic purposes he categorized actors into groups that were either for military regimes or opposed to them. Transitions to democracy began when softliners within military regimes won over hardliners. Moderate centrists within the opposition then found a way to accommodate softliners, capitalize on their weaknesses and expand political liberalization into democratization. Their willingness not to raise socio-economic issues went a long way to calm the fears of the supporters of military government.

By the 1990s, O'Donnell (1992) became disenchanted with the analysis of transitions to democracy. He felt the focus should shift to the study of the type of democracies that were emerging in Latin America. Too many had features that restricted contestation for power and the participation of subordinate social groups in policy-making. This was true for Chile after Pinochet, Argentina under Menem, Brazil since 1985 and Peru since 1979, to name but a few cases. In these analyses, O'Donnell blended modernization-inspired theorizing with political economy. In his view, democratic states had once again to address socio-economic issues if they were to survive. This conclusion certainly fitted with either theoretical approach. However, he drew more from political economy when he analysed the use that upper-class groups and political elites made of political institutions that restricted the organization and political impact of lower-class groups. Here, O'Donnell joined a number of scholars who were making the same points from a more explicitly political economy perspective (Smith *et al.*, 1994; Collier and Levitsky, 1997).

Regardless of whether one favors modernization-inspired or political-economy-inspired approaches, there is one element of heavily criticized dependency theory that is worth rescuing: the analytical importance of the region's historical specificity, that is, an understanding of how the region's unique economic, social, political history and the region's particular insertion in the world economy affect outcomes. Theory is supposed to be generalizing. Therefore, it is perfectly acceptable to focus first on a few general core variables (or cluster of variables), such as the political culture of elites, the structure of political institutions, the relative autonomy of the state, class structure and the impact of the international economy and politics. These variables, however, are perhaps best thought of as guideposts. How they are constructed, the meaning actors ascribe to them, the situation of countries internationally – all of these are region-specific. Latin America's history is different from that of Asia or Africa (not to mention Europe).

In this respect, rich, historically textured comparisons of carefully selected cases probably shed more light on the prospects for democratic consolidation than do multivariate analyses of large numbers of cases drawn from all over the globe. The latter may establish some plausible correlations, but the former will provide more explanation for why specific countries have democratic or authoritarian states. The same applies to regional and cross-regional analyses. The economic and social challenges facing state actors – as well as the institutions, groups and traditions involved – are region-specific. Analysts ignore that specificity at their peril as they interpret trends and offer options for pressing problems.

FURTHER READING

Chalmers, D. A., Vilas, C. M., Hite, C. (eds) 1997 *The new politics of inequality in Latin America: rethinking participation and representation*. Oxford University Press, Oxford. The issues of inequality, participation and representation have been at the heart of tensions that contributed to the breakdown of democracy in the past. Therefore, how they are dealt with today is vital for democratic consolidation as discussed in this book.

Drake, P. W. 1996 *Labor movements and dictatorships: the Southern Cone in comparative perspective*. The Johns Hopkins University Press, Baltimore, MD. This book analyses the role of organized labour in democratic breakdowns, authoritarian regimes and transitions to democracy. This social group is a crucial component of organized civil society whose condition affects patterns of democratic participation and representation.

Huber, E., Safford, F. (eds) 1995 *Agrarian structure and political power: landlord and peasant in the making of modern Latin America*. Pittsburgh University Press, Pittsburgh, PA. This volume explores how the historical legacies of large landholdings and agrarian class relations have affected democratic versus authoritarian trajectories in Latin America.

Kincaid, D., Portes, A. (eds) 1994 *Comparative national development: society and economy in the new global order*. University of North Carolina Press, Chapel Hill, NC. Economic development affects political change. Thus, we must understand how globalization influences economic modernization. Drawing on past debates in development economics, this book offers an excellent introduction to those themes.

Lijphart, A. 1984 *Democracies: patterns of majoritarian and consensus government in twenty-one countries*. Yale University Press, New Haven, CT. Although not focused on Latin America, this is one of several works by Lijphart that have heavily influenced Latin Americanists. It is a well-constructed analysis of how the structure of political institutions affect democratic stability.

Smith, P. H. 1995 *Latin America in comparative perspective: new approaches to methods and analyses*. Westview Press, Boulder, CO. Understanding Latin America involves more than analysing what is going on. It also requires the rigorous application of new concepts and methods. This book is an exciting and highly readable introduction to comparative method, concept formation and research ethics.

Whitehead, L. (ed.) 1996 *The international dimensions of democratization: Europe and the Americas*. Oxford University Press, Oxford. Transitions to democracy are deeply influenced by the international context. This volume explores the role of international relations in the formation of democratic states in Latin America and Europe.

3

THE NEW POLITICAL ORDER IN LATIN AMERICA: TOWARDS TECHNOCRATIC DEMOCRACIES?

PATRICIO SILVA

This chapter focuses on what seem to have become permanent traits of the new democratic order which has been emerging in Latin America since the early 1980s. It represents an attempt to go beyond the 'transitional' scope followed by most democratization studies so far. Indeed, until now most attention has been centered around the specific ways in which the transition from authoritarian rule to democratic restoration has taken place. However, little has been done in attempting to characterize the chief features of the new democratic order emerging from that process.

When dealing with the current modernization process in Latin America it is essential not to overlook the conspicuous fragmented nature of this phenomenon, particularly with respect to its geographical, social and cultural dimensions. For instance, huge contrasts can be found in degrees of modernity achieved not only between countries but also between regions within a single nation. In addition, the extreme social disparities existing in most Latin American countries (between 'the haves' and 'the have nots'), have led to very unequal levels of access to the fruits of modernization for the several social strata. Furthermore, the modernization of society has been mainly concentrated in the areas of production and services (use of modern technologies and access to modern communication facilities). In sharp contrast to this, scarce transformations have yet taken place in patterns of income distribution, more equal access to basic services, more respect and guarantee for citizenship and other civil rights for marginal groups and ethnic minorities, and so forth. As I try to show in this chapter, this fragmentation has led to a quite differentiated impact of the modernization process on the different segments of the population, resulting also in different social and political responses to the new democratic order.

Neoliberalism, Modernization and Democracy

Since the restoration of democratic rule in the region, the relation between neoliberalism, modernization and democracy has been an uneasy one, being plagued by continuous tensions. The inability of most of the new democratic governments so far in successfully legitimating both economic neoliberalism and its related discourse of modernization has largely been the result of three factors.

First, we must not forget that both the neoliberal economic project as well as its accompanying discourse of modernization were applied in countries such as Chile, Argentina and Uruguay by military regimes, that is, by forces opposed to democracy. The *ideé force* behind the application of the neoliberal policies during those authoritarian regimes was the achievement of the '*modernization* of society'. This modernization project was translated in practical terms as the achievement of macro-economic goals such as the privatization of the economy, the reduction of state bureaucracy, the liberalization of the markets to foreign competition and the strengthening of the export-orientated economic sectors. What we observe today is that, in fact, most of the current democratic governments have largely continued to visualize modernity along the same macro-economic lines as did the former military governments.

This element of continuity from the previous regime makes it now extremely difficult for the new democratic authorities to legitimate neoliberal policies. The fact is that the implicit call of the new democratic governments for accepting the neoliberal economic guidelines is now unpalatable for the social and political forces who fought against the military regimes. They experience this reality as an inadmissible legacy of the past, and consequently they resist accepting the neoliberal model as being the economic engine which must sustain the democratic fabric for the coming years.

Second, an important obstacle in the attempts to legitimate neoliberal policies in the region is that in many countries (such as Costa Rica, Venezuela and Ecuador) the adoption of neoliberal economic policies and its correlated modernization discourse have been regarded by broad sectors of the population as being externally imposed. International financial institutions such as the International Monetary Fund (IMF) and the World Bank have been constantly denounced by several Latin American political forces and social movements for asserting since the early 1980s an unprecedented pressure on Latin American governments to apply structural adjustment policies (Petras and Morley, 1992). These 'dictates' from the North should have left almost no room of maneuver for the new democratic governments in the formulation and implementation of their own socio-economic agendas (Green, 1995). Paradoxically, this thesis (and the resulting rejection of the neoliberal agenda) has found support in Latin America among quite heterogeneous political and social groups; these include the left, certain religious groups, some nationalist sectors (both inside and outside the army) and some entrepreneurial circles (afraid of free-market policies and especially of foreign competition).

The lack of popularity and support for neoliberal policies in many countries of the region has led to a quite peculiar situation. So while only very few political forces in Latin America dare openly to express their unconditional support for neoliberalism, neoliberal economic policies have become *de facto* dominant in the continent. In other words, almost no political force is seriously trying to elaborate an ideological legitimation for the new neoliberal order. One of the few exceptions in this respect can be found in Chile, where neoliberalism is not only enthusiastically defended by the right-wing parties and entrepreneurs but even, though in more implicit terms, by some sectors of the left (Moulian, 1997).

Finally, another major difficulty faced by the new democratic governments in getting their neoliberal economic policies accepted by the people is related to the great social expectations generated by the restoration of democracy in the region. For many Latin Americans, democracy still means in the first place the existence of a government which really cares for the needs of the majority (that is, the masses) and which has to be actively and genuinely engaged in the struggle against poverty and social injustice. From this follows the expectation that in the new democratic era an active role for the state is required to meet the consequence of the immense 'social debt' left by the former authoritarian regimes. What many people have experienced instead is that the restoration of democracy has been accompanied by a further abandonment by the state of its traditional social tasks, resulting during the past decade in a dramatic deepening of the social inequalities in the large majority of Latin American countries. This lack of 'social dimension' in most of the neoliberal economic programs applied in the region has been severely criticized by leading intellectuals, as well as by the Church and nongovernmental organizations (NGOs).

From the very beginning, the restoration of democratic rule since the early 1980s became strongly conditioned by the implementation of structural adjustment programs by an increasing number of countries, following Mexico's dramatic announcement in August of 1982 that the country was unable to keep its international financial commitments on the repayment of its foreign debt. This marked in the entire region the initiation of a profound shift in development strategies. Following the outset of the debt crisis the traditional pattern of import substitution industrialization (ISI) was strongly criticized by domestic and international actors who demanded the adoption of market-orientated reforms. Initially, many Latin American countries decided to apply nonorthodox stabilization program (for example the Austral Plan in Argentina, the Cruzado Plan in Brazil, the Inti Plan in Peru) in an attempt to diminish the social costs of these austerity policies. By the late 1980s, however, it had became clear that these and other stabilization programs had failed to put an end to the crisis and to provide the expected economic recovery. As Green (1995: 69) indicates,

> once 'easy' heterodox solutions had been discredited, neoliberalism spread rapidly across the region; these were the years when the long-term structural adjustment of Latin America's economy gathered pace. Trade liberalization, government cutbacks, privatization and deregulation have since then become the norm in almost every country.

While a general consensus emerged among most of the Latin American governments about the need to abandon the traditional pattern of state-led industrialization and to modernize their economies by adopting neoliberal free-market policies, they have had immense difficulties in translating the goal of modernization into *political* terms. For instance, rarely has an attempt been made to provide a clear explanation about how modernity is related to democracy or to indicate the concrete differences existing between the modernization project of the former authoritarian regimes and the new democratic authorities. Both the pre-eminent role of structural adjustment policies under the new democracies and the lack of a clear political project have resulted in the *invasion* of neoliberal economicism into the political realm. In this manner, official politics in contemporary Latin America has tended to lose its own dynamic, having been in many cases reduced to a functional mechanism for the implementation of the neoliberal economic project.

THE DEPOLITICIZATION OF SOCIETY

Contrary to what was broadly expected, in most Latin American countries the restoration of democratic rule was not followed by a strong 'political resurrection' of civil society. On the contrary, one of the most striking features of the new democracies has been the growing depoliticization of society and the marked absence of national political debates. This phenomenon of increasing depoliticization of Latin American societies has been the product of a complex blend of past and current political experiences faced by the people, as well as a consequence of the neoliberal modernization project and its ideological impact on the people's political behavior.

The origins of this process of political deactivation have to be sought during the former authoritarian regimes, when systematic repression against any independent political expressions inaugurated a dark period of 'forced depoliticization'. Paradoxically, state repression eventually led to the emergence of a firm response from certain sectors of civil society. This was expressed in the germination of active social movements and the creation of many NGOs that defended human rights and attempted to ameliorate the social conditions of the rural and urban poor.

At the same time as the use of physical repression, military governments in countries such as Chile, Argentina and Uruguay attempted to convince the population that 'politics' was synonymous with subversion, chaos, decadence and corruption. For this purpose, the military simply exalted and manipulated the feelings of discontent with politics and politicians, which were already entrenched among some parts of the population as a result of the general political and economic crisis which preceded the arrival of the military to power.

The extremely repressive nature of military governments also convinced many individuals that to become involved in politics could lead to big problems, as one risked not only one's own life but also the physical integrity of family members and friends. Other people, after so many years of official anti-political indoctrination (through mass media and education), became finally convinced that politics was 'indeed' intrinsically perverse. So, although at the end many people repudiated the systematic violation of human rights by the military and demanded an immediate re-establishment of the rule of law in these countries, the restoration of democracy did not completely eliminate the deep apprehensions and mistrust that had been engendered in the previous decades against politics in general, and political parties and politicians in particular.

Another factor contributing to the further depoliticization of Latin American societies has been the negotiated nature of most transitions, as restoration of democratic rule was achieved following a series of bargains between the democratic forces and the military authorities (Casper and Taylor, 1996). During these negotiations the opposition forces involved tacitly or explicitly agreed in not actively encouraging political effervescence among the people. Furthermore, several democratic political leaders came to regard the continuation of the political demobilization of the masses as a prerequisite for achieving an ordered and peaceful democratic transition and to guarantee governability under the new democratic scenario.

Strong calls for political moderation also came from academic circles. A group of prestigious political scientists, for instance, offered in an influential four-volume study detailed practical advice to Latin American civilian leaders about how to minimize the levels of political instability which usually accompany transitional processes. In the last volume, O'Donnell and Schmitter recommended, among other things, to provide the armed forces with an honourable role in accomplishing national goals. Meanwhile, political parties were asked to leave behind their traditional role as agents of mobilization to become instruments of social and political control of the population (O'Donnell et al., 1986: 32, 58). They furthermore recommended that democratic leaders should not affect the interests of the dominant groups, and not threaten the institutional existence, assets and hierarchy of the armed forces. With respect to the left-wing forces, they advised them to accept the political restrictions of the transition process, leaving them only 'to hope that somehow in the future more attractive opportunities will open up' (69).

The 'cupular' or top-down nature of the democratization process, which has been often portrayed as an 'elite settlement' (Higley and Gunther, 1992), produced a deep disillusion (and a subsequent demobilization) among many people, such as supporters of left-wing parties and members of social movements who had expected a more participatory process of democratic reconstruction. This, together with the conscious decision of political leaders of maintaining the legacy of political demobilization, has certainly played an important role in the further depoliticization of society following the restoration of democracy in the region.

Following the restoration of democratic rule, the expectation among popular organizations was that the new authorities would bring to justice those responsible for human rights abuses during the military governments. However, the specific way in which the new democratic governments finally dealt with the highly controversial question of human rights' abuses also contributed to political demobilization of the population, as a consequence of considerable disappointment at the results. In Brazil, for instance, there was an implicit agreement between politicians and the military not to make the prosecution of human rights violators a political issue. In Argentina, the Alfonsín government passed in 1987 a *punto final* amnesty law by which prosecutions of most lower-rank military for human rights abuses were declared ended on the grounds that they had simply carried out orders. In Chile, the Truth and Reconciliation Commission, established by the Aylwin government in 1990, brought a detailed report about the human rights abuses during the Pinochet regime. However, the existence of the self-decreed amnesty law of 1978 protected the military from possible trials. An amnesty law was passed in El Salvador in 1993 as a part of the peace agreements between the government and the guerrillas. In countries such as Guatemala and Peru, the human rights issue remained largely unresolved (Jelin and Hershberg, 1996). The practical impunity obtained by the military in most of the countries produced deep disappointment among significant sectors of society, resulting very often in indifference towards the government and politics in general.

Another factor which has helped to reinforce apoliticism among the Latin American population has been the traumatic effects of the hyperpoliticization experienced in the past. The high degree of social and political confrontation preceding the military coups in the Southern Cone countries, together with the institutionalization of fear and repression during the military regimes, has now generated a kind of 'political exhaustion' among the generations who consciously and actively lived those years. The same is true for people who for years have suffered from political violence resulting from open or disguised civil wars such as in El Salvador, Guatemala, Nicaragua and Peru. The endless effervescence, the destabilization of daily life, produced in the long term a distressing psychological situation and an unbearable (political) fatigue. Under these circumstances, the disconnection with politics has become for many people a kind of personal 'survival strategy'.

The increasing 'social democratization' experienced during past years by broad sectors of the Latin American left is certainly another factor in contributing to the political deactivation of civil society. Most left-wing parties have explicitly abandoned the objective of revolution to replace it with the search for gradual and consensual changes in the struggle against poverty and social inequalities (cf. Vellinga, 1993). This has been the case with parties such as the Brazilian Social Democratic Party (PSDB), Mexico's Party of the Democratic Revolution (PRD), the Chilean Socialist Party (PSCh), and Venezuela's Movement to Socialism (MAS). As Castañeda concludes, 'As left-of-centre reformism in Latin America is pushed to the fore and transformed into a meeting point for many other historical currents of the Latin American left, it is also being driven to the centre' (1994: 174).

In the new democratic order, political parties have, in general, dramatically lost their appeal amidst the population. This has been in part another aspect of the heritage from the authoritarian era, as repression severely hit and dislocated political organizations. This resulted, in many countries, in the establishment of a substantial distance and mutual alienation between political parties (most operating clandestinely) and civil society, characteristics that continued after the restoration of democracy. The rendezvous between parties and the electorate after so many years of disconnection has been full of surprises. For instance, many parties had experienced dramatic ideological changes during the authoritarian period and had become almost unrecognizable to the population in the democratic transition. In contrast, other organizations maintained their pre-coup postulates almost unaltered, and hence became anachronistic in countries which in the meantime had experienced profound transformations in their demographic, social and economic structures.

The depoliticization of society has led to the breakdown of historical party loyalties as during the past years a significant segment of the Latin American electorate has not voted for political parties or political projects but for specific *individuals*. The presidential victories of political outsiders such as Collor de Mello in Brazil and Alberto Fujimori in Peru are cases in point. As Little (1997: 191) points out,

> both . . . came out of complete obscurity as self-proclaimed saviours of the nation and their promises of a clean sweep of the political stables clearly struck a chord with the electorate. Collor . . . [and] Fujimori . . . symbolize a popular distance from the old party politics.

Both politicians fully compensated for the lack of firm party structures with the effective use of television. In current Latin American politics, television has replaced parties, unions and 'the streets' as the most important instrument for creating (and destroying) the public 'images' of the members of the political class. Political leaders no longer speak to the crowd, but address directly the atomized community of millions of television spectators. Paradoxically, this modern medium could help to generate a revisited version of old-fashioned personalistic politics. Fujimori, for instance, has been able to combine a smart use of the modern media (as he did during the hostage crisis of the Japanese embassy in Lima) with old-fashioned populist methods (such as frequent visits to urban shantytowns and rural communities, distributing presents and promising solutions for people's problems).

POLITICAL LEGITIMATION, CONSUMERISM AND MACRO-ECONOMIC STABILITY

Most of the new democratic governments have attempted to counterbalance their lack of a political project by putting special effort into expanding the level of consumption among certain segments of the population. One of the main arguments used by the authorities in legitimating the application of policies directed to market liberalization was the promise of full access to better and cheaper foreign products. Indeed, after many years of applying neoliberal economic policies, the upper and middle classes in countries such as Mexico, Argentina and Chile have acquired very sophisticated patterns of consumption which has resulted in the configuration of veritable 'consumer societies'. The increasing presence of foreign consumer goods in most Latin American capital cities has in many cases radically transformed the physiognomy of the streets, as thousands of new cars have invaded the roads and a large number of giant shopping centres (malls) sell products from all over the world.

Although the main beneficiaries of this pattern of increasing consumption have certainly been the high-income groups, one cannot underestimate the extent to which the rest of the population has also participated in the consumption of foreign goods. Many people who actually cannot afford these products have obtained access to these goods by contracting consumer credits, or more often by making use of payment facilities (in monthly terms) offered by most large shops and stores. While the high-income sectors obtained access to modern European cars, to very sophisticated US or Japanese electronic products and to tourism abroad, the poorer segments of the population acquired radios, television sets, battery watches, foreign clothes and trainers, products to which they previously had less access.

In fact the idea of replacing 'politics' by increasing consumption constituted, together with repression, one of the central mechanisms used by the former military regimes to depoliticize society. In the authoritarian conception of modernity 'liberty to consume' was intended to replace political liberty in an effort to deactivate civil society politically and to obtain the required civilian support for the military rulers. Chile is a clear case in point.

The attempt by the military governments to redefine Chileans as consumers instead of citizens was mainly directed to privatize the nature of the social relations within civil society. For this purpose, the regime tried to destroy all kinds of collective identities existing in Chilean society such as party and neighborhood loyalties and social solidarity with the needy, which were officially seen as unwanted heritages of a 'socialistic' past. As a substitute for the search for *collective* goals, the military government offered a neoliberal ideology which was entirely directed to the achievement of *individual* ambitions. In this manner individual freedom was redefined as representing the free access to open markets, while the 'pleasure of consumption' was presented as an instrument to express social differentiation and as a way to obtain personal rewards. From this perspective, the regime's ideologues pointed out that social mobility was in fact mainly a question of personal achievement.

According to this conception of modernization, to be up-to-date in terms of the acquisition of consumer goods represented the single most important criterion for modernity. Moreover, the imitation of lifestyles and values imported from the industrial world as a result of the free-market policies became in fact the only way to participate in the experience of modernity; in other words 'to be modern'. As Brunner indicates, the market is unable by itself to produce normative consensus among the population or to generate social identities. Together with this, the market does not accept the constitution of solidarity bonds and rejects any behavior that is not based on rational calculations. At most, the market can only create lifestyles that are crystallized in the consumption of particular goods (1988: 97, 119). At the end, the expansion of consumerist behavior in Chile generated a kind of passive conformism among the population, who eventually accepted the individualistic tenets of the neoliberal economic model based on the search for private satisfactions (Silva, 1995).

The increasing internationalization of the Latin American economies has not only strengthened consumerism in the local culture but also led to the adoption of values, beliefs, ideas and even patterns of behavior and cultural orientations which also resemble those of the core countries. Many people have embraced a system of meritocratic and individualistic mobility, replacing the old system in which one had to be part of a group (mainly political parties) and in which mobility was conditioned mainly by the capability of the group to exert political pressure on the state. More generally, the 'discovery' of this new world of consumption convinced several social forces that they have a direct advantage in the continuation and deepening of the ongoing process of transnationalization of their societies.

Together with consumerism, the new democratic governments have also tried to secure the political support of the population by attempting to obtain (and maintain) macro-economic stability. Today in Latin America it is no longer rhetoric but real socio-economic and financial achievements that have become the main measuring criteria to evaluate the quality of governments. After so many years of neoliberal economic rationalism many people in Latin America have learned to evaluate government performances almost exclusively on the basis of economic success. Variables such as the rate of inflation, the dollar rate, volume of exports and the balance of payments are the main evaluation parameters. It seems that after many decades of overideologized discourses an increasing part of the population has become extremely conscious of choosing tangible economic benefits. As the Argentine case has shown, the ability of the Menem government to defeat inflation and to achieve macro-economic stability became decisive for his re-election. Because of the increasing importance of economic stability in current Latin American politics, the possession of a team of prestigious economists in charge of economic and financial policies has become a condition *sine qua non* not only for keeping the confidence of the business community but also that of the wider electorate.

THE TECHNOCRATIZATION OF DECISION-MAKING

Following the restoration of democratic rule, technocrats[1] have acquired in many countries a clear public presence and a higher degree of acceptance and legitimacy among the political class and the population than in the recent past. This is reflected among other things in the fact that the leaders of technocratic-orientated economic teams, such as the ministers of finance Fernando Henrique Cardoso in Brazil, Domingo Cavallo in

Argentina and Alejandro Foxley in Chile, have obtained significant popularity amongst the population. This situation, however, cannot be solely explained by referring to the technocrats' central role in the application of the recent stabilization programs or to the fact that they are now operating in a legitimate democratic environment. In my opinion an even more important factor for the consolidation of technocratic politics has been the dramatic weakening and – in some countries – a virtual disappearance of the forces which traditionally resisted technocracy in the past, such as left-wing parties, trade unions, student movements and so forth. It is the latter factor which also helps to explain the reason why, since the restoration of democratic rule, most countries have adopted (or maintained) neoliberal economic policies.

One must remember that the existence of technocratic economic teams and the application of neoliberal economic policies constituted one of the main features of the former military regimes. Thus it is remarkable that, following the democratic restoration, electoral formulas which openly or implicitly support neoliberal policies (such as those of Fujimori in Peru and of Menem in Argentina during his re-election) have become successful. This certainly has to do with the global hegemony achieved by neoliberal ideology, the pressure of international financial organizations, the perceived lack of economic alternatives and the increasing apoliticism of the Latin American population (Espinal, 1992). However, I think the more profound reasons why large segments of civil society are beginning to accept this new technocratic and neoliberal reality lie in quite traumatic events of the recent political past.

The military dictatorships inaugurated in the 1970s inflicted a major blow to the politics of populism, based on clientelistic relations between the state and civil society. Indeed, populism suffered a significant psychological defeat as many people, right or wrong, internalized the view that populism had been one of the main causes for the economic and political crisis that had preceded the breakdown of democracy. This is partly the reason why in most countries in the region the electorate was not inclined to support the adoption of populistic policies after the departure of the military.

The adoption of orthodox adjustment programs has almost always been accompanied by the appointment of technocratic-orientated neoliberal economists in strategic governmental positions (ministries of economic affairs, finance, central banks, planning agencies, etc.) who have become responsible for the formulation and application of these new economic guidelines. The extreme visibility which many governments have consciously given to the economic teams is related to their efforts to send the right signals both to the domestic and to the international business community (Schneider, 1998). Moreover, Latin American technocrats have felt quite confident in the policies adopted, as the policies encompass completely the neoliberal economic thinking which – since the early 1980s – began to achieve an almost uncontested hegemony. As Stallings (1992: 84) points out,

> technocrats who had long argued for more open economies and a bigger role for the private sectors suddenly found increased backing from the outside. They could count on political support from the United States and other advanced industrial countries, intellectual reinforcement from the IMF and World Bank, and empirical evidence of successful performance from countries that had followed an open-economy model.

These technocrats have also played a strategic role in conducting negotiations with industrialized countries as a means to reschedule existing debts and to obtain new credits and financial aid. As Kaufman (1979: 189–90) indicates, these technocrats are more than simply the principal architects of economic policy: they [are] the intellectual brokers between their governments and international capital, and symbols of the government's determination to rationalize its rule primarily in terms of economic objectives... Cooperation with international business, a fuller integration into the world economy, and a strictly secular willingness to adopt the prevailing tenets of international economic orthodoxy, all [form] a... set of intellectual parameters within which the technocrats could then 'pragmatically' pursue the requirements of stabilization and expansion.

In this manner, local neoliberal technocrats have become the national counterparts of foreign financial experts from lending institutions who assess the performance of the Latin American economies that are currently executing adjustment programmes. As Centeno (1993: 325–6) points out, the communication between the foreign financial experts and the local technocrats has been clearly facilitated by their common academic backgrounds; they

> not only share the same economic perspectives, but perhaps most importantly, speak the same language, both literally and metaphorically... The technocrats do not necessarily have to represent one ideological niche or the other, they simply share a familiarity with a certain language and rationale... The graduate degrees from U.S. universities... enable these persons to present arguments that their fellow alumni at the World Bank... understand and consider legitimate.

Although the above-mentioned international political and economic factors have certainly played a decisive role in legitimating and consolidating the position of technocrats within the political elite, any attempt at explaining the technocratic ascendancy as being exclusively the result of external influences would be inaccurate. As the Chilean case shows, the ascendancy of neoliberal technocrats and the adoption since the mid-1970s of severe economic stabilization policy reform in that country was primarily a product of domestic political and ideological struggles resulting in a distinctive balance of power in favor of neoliberalism (Valdés, 1995).

The increasing technocratization of decision-making in most countries in the region has resulted in a general trend of governments trying to 'technify' social and political problems. The problem of poverty has been mainly approached in technical terms, and its solution is posed in terms of adopting the 'technically correct' social policies. By doing this, poverty and social inequalities have been consciously 'filtered' from their political, economic and social dimensions. This has resulted in countries such as Bolivia, Chile and Costa Rica adopting anti-poverty strategies with a strong *asistencialista* orientation, intended to alleviate to some extent the hardship of the neoliberal policies. For this purpose, Latin American governments have obtained technical and financial assistance from regional institutions such as the UN Economic Commission for Latin America (ECLA) and the Inter-American Development Bank (IADB), as well as from the World Bank and from developed nations.

THE DISENCHANTMENT

Today in Latin America a generalized mood of disenchantment with the accomplishments of the new democracies can be perceived. Some people are deeply disappointed by the inability of many governments to improve the social conditions of the less privileged segments of the population. Instead, and as a result of the application of neoliberal policies, the breach between rich and poor has increased in most Latin American countries. The restoration of democracy has brought a clear improvement in the human rights records in many countries – and in the general macro-economic situation. Nevertheless, time and again the great inequality existing in the distribution of the fruits of modernization and economic growth among the different socio-economic and ethnic segments of society becomes manifest. This is partly inherent in the nature of neoliberal policies as their emphasis lies in generating economic growth and not in producing a better income distribution. Most governments have abstained from playing an active role in the area of income distribution, as that option has been regarded as a step back to the old interventionist state.

The increasing impact of the process of globalization in the national economic and political agendas has also discouraged many people about the real possibilities they have to influence decision-making in their countries. Today, an important part of what is happening to their economies is the result of regional trade agreements (such as Mercosur) and other arrangements and processes which go beyond the national borders. The adoption of neoliberal adjustment programs has been seen by many in countries such as Nicaragua, Venezuela and Costa Rica as a clear loss of national sovereignty as they, rightly or wrongly, perceive these policies to be the result of foreign imposition. This has negatively affected not only the degree of legitimacy of these policies but also the status of democratic governments as they appear as being weak *vis-à-vis* the international financial institutions.

There is also a growing discontent among the indigenous people, as in many countries the democratic authorities have not been able to satisfactorily protect their civil rights and to guarantee respect for and acknowledgement of their cultural contribution to national identity. In this respect, the commemoration in 1992 of the 500 years since the discovery of America was chosen by the indigenous movement all over the continent to protest firmly against the factual maintenance of their status as second-class citizens. In countries such as Ecuador, Bolivia and Guatemala significant legal steps (through amendments in the constitution and the adoption of special indigenous legislation) have been taken in recent years, directed at the elimination of formal discrimination and tutelage towards the indigenous population. Notwithstanding the willingness existing among several governments in order to accomplish some tangible improvements in the general condition of the indigenous people, very little has been achieved yet (Van Scott, 1995; Díaz Polanco, 1997).

Democracy has also experienced a loss of prestige through the high degree of political corruption existing in many countries in the region. This has been the case in Peru where the initially very promising government of president Alan García ended in 1983 amidst a situation in which corruption had penetrated the entire structure of power, including the president himself who finally fled the country. Unfortunately this has not been an isolated case. Presidents Fernando Collor de Mello in Brazil and Carlos Andrés Pérez in Venezuela were forced to abandon power in 1992 and 1993, respectively, following accusations of corruption (Little and Posada-Carbó, 1996). Political corruption and the spread of nepotism were also among the main causes that provoked the destitution of president Abdalá Bucaram in Ecuador in 1996. During the past years, the growing influence of capital coming from drug-trafficking activities has further reduced the already low levels of probity characterizing state institutions in almost all Latin

American countries. In countries such as Honduras and Guatemala, corruption related to *narcotráfico* has already penetrated the judicial system, army and police forces, mass media and almost all spheres of public life.

Increasing poverty, drug-trafficking, corruption and the material and institutional weakness of police forces have led in recent years to a dramatic increase in the levels of violence and delinquency in most Latin American countries. As a result of this, there are even people who feel that the levels of security have decreased under the new democracies rather than increased in comparison with the situation under the former authoritarian regimes. Opinion polls of inhabitants of the largest Latin American cities continually indicate 'criminality' as the most urgent problem the government should resolve. The question is, of course, whether the growing levels of delinquency are the direct result of the existence of democracy *per se*. Most probably this phenomenon has more to do with the ongoing process of modernization which has resulted in a gradual disintegration of traditional norms and values, mechanisms of social control and so forth. Similar processes (with similar results) can be observed in industrialized nations. However, as a result of the recent nature of democratic rule some people in Latin America constantly compare the current democracies with the former military regimes, being often inclined to blame democracy for almost all the problems affecting their societies. Nevertheless, the militaristic option has today no appeal whatsoever among the population, so the only viable alternative to the current democracy, they perceive, is not authoritarianism but an improved democracy.

Paradoxically, the several deficiencies of current Latin American democracies have stimulated political indifference among the population, as has been the case in most Central American countries. Even in those few countries in which the restoration of democratic rule has led to manifest improvements in the general socio-economic and political conditions of the population, political apathy has also become generalized. Chile is a case in point. Since the restoration of democracy in 1990 this country has achieved an excellent performance in terms of socio-economic development, as the living standards of the entire population have dramatically improved since then. Besides, the political situation in this country has been very stable as the ruling *Concertación* coalition has been able to concert workable agreements with the opposition, permitting a successful consolidation of democratic rule in the country. Nevertheless, a strong apathy for politics has emerged among the population. During the parliamentary elections of December 1997 (voting is compulsory), almost a third of the people who attended voting stations did not vote for any candidate (by leaving their vote void or by invalidating it). Moreover, there are more than 1.5 million youngsters, and 0.5 million adults who have not even enlisted themselves on the electoral register. Although the Chilean political class has become alarmed by the increased political indifference among the population, there are observers who explain this phenomenon as just being the consequence of the high levels of political and economic stability existing in the country, as is the case in most Western democracies. What we generally can observe in Latin America is that the disenchantment with democracy has not resulted in the adoption of anti-system attitudes, but has rather strengthened the general mood of apathy and depoliticism.

THE FUTURE OF DEMOCRACY IN LATIN AMERICA

At the beginning of the process of democratic transition in the region several political scientists expressed their doubts about the chances for firm consolidation of democratic rule in the continent. Scholars such as James Malloy and Mitchell Seligson (1987), for instance, openly declared their scepticism about the prospects for democracy. They referred to the historical inability shown by Latin American countries in the recent past of maintaining democratic rule for a long period of time. Until now, the presence of democracy in the region has been characterized by a cyclical pattern in which democratic and authoritarian 'moments' have been constantly alternating with each other. In other words, the current democratic period could simply be just another democratic *intermezzo* that after a certain period of time could be followed by another wave of authoritarian rule.

Over a decade after the publication of the volume edited by Malloy and Seligson one can establish that definitely the new democratic governments have been confronting very serious social, economic and political problems and that generally their performance has not been very satisfactory. But despite all the difficulties they have faced so far, the probability of an authoritarian regression in the region has been substantially reduced. The reasons for this, however, have not been always a direct accomplishment of the new democracies themselves. So, for instance, the almost complete disappearance of the revolutionary left has eliminated one of the main grounds on which the armed forces in the past attempted to legitimate their military coups. While a large part of the left has adopted social democracy and showed a strong commitment to the preservation of democratic rule, the military has consciously chosen to stay away from contingent politics and to concentrate its attention on the further modernization of its institutions (Millet and Gold-Biss, 1996). In addition, following the end of the Cold War the United States is no longer disposed to support military coups in the region, as they made clear to the military who deposed President Aristide in Haiti and to seditious officers threatening Paraguay's fragile democracy. Since the late 1980s US foreign policy towards Latin America has begun to emphasize countries' human rights performances, the achievement of good government and the consolidation of democratic rule (Wiarda, 1990).

Nevertheless, it is still too early to say that the armed forces have fully accepted their institutional subordination *vis-à-vis* the civilian authorities in their countries. It has been said, for instance, that today the Peruvian army co-govern the country. In Chile, the armed forces still maintain a strong presence in strategic political institutions such as the state security council and even in the Senate, where they have their own 'bench' (the so-called *bancada militar*) with Pinochet as a senator-for-life. Also, in countries such as Guatemala and Paraguay the armies still have a lot to say in national political events, while in Venezuela a possible intervention by the military in political events is not completely dissipated. In short, most of the new democracies have so far been unable to eliminate fully the 'authoritarian enclaves' left by the former authoritarian regimes (in the form of nondemocratic legislation and special attributions for the armed forces).

In the coming years, poor management of the country's affairs (*desgobierno*) and conspicuous corruption will remain serious threats for the new democracies, as the destitution of president Bucaram in Ecuador has shown. In that case, the army played a decisive role in forcing him to abandon power and, in that critical moment, a *coup d'état* was clearly among the possible outcomes of the crisis. The legitimacy provided by democratic procedures towards elected authorities is certainly not enough to guarantee the support of the people under any circumstance. What many democracies still have to prove towards the citizenry is that this form of government is more efficient and successful in economic terms, and more sensible from a perspective of social justice than was the case during the previous authoritarian regimes.

In the years ahead, further technocratization of the democratic governments can be expected. In my view the current technocratization of decision-making in government circles does not represent a temporary phenomenon, associated with the political and economic requirements of the democratic transition and the application of structural adjustment programmes; it has rather become an integral feature of many Latin American democracies.

Since the restoration of democratic rule in the region one can argue that the concept of democracy has lost much of its participatory implications in relation to the decision-making process. Instead, the Schumpeterian view of democracy is beginning to be tacitly accepted in practice. In this manner democracy is merely conceived as a *method* for arriving at political decisions and in which citizens reserve the right to decide by whom they will be governed through elections in which various elites compete for the electorate's vote. Moreover, we see that today the use of traditional methods of civil pressure and protest (such as property seizures, unauthorized street protests, politically motivated strikes, etc.) are generally considered by political elites as illegitimate acts.

Despite the fact that technocratically-orientated groups have become key actors in the new political landscape of Latin America, in many cases local public opinion is not very aware of the existence of this phenomenon. As I have attempted to show in this chapter, negative collective memories about the populist past, profound changes in the political culture of major left-wing sectors, together with the recent neoliberal transformations have allowed the almost unchallenged ascendancy of technocrats in the continent.

Although it is quite probable that in the coming years a further depoliticization of Latin American societies will take place, this certainly does not mean that the continent will be spared social effervescence or even violent political confrontations. The modernization process must not be reduced to the financial and economic spheres, as has been the case so far in many Latin American countries. Profound changes must also take place in the social and cultural realms. The goal of achieving the modernization of society – as it has been repeatedly stated by most democratic governments in the region – also demands huge efforts to obtain a substantial reduction of the existing high levels of poverty among the population. In addition, real efforts have to be made to expand access to education, health and housing to those social and ethnic groups which are today marginalized from the fruits of modernization. The existence of real tolerance on the part of the political establishment towards people and political organizations who do not agree with the neoliberal postulates and exercise their democratic rights to fight for a possible alternative social order can still not be guaranteed in most countries in the area. In short, the persistence of extreme poverty, open discrimination against the indigenous and black populations, political corruption and excessive isolation of government technocracy will remain serious obstacles for a definitive consolidation of the democratic order in Latin America.

ENDNOTES

1 Defined here as 'individuals with a high level of specialized academic training which serves as a principal criterion on the basis of which they are selected to occupy key decisionmaking or advisory roles in large, complex organizations – both public and private' (Collier, 1979: 403).

FURTHER READING

Haggard, S., Kaufman, R. R. 1996 *The political economy of democratic transitions*. Princeton University Press, Princeton, NJ. An excellent inquiry into the relationship between economic crisis and the breakdown of authoritarian regimes in a series of Latin American and Asian countries. It also provides a forceful assessment of the economic problems faced by new democracies and their impact on democratic consolidation.

Higley, J., Gunther, R. (eds) 1992 *Elites and democratic consolidation in Latin America and Southern Europe*. Cambridge University Press, Cambridge. This volume explores a particular dimension which has often been ignored in the democratization literature: the critical role played by elites in creating and consolidating democratic regimes in various Latin American and South European countries. Special emphasis is placed on the ways elites negotiate compromises and make tactical decisions about the transition and beyond.

Mainwaring, S., O'Donnell, G. and Valenzuela, J. S. (eds) 1992 *Issues in democratic consolidation: the new South American democracies in comparative perspective*. University of Notre Dame Press, Notre Dame, IN. Discusses the varied modalities leading to democratic transition from a comparative perspective. Of particular interest is the attention given to elements of continuity between the new democracies and the old authoritarian regimes.

O'Donnell, G., Schmitter, P. C., Whitehead, L. (eds) 1986 *Transitions from Authoritarian Rule*. The Johns Hopkins University Press, Baltimore, MD, and London. This four-volume study inaugurated the Latin American debate on democratic transitions. It has been very influential and soon became an obligatory starting point for later studies on the subject.

Peeler, J. 1998 *Building democracy in Latin America*. Lynne Rienner Publishers, Boulder, CO, and London. A thoughtful analysis of the Latin American experiences with democratic rule from the nineteenth century to the present. It provides a very useful introduction to basic issues of general democratic theory and contrasts it to the Latin American reality. A strong plea is made for deepening the current democratic systems by expanding popular participation.

Part 3

Globalization and economic transformations

4

GLOBALIZATION, NEOLIBERALISM AND ECONOMIC CHANGE IN SOUTH AMERICA AND MEXICO

ROBERT N. GWYNNE

At the turn of the century, the countries of South America and Mexico are becoming increasingly integrated into the global economy. This is not only because of changes at the global scale but also because of change at the continental scale. At the global scale, the world economy has become increasingly liberalized through the latter half of the twentieth century and costs of interaction (or friction of distance) within that global economy have been dramatically reduced. This has created the conditions for global corporations to thrive in all economic sectors, in services as well as manufacturing and resource development. Meanwhile in South America and Mexico a range of economic reforms have been introduced since the mid-1980s which have made these economies more closely integrated into the world economy – particularly through cross-border flows of trade, investment, financial capital and technology. These reforms have often been labelled neoliberal because of their emphasis on increasing the influence of markets and decreasing the significance of government in economic decision-making.

The nature and impact of these economic reforms in South America and Mexico are the central questions being addressed in this chapter. In Part 3, South America and Mexico are examined separately from Middle America (see Chapter 5). The main reason for this is that Middle America tends to be characterized by small countries that have traditionally needed to be much more closely integrated into the global economy in order to survive. In contrast, South America is characterized by one very large economy (Brazil) and by a large number of middle-sized economies (see Table 4.1); Mexico constitutes the second largest Latin American economy after Brazil.

Furthermore, the economic, social and cultural development of the small countries of Middle America (together with Mexico) tends to be much more dominated by their powerful Anglo-American neighbour, the United States, partly as a result of their geographical proximity; the only exception here is Cuba which has maintained a resolutely anti-US position in its external affairs for 40 years. In contrast in South America, the geopolitical context is less heavily influenced by the United States. Indeed, at present, the development of Mercosur (including a population of 216 million in terms of both full and associate members) around the regional axis of Brazil and Argentina is creating a distinct counterweight at the continental level to the United States and the North American Free Trade Agreement (NAFTA) countries (combined population of 385 million in the United States, Canada and Mexico). Brazil, with a 1995 gross national product (GNP) of US$579.5 billion (see Table 4.1) is one of the world's top ten economies in terms of size, and Mercosur could be described as the third most significant trading bloc in terms of production and population after the European Union (EU) and NAFTA.

The South American economies can be divided into those that are part of Mercosur (Brazil, Argentina, Uruguay, Paraguay, Chile) and the Andean Community (Colombia, Venezuela, Peru, Ecuador and Bolivia). There are considerable variations between the national averages of income per capita, as measured through purchasing power parity (see Table 4.1). According to these data, the average Chilean had an average income one third that of the average US citizen, whilst inhabitants of Bolivia had an average income only quarter of the Chilean. Of course, such data hide significant variations in income distribution within countries, but we will return to this issue later in the chapter.

Since the late 1980s, most South American countries and Mexico have shifted to more export-orientated economies. Table 4.1 shows that all countries demonstrated a rapid annual growth rate in exports in the first half of the 1990s (between 4.4 per cent and 13.8 per cent a year); normally these rates were much higher than those recorded in the 1980s. Exports have as a result become an increasingly important percentage of gross domestic product (GDP), particularly in the small and medium-sized countries; however, even in Mexico export activity accounts for 25 per cent of GDP (up from 11 per cent in 1980). Only in Brazil and Argentina, partly because of their large size and relatively

TABLE 4.1 Population and economic indicators for South American Countries and Mexico

Country	Population (millions, 1995)	GNP (US$ billions, 1995)	Export growth rate, 1990–95 (annual %)	Exports as percentage of GDP, 1995	Purchasing power parity of GNP per capita, 1995 (US$)[a]
Brazil	159.2	579.5	7.4	7	5400
Mexico	91.8	304.8	6.8	25	6400
Argentina	34.7	278.6	6.9	9	8310
Colombia	36.8	70.3	7.2	15	6130
Venezuela	21.7	65.5	4.9	27	7900
Chile	14.2	59.1	9.2	29	9520
Peru	23.8	55.0	8.3	12	3770
Uruguay	3.2	16.5	4.4	19	6630
Ecuador	11.5	16.0	7.4	29	4220
Paraguay	4.8	8.1	13.8	36	3650
Bolivia	7.4	5.9	6.7	20	2540

Source: World Bank (1997).
[a] Calculated by converting gross national product (GNP) to US dollars using purchasing power parity instead of exchange rates as conversion factors.

recent conversion to more outward-orientated policies, do exports account for a low proportion of GDP.

The nature and impacts of this economic shift to outward orientation, combined with the economic justifications for neoliberal reform, will be the main focus of this chapter. In order to contextualize these shifts and examine their impacts, the chapter will be divided into four sections:

- the historical context of the shift to economic reform;
- the nature of economic policy change, with particular reference to macro-economic and trade reform and the significance of nontraditional exports;
- the impacts and problems of neoliberal reform, particularly in terms of economic growth, employment, income distribution and poverty;
- the future of more market-orientated and outward-orientated policies in the light of the evolution of regionally based trading blocs.

Historical contexts

Each age has its own visions of the future, whether this be economic, cultural or political. The present period in South America is characterized by a distinctive economic philosophy based on outward orientation, freer markets and a reduction of state influence in the economy. However, this paradigm has replaced a previous economic paradigm that had been in place in Latin America since the 1930s. The previous paradigm was based more on inward orientation and a greater intervention of the state in matters economic. Why did this economic paradigm of inward orientation hold sway for such a long period of time and why is it a paradigm that is now so criticized?

Inward orientation – why did it last so long?

In many ways the answer to the first question lies in the continent's relations with the wider world economy and its reaction to the crisis of the World Depression of 1929–33. Before then, Latin American governments had generally supported free trade and close integration with the world economy (Bulmer-Thomas, 1994). Economic growth in Latin America in the nineteenth and early twentieth centuries had been closely linked to the rapid expansion in world trade and to the increasing flow of investment capital from such core economies as Britain. As a result, Latin American governments encouraged open economies, inward investment and export-led growth. Generally, Latin American countries exported raw materials (mineral and agricultural) and imported manufactured goods. However, World War One and its aftermath, in which European countries engaged in protectionism, dampened the enthusiasm for outward-orientated policies in Latin America. However, it was not until the Depression and the more than halving of world trade that Latin American governments reacted with dramatic changes in economic policies.

Between 1928 and 1933, Latin American exports declined from about US$5 billion to US$1.5 billion (Bulmer-Thomas, 1994). This was partly a result of a decline in volume. As Table 4.2 reveals, export volumes declined by 22 per cent in Latin America as a whole between 1928 and 1932, although some countries fared much worse; the volume of Chile's exports (dominated by copper) fell by nearly 70 per cent. In addition, export

TABLE 4.2 Percentage price and quantity changes for exports, net barter terms of trade and export purchasing power, 1932 (1928 = 100)

Country	Export prices	Export terms volumes	Net barter of trade	Purchasing power of exports
Argentina	37	88	68	60
Bolivia	79	48	na	na
Brazil	43	86	65	56
Chile	47	31	57	17
Colombia	48	102	63	65
Ecuador	51	83	74	60
Mexico	49	58	64	37
Peru	39	76	62	43
Venezuela	81	100	101	100
Latin America	36	78	56	43

Source: Bulmer-Thomas (1994: 197).

prices were in free fall, declining by two-thirds between 1928 and 1932 (see Table 4.2). Of course, the prices of manufactured imports also declined. However, the net barter terms of trade (NBTT) measures the behavior of export prices in relation to import prices. The overall effect in Latin America was of a 44 per cent reduction in terms of trade in four years. An even more dramatic indication of the effect of the Depression on Latin America's export-orientated economies is found through the purchasing power of exports (where the NBTT is adjusted for changes in the volume of exports). The purchasing power of exports fell by 57 per cent in Latin America overall but in mineral-exporting countries such as Chile and Mexico the fall was even greater – 83 per cent and 63 per cent, respectively.

Such a statistical analysis gives an indication of the scale of the huge economic crisis that affected Latin American countries. The severe crisis in the world economy and the high dependency of Latin American countries on that economy suddenly demonstrated the extreme vulnerability of their economies. Through the 1930s and subsequent decades, Latin America tried to restrict interaction with the world economy, particularly in terms of trade, investment and the transfer of technology. Tariffs, quotas and exchange controls provided protection from foreign competitors by making the entry of foreign goods expensive or impossible. Latin American entrepreneurs observing the scarcity of goods and the level of protection began to produce or increase the production of goods previously imported. Industrial production and employment increased as a result in most South American countries.

Indeed, industrialization became a key theme of the subsequent period of inward orientation. It was argued that all developed countries had industrialized behind high protective tariffs and that it was only after a country had developed a more mature industrial structure that it could become involved in the freer trading of goods (Prebisch, 1950). In addition, and in order for South American countries to achieve a more mature industrial structure, the political consensus was that governments should actively intervene not only through the elaboration of industrial policy but also through the creation of state-owned development corporations. Governments drew up strategic plans for industrial sectors and facilitated investment in key feedstock industries, such as steel, where it was thought that national private investors might be unwilling to venture. The modernization of the state through industrialization became a key theme of the inward-orientated period. In Brazil, for example, the crucial role of industrialization in the rapid accumulation of capital and in improving national technological capabilities, and the pivotal role of the state in facilitating such development, were central tenets of governments stretching from the 1930s to 1970s – most notably in the regimes of Presidents Vargas, Kubitschek and Geisel.

Governments also became more actively involved in economic development because of the way in which the world economic crisis had so seriously affected their economies. Governments were forced to make some response to the two pressures that the world crisis inflicted on the economies of their countries. First of all, there was the external imbalance, the collapse of national earnings from exports and the drying-up of capital inflows from international sources. Second, and even more serious for government, was the internal imbalance – the decline in government revenues as a result of the critical falls in the volume of export and import taxes. In this sense, state intervention in the economy and the increased rate of import taxes (through tariffs) can be seen as part of governments' attempts to solve their own huge financial problems. Questionable policies, such as multiple exchange rates and the printing of excess money, can often be attributed to governments reacting to serious financial crises. These policies tended to become embedded in the history of inward orientation. Edwards (1995: 83) measured the levels of seigniorage (the change in base money relative to total government revenue) between 1971 and 1982 and found that 'in some countries money creation accounted, on average, for almost one-fourth of government revenues'.

Further justification for strong government involvement in the economy came from the core economies in the form of Roosevelt's New Deal policies and Keynesian economics. Meanwhile, the main theories of political economy emerging from Latin America were those of structuralism and dependency (Kay, 1989). These contextualized the problems of Latin American countries developing within a world economy characterized by highly asymmetrical relations, especially in terms of trade. The recommended policies for Latin American countries were to shift their trading patterns away from a reliance on primary commodities, and industrialize within the framework of a strong state – incidentally the basic policy statement of the East Asian 'tigers' whose development paths have received such international praise – at least until the crises of 1997.

Indeed, it must be emphasized that the inward-orientated period of Latin American economic history did bring high rates of economic growth and a rapid expansion of the manufacturing product. This was particularly apparent in the 1950s, 1960s and 1970s. Between 1950 and 1978, manufacturing GDP in Brazil expanded 10 times in real terms (Gwynne, 1985: 36) – equivalent to an average annual growth rate of 8.5 per cent over the three decades. Although Brazil proved the most successful at the industrializing strategy, manufacturing GDP also expanded by a factor of seven in Mexico, by a factor of five in the six Andean countries and by a factor of three in the three River Plate countries (Gwynne, 1985: 36–38). It is misleading to characterize the inward-orientated phase as one of low economic growth. Indeed, growth rates in the 1960s were higher than those of the 1990s for many countries in Latin America.

Inward orientation – why is it now so criticized?

Why then has the paradigm of inward orientation been so criticized? There are perhaps two basic issues involved in that question – trade and inflation. The period of inward orientation brought a major decline in Latin America's participation in world trade. Table 4.3 shows that total Latin American exports as a percentage of total world exports fell from 13.5 per cent in 1946 to only 4.4 per cent in 1975. The larger countries had a proportionally higher fall than the smaller countries of Latin America (Bulmer-Thomas, 1994). This can be exemplified by the case of Argentina; its share of total Latin American exports declined from a substantial 25.5 per cent in 1946 (when it was South America's major exporter) to only 8.2 per cent in 1975.

This serious decline in the global importance of South American trade was a result of inward-orientated policies. Exchange rate policies, in particular, created an overvalued exchange rate for most national currencies. As a result, exports became more expensive than they should have been, making it more difficult for entrepreneurs to expand price-sensitive exports, such as manufacturing or agricultural products. Imports were of course theoretically cheap, but their entry was prevented by high tariffs and quota restrictions. Expensive exports and restricted imports were two powerful reasons behind the stagnation of South American trade.

The resulting export profile of most South American countries proved highly vulnerable. Exports were mainly in areas that were not too price-sensitive in terms of demand on world markets, such as oil; indeed, Latin America's main oil exporter of the period, Venezuela, was the country with the most sustained growth in exports in the post-war period (see Table 4.3). Most South American countries came to rely on exporting just one or two primary commodities; for example in 1974, about 80 per cent of Chilean exports corresponded to copper, and 95 per cent of Venezuela's to oil. Because these primary commodities were characterized by volatile international prices, the external position of most South American economies was very precarious, leading to periods of boom and bust (Auty, 1993).

Furthermore, the partial delinking from the world economy meant that the potential for economic growth was reduced. Most manufacturing producers geared their opera-

TABLE 4.3 Latin America's percentage shares of world and regional exports, 1946–75

Year	Share of total world exports			Country shares of total Latin American exports		
	Total Latin America	Major countries[a]	All other republics	Argentina	Brazil	Venezuela
1946	13.5	8.9	4.6	25.5	21.2	11.1
1950	10.7	6.7	4.0	18.4	21.2	14.5
1955	8.9	4.9	4.0	11.8	18.0	23.0
1960	7.0	3.5	3.5	12.8	15.0	27.2
1965	6.2	3.2	3.0	13.9	14.9	22.8
1970	5.1	2.8	2.3	12.0	18.5	17.7
1975	4.4	2.2	2.2	8.2	24.0	24.3

Source: Bulmer-Thomas (1994).
[a] Includes Argentina, Brazil, Chile, Colombia, Mexico and Uruguay.

tions to the domestic market with the consequent problems of low economies of scale that this produced (Gwynne, 1985); for example, in 1973, a Citroen 2CV produced in Chile was three times the cost of one produced in France (Gwynne, 1978). The only exception was Brazil, where the large domestic market allowed manufacturers to benefit from some economies of scale (Dickenson, 1978). Furthermore, after the late 1960s, Brazil started to promote manufactured exports through a complex range of subsidies – which took Brazil's share of total Latin American exports up from 14.9 per cent in 1965 to 24.0 per cent in 1975, thus making it Latin America's most dynamic exporter within the inward-orientated model.

Inward orientation had another serious failing – high inflation (annual price rises of between 10 per cent and 50 per cent) and hyperinflation (above 50 per cent). Table 4.4 shows that only two countries (Paraguay and Colombia) in South America have not suffered from hyperinflation since 1970. Four countries recorded annual inflation levels of over 1000 per cent, with two countries (Bolivia and Peru) recording figures around the 8000 per cent mark – when prices increase by over one fifth each day of the year! It is very difficult for producers and consumers to live through such periods. Savings rapidly become worthless and speculation becomes rife. Very high price rises create economic (and often political) instability. There are perhaps three significant areas in which high inflation can cause economic damage.

- Increasing uncertainty: uncertainty increases for consumers and producers alike as major shortages of basic products can occur. Producers linked to international markets, such as through component supply, face problems as national prices rise faster than international ones.
- The implications of boom and bust: producers for the national market can do well in periods of inflationary growth. However, if inflation is to be significantly reduced, a period of major austerity (through stabilization programmes) must inevitably occur which will significantly reduce domestic demand; for producers strongly geared to

TABLE 4.4 South American countries: peak inflation years since 1970 and annual inflation average over the period 1984–93

Country	Peak inflation (%)[a]	Average (%)
Bolivia	8 170.5 (1985)	1 051.6
Peru	7 649.6 (1990)	1 283.7
Argentina	4 923.6 (1989)	811.5
Brazil	2 500.0 (1993)	944.8
Chile	650.0 (1973)	19.5
Mexico	159.2 (1987)	52.9
Uruguay	129.0 (1990)	75.5
Ecuador	85.7 (1988)	44.5
Venezuela	81.0 (1989)	34.0
Paraguay	44.1 (1990)	24.5
Colombia	32.4 (1990)	24.8

Sources: Edwards (1995) and Gwynne (1976).
[a] The year in which peak occurred is shown in parentheses.

the domestic market, their operations will have to be cut back severely at some stage. One adverse implication of this is low investment. In boom periods of inflationary growth, there is little incentive for entrepreneurs to make long-term investments in expanding capacity as in the downturn there will be spare capacity. Instead, in the boom periods, producers charge high prices (adding to inflation) in order to dampen demand and make high profits. These high profits during the boom period will compensate the poor returns during the downturn when prices will have to be kept low to encourage demand.

- High consumption and low savings: inflation rates can be higher than interest rates, particularly if governments wish to bolster demand in the short term. Individuals therefore see themselves as effectively losing money if they save as opposed to maximizing the return on their money if they spend (and even borrow). Because of the predilection to spend rather than save, periods of high inflation are associated with low national savings ratios. There is also a problem of capital flight (Mahon, 1996). This is because the affluent wish to conserve their savings against high rates of national inflation and find ways of sending these savings overseas, preferably to US or offshore dollar bank accounts. This reduces the national savings ratio even more, particularly during periods of crisis.

The damaging repercussions of long periods of high inflation can be seen as a significant problem of the inward-orientated phase of Latin America's development. High inflation has also been the site of the ideological battle between the theorists of structuralism and neoliberalism. If structuralism (and the related theories of dependency) constituted the theoretical framework for the political economy of inward orientation, then neoliberalism (and the related economic theories of monetarism) have been the theoretical inspiration for the political economy of outward orientation. Structuralism and neoliberalism are wide-ranging theories (Kay, 1989: 47–57), but one crucial debate has been over explaining inflation. Structuralists have traditionally taken a very wide view of inflation, seeing it as arising from the socio-political tensions, sectoral imbalances and expectations generated by the process of development itself. Structures such as an unequal landholding system (meaning that demand for food outstripped supply), the foreign trade gap and an inefficient and regressive tax system were seen to be at the root of the inflationary problem. In other words, inflation was a symptom of the problem and not the main problem. In contrast, the neoliberal position is that inflation itself constitutes the main problem and that its reduction must be the main economic priority of governments. The neoliberal interpretation of inflation has become the consensual one during the 1990s.

The legacy of inflation within inward orientation left countries short on national savings and increasingly reliant on external finance. Indeed the 1970s and early 1980s became a notorious period of what could be termed debt-led (and inflationary) growth in Latin America. The end of this period came in another dramatic crisis for the continent – the debt crisis that started in August 1982 after Mexico declared a moratorium on its debt repayments.

The impact of the debt crisis

In many ways, the paradigmatic shift in political economy can be attributed to another crisis and its aftermath – the debt crisis that started in 1982. The confused and messy aftermath of this crisis was instrumental in causing Latin America as a whole to suffer a decade of stagnation and policy turmoil in the 1980s. The global capital markets that had appeared so benign during the 1970s turned against Latin America in the 1980s. As a result, the reliance of Latin American countries on external finance (because of low domestic savings, partly a result of the legacy of high inflation) became the Achilles heel of their efforts at economic growth in the 1980s.

In general, it took time for Latin American governments to come to terms with the realities of the new world economy. Since the 1930s, they had tried to protect their economies from the vagaries of the global trading system and from the economic influences of resource and manufacturing multinationals in their economies. They thought that the dependence of their countries on the global economy would thereby be reduced. However, it could be argued that during the 1970s policies of inward orientation and strict controls on inward investment survived only because of the huge inflow of capital from international banks, recycling the capital surpluses of the oil-rich countries. These flows peaked at nearly US$22 billion in 1978, more than 10 times greater than at the beginning of the decade. Latin American governments favored these borrowings as there were no strings attached, such as macro-economic policy recommendations, as normally came with lending from the International Monetary Fund (IMF) and other multilateral organizations.

In retrospect, Latin American economies were becoming more closely linked to the global economy – albeit in new ways. Governments did not realize that international bank lending, although not strongly regulated at that time, was intimately linked to business confidence. When business confidence in Latin America collapsed in August 1982, bank lending, the lubricator of Latin American economies for nearly a decade, dried up virtually immediately. Multinational banks tried to reduce drastically their exposure in Latin America, with policies of no new lending, tight renegotiations and even insistence on the socialization of private debt – where Latin American governments were forced to take responsibility for bad private debts in their country (for which they were not legally bound) as well as the bad public debts for which they were legally answerable (Congdon, 1988).

In similar ways to the World Depression, during this time Latin American governments faced severe imbalances both on the external and on the internal front. Their external accounts were characterized by large current account deficits, partly because of high interest payments and partly because of the drying up of new funds. Interest payments increased because of high world interest rates in the early 1980s and the fact that most international bank lending in Latin America had been in the form of roll-over credits, where the credit is rolled over periodically for interest rate review (such as every six months); in this way the risks of interest-rate movements had been passed to the borrower. In addition, most Latin American countries faced the problem of capital flight. During the 1980s, the more affluent Argentines, Mexicans and Venezuelans held dollar deposits and other deposits abroad worth nearly as much as their country's debt (Mahon, 1996). On the internal side, Latin American treasury departments were not only having to deal with big increases in interest payments on their loans and extreme difficulties in finding new funds but also with the problems of increased expenditure as economic activity declined, state firms declared increasing losses and as unemployment and other social problems rose.

Latin American governments tried to negotiate with the US government for a 1980s version of the Marshall Aid Plan, but US government finances in the 1980s were themselves in deficit and very weak compared with their huge strength after World War Two. With opportunities for bilateral loans limited, the only possible way to receive some external financing was through loans from multilateral agencies such as the IMF and World Bank. However, this came with strings attached; in other words, Latin American governments would have to impose major economic reforms, even if they were only short-term, in order to receive external financing from these organizations. Between 1980 and 1986, net capital flows into the continent declined by around 40 per cent, but private net flows from international banks declined by an astonishing 80 per cent, demonstrating the impact of the change in perception of Latin America from the international banks; meanwhile, the net flows from bilateral loans were halved in value (World Bank, 1992). With other financial flows static, the only significant increase came from official loans from multilateral agencies.

The nature of these economic reforms will be discussed in the next section. The important point to be made here is that most Latin American countries had little alternative but to impose these economic reforms in return for assistance to alleviate their serious financial dilemmas. Latin American countries had a great need for foreign finance, because of low internal savings as well as the great need and potential for investment. But there was also a general lack of effective regulation and surveillance of foreign finance, which led to greater possibilities for fraud, mismanagement and corruption. These financial dilemmas demonstrated that Latin America still had a dependent relationship with the world economy; no longer was this just in relation to trade but also to foreign finance, where shifts in bank confidence and interest rates could have such a huge impact on the management of national economies.

Crises often reveal the true nature of economic and social relations. In Latin America they have revealed that economic performance in Latin America is highly dependent on the world economy, in terms of both growth and decline. In the crisis of the 1930s, Latin American governments understood this and attempted to protect themselves from the extremes of the world economy. After the crisis of the 1980s, it would appear that governments have responded by becoming more closely connected to the world economy (see next section). Crises have been important in changing the nature of the prevailing economic paradigm. Paradigmatic shifts have not necessarily been guided by ideology – although the Chilean shift during the 1970s was more ideologically driven (through the Chicago boys). Rather, the aftermath of crises could be represented as Latin American governments deciding (with different chronologies) to respond in highly pragmatic ways to the contingencies of the crisis. It must be remembered that the inward-orientated paradigm brought at least three decades of economic growth. Will the outward-orientated paradigm prove to have such success in the long term?

Economic Policy Change: Moves to a New Consensus

By placing the current paradigm of neoliberal economic reform in a historical perspective, it can be argued that the paradigm shift has been as much to do with pragmatic considerations of coming to terms with economic crisis and the deficiencies of inward orientation as with the theoretical benefits and ideological justifications of closer integration with the world economy and market-led economics. During the aftermath of the debt crisis in the 1980s, Latin American countries suffered a severe lack of capital, which exacerbated the inherited problems of inward orientation – low domestic savings, reliance on external financing, high inflation, low investment rates and stagnant trade. This section will thus examine the nature of this paradigmatic shift in three parts:

- searching for ways out of the crisis;
- the opening up of Latin America;
- theoretical and ideological elements of the new consensus.

Searching for ways out of the crisis

It took a long time to work out a coherent strategy to get out of the debt crisis. Edwards (1995: 69) points out that multilateral institutions and leading analysts saw the debt crisis as a short-term problem that most Latin American countries could 'grow' out of. It was not until 1987 that creditors and debtors began to acknowledge the structural nature of the problem and the need to implement deep measures to regain macroeconomic equilibrium. One issue here was the need to set clear economic priorities as there were many trade-offs between policy reforms. In the crucial area of exchange rate policy, there was a key dilemma between keeping the exchange rate high in order to guide inflation downward and pushing it down in order to promote export-led growth. Overall, the scale of the debt crisis was not readily identified in the 1980s and many different policies were set in motion in different countries.

Multilateral institutions exerted considerable influence in this search for ways out of the debt crisis. As the IMF and World Bank became the main sources of new funds for the debt-laden countries of Latin America they had the leverage to release funds on condition that those countries implemented basic reforms. These conditions covered highly diverse areas (Table 4.5) and sometimes the package could include contradictory recommendations. However, the emphasis was on achieving export-led growth (trade liberalization, exchange rate action), improved domestic capital formation (tax reform, financial reform) and a reduction in government intervention in the economy (Table 4.5). During the 1980s, many governments resisted the implementation of conditional reforms, but others moved faster than was required by the multilateral agencies (Chile, Mexico and Colombia, for example).

TABLE 4.5 Conditionality contents of International Monetary Fund (IMF) and World Bank programs (percentage of programs with particular conditions)

Condition	IMF, 1983–85	World Bank, 1982–89
Financial reform	44	51
Public enterprises reform and privatization	59	65
Trade liberalization	35	79
Exchange rate action	79	45
Tax reform	59	67

Source: Edwards (1995: 57).

The IMF and World Bank were thus attempting to coordinate the international response to Latin America's debt crisis and introducing their own outward-orientated and market-led solutions to the problem. The international banking community endorsed this view and strongly urged for the burden of new financing to be placed on multilateral institutions. Debt-restructuring operations, IMF-sponsored programs, and World Bank structural adjustment loans were the most important elements of this strategy; between 1983 and 1988 Latin American countries engaged in 29 debt-restructuring operations with the private banks. Nevertheless, Latin American countries remained starved of finance as international banks insisted on debt repayments without new money. It was not until 1989 that a breakthrough occurred in the approach to the debt crisis when the international creditors and multilateral institutions recognized that providing some debt forgiveness could be in everyone's interest.

The outcome was the Brady Plan, which encouraged creditors to enter into voluntary debt-agreements with the debtor countries. There were two basic mechanisms for alleviating the debt burden. First, the use of debt-reduction schemes based on secondary market operations was actively encouraged. This technique acquired special momentum after 1988, when, in a number of countries, debt–equity swaps became an important mode for attracting new investment from multinational companies and privatizing state-owned enterprises (Box 4.1). Table 4.6 shows the behavior of the secondary market for Latin American debt. It shows that not only were there huge discounts available for prospective investors but also that the secondary market firmed up considerably between 1989 and 1991. Second, direct debt-reduction agreements between the international creditor banks and individual countries became more common after the introduction of the Brady Plan; between 1989 and 1992, Argentina, Brazil, Mexico, Uruguay and Venezuela reached agreements with their creditors to reduce their debt burdens.

TABLE 4.6 Latin America: prices of external debt paper on the secondary market (percentage of nominal value)

	1989		1990		1991	
	January	June	January	June	January	June
Argentina	20	13	12	13	19	25
Brazil	37	31	25	24	23	33
Colombia	56	57	60	64	64	73
Chile	60	61	62	65	75	88
Ecuador	13	12	14	16	20	22
Mexico	40	40	37	45	45	55
Peru	5	3	6	4	3	7
Venezuela	38	37	35	46	50	60
Weighted average	35	32	30	33	34	43

Source: United Nations (various years).

BOX 4.1: Debt–equity swaps and multinational investment

The interesting concept behind debt–equity swaps was to change debt into investment through use of the secondary market. In the late 1980s, many international banks, particularly smaller ones, wanted to get rid of their foreign loans and were willing to do so at big discounts using the secondary market (Table 4.6). Chile was one country that utilized the debt–equity swap to good effect. A hypothetical example can show how it worked, and how most participants benefited.

A New Zealand timber firm, South Island Forestry (SIF) wanted to invest in Chile's booming forestry sector in 1989 and realized that substantial discounts could be achieved in locating a US$100 million plant there. Contact was made with a large international bank as broker, such as Citicorp. Citicorp knew of a small Mid-West bank that wanted to get rid of a US$100 million Chilean loan on its books. The secondary market showed that Chilean loans were trading at 60 per cent of their true value in early 1989 (Table 4.6). Arrangements were made. The Mid-West bank received US$60 million (60 per cent of US$100 million) and SIF paid this plus the commission to Citicorp. After receiving permission to invest from the Chilean government, SIF, with US$100 million of loans, exchanged these in the Chilean Central Bank and received US$100 million worth of pesos at the official exchange rate. SIF started to build.

The loser in this debt–equity swap was the Mid-West bank which traded loans with a book value of US$100 million for US$60 million in cash. However, its debt exposure in Latin America was considerably reduced and its shareholders content as accounting provision had already been made. The winners in the swap were the broker (because of the commission) and SIF (receiving a 40 per cent discount on investment). Meanwhile, Chile not only received US$100 million of investment but also no longer had to pay interest on this amount. Thus the debt–equity scheme acted to increase investment (at a time when it was low) and enabled the national economy to grow out of its heavy indebtedness.

For the links with the reality of debt–equity investment in the Chilean forestry sector, please see Gwynne (1996a).

Figure 4.1 Loading logs onto a truck of the forestry company, Colcura, to supply the cellulose plant of the Santa Fe corporation. Both companies were owned in the early 1990s by a multinational consortium of Royal Dutch Shell (60 per cent), Scott Paper (20 per cent) and Citibank (20 per cent). Considerable discounts on invested capital were achieved through debt–equity swap arrangements (for further details, see Gwynne, 1996a). Photograph by R.N. Gwynne.

The opening up of Latin America

One of the key elements of Latin American economic reform during the late 1980s and 1990s has been that of trade liberalization. It is this reform that most distinguishes the neoliberal paradigm from that of inward orientation. Throughout Latin America, tariff and nontariff trade restrictions have been reduced and controls on foreign exchange markets lifted – particularly in the smaller countries. Although this process has been primarily the result of the decisions of the governments of Latin American states, the multilateral Uruguay Round negotiations (1986 93) and the wave of regional trade agreements during the 1990s have also played a role.

The impact of the debt crisis and its aftermath in the early 1980s caused this fundamental change in policy. Economic reform allied to trade liberalization offered the possibility for exchange-rate devaluation, increased exports and higher trade surpluses, these surpluses providing a valuable source of finance to start balancing the current and capital accounts of the indebted Latin American nations. Practically all Latin American countries under review began significant programs to liberalize their trade regimes between 1985 and 1991 (Table 4.7).

According to the Inter-American Development Bank (IADB, 1996: 95), Latin American countries now have the most liberalized trade and foreign exchange system they have had since the 1920s. There has been a sharp reduction in the levels of tariff protection on imports. Taking the region as a whole average tariffs declined from 44.6 per cent in the pre-reform years to 13.1 per cent in 1995 (IADB, 1996: 98). Another important feature of the liberalization process has been the gradual adoption of more uniform tariff structures. These provide advantages in terms of administration and transparency, preventing tariff policy from being manipulated by interest groups capable of applying pressure on government policy. Despite this process of lowering and levelling tariff structures, there are still significant differences in protection levels by the major sectors of production (Fig. 4.2). Interestingly, Mexico maintains high tariffs on consumer imports and Brazil on capital goods imports compared with the Latin American average; only Chile and Bolivia have both low and uniform rates.

Meanwhile, quantitative limits on imports were gradually eliminated both unilaterally and multilaterally (through the Uruguay Round). Some countries, such as Chile, were quick to remove quotas on imports, whilst the more industrialized countries (such as Mexico and Brazil) have been much slower in this respect. Overall in Latin America, nontariff measures that affected 33.8 per cent of imports in the pre-reform period

TABLE 4.7 Chronology of trade liberalization in Latin America, 1985–91

Year	Country
1985	Chile, Mexico
1986	Bolivia, Costa Rica
1988	Guatemala
1989	Argentina, El Salvador, Paraguay, Venezuela
1990	Brazil, Ecuador, Honduras, Peru
1991	Uruguay, Colombia

Source: IADB (1996).

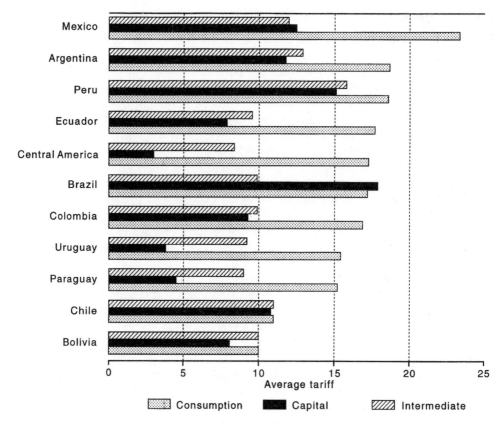

FIGURE 4.2 Average percentage tariff by type of good, 1995. Source: IADB (1996).

subsequently declined to 11.4 per cent in 1995; the number of tariff lines affected by these measures fell from 29.6 per cent to only 1.6 per cent (IADB, 1996: 99).

Trade liberalization has normally been associated with crucial changes in exchange rates. After the debt crisis, many Latin American currencies were devalued and subsequently many have been classified as either effectively-valued (value of currency broadly in line with the real market rate) or undervalued (value of currency lower than the real market rate). This change has made exports from most Latin American countries much more competitive and has been one of the main causes of rapid export growth in many countries. According to the IADB (1996: 105), a new era of exchange-rate systems began in 1991–92, with the return of substantial external capital flows. Liberalization of the foreign exchange market has had a significant impact on reducing the distorting effect of the exchange-rate premiums; this averaged 72 per cent in 1989, but only 2 per cent in 1995 (IADB, 1996: 104). There is still considerable debate about the details of exchange-rate management. Some countries have fixed exchange rates (such as Argentina, in order to reduce inflation), whereas most countries still opt for flexible, floating rates (as with Chile, in order to maintain a competitive rate and promote exports).

Theoretical and ideological elements of the new consensus

> I find it ironic that some critics have condemned the Washington consensus as a neo-conservative tract. I regard it rather as embodying the common core of wisdom embraced by all serious economists (Williamson, 1993: 18).

It has already been noted that Washington-based multilateral organizations such as the World Bank, the IMF and the IADB were highly influential both as lenders and as policy reform advisers in the 1980s in Latin America. However, in 1989, John Williamson, a fellow at the Institute for International Economics, listed 10 policy reforms being implemented in Latin America which, he thought, the international financial institutions would applaud. The idea, no doubt, was to give the idea of a bottom-up approach to policy formulation in Latin America rather than a top-down trajectory (the international institutions demanding policy reforms). These 10 policy reforms had been dear to the heart of Washington-based institutions for many years, and hence the package became known as the Washington consensus.

The Washington consensus had three main ideological thrusts:

- the opening up of Latin American markets to the world economy through trade liberalization and easier foreign direct investment (see previous section);
- the reduction of direct government intervention in the economy through privatization as well as increasing the professional role of economic ministries – through fiscal discipline, balanced budgets and tax reform;
- increasing the significance of the market in the allocation of resources and making the private sector the main instrument of economic growth through deregulation, secure property rights and financial liberalization.

Such ideological principles were not too different from those of the liberal economists of the eighteenth and nineteenth centuries, such as Adam Smith and David Ricardo – hence they have also been termed the 'neoliberal manifesto' and are associated with conservative political forces.

However, as Williamson refers to in the above quote and Edwards (1995) implies in his book, such policies can also be seen as the common core of wisdom of the economics profession – at least in those sectors involved in policy formulation. In 1996, additions were made by Williamson to the policy reforms of the Washington consensus at the annual IADB Conference in order to make them more appealing to the political centre ground. These changes included the following.

- Government was to build more solid and professional institutions, particularly independent central banks and stronger budget offices;
- Although maintaining the theme of reducing direct government spending in economic matters, the need to increase public expenditure in social areas was highlighted, particularly education and health. Education was seen as a high priority for increased spending as this would develop the human resources so essential for economic development in the long term.
- The importance of strengthening bank supervision, combining financial deregulation with stronger oversight, was emphasized in the wake of the Mexican financial crisis of 1994–95.
- The need to create more competitive economies not only through privatization and deregulation but also through investment in institutions and human resources was highlighted.

Such changes have emphasized the consensual nature of economic reforms. Indeed, they have been taken on board by center–left administrations, such as those of the *Concertación* in Chile under Aylwin and Frei and by the Cardoso government in Brazil (Chapter 1).

IMPACTS AND PROBLEMS OF NEOLIBERAL REFORM

Economic growth and inflation

The theme of the book by Edwards (1995), the World Bank's chief economist for Latin America at the time, is that the package of market-orientated economic reforms introduced in the late 1980s and early 1990s has transformed Latin America from a continent of economic despair to one of hope. He notes, however, that the reform process needs to be consolidated and that this requires the majority of each country's population 'to recognize that the modernization effort will generate sustainable and solid results in the form of rapid growth and improved social conditions' (Edwards, 1995: 303). What then has been the economic record of neoliberal reform? Table 4.1 demonstrated the record of export growth. Table 4.8 points to the record of economic growth and inflation for South America and Mexico as the 1990s draw to a close.

The record of economic growth has been distinctly varied ranging from a high 7.3 per cent average annual growth rate for Chile to a low 1.9 per cent rate for Mexico in the 1990–97 period. The Mexican performance was adversely affected by the severe financial crisis of 1994/95; economic growth actually averaged 7.3 per cent in the two years 1996 and 1997. World Bank economists would point to the economic success of those countries that have been most committed to neoliberal reform in the late 1980s and 1990s (Chile, Argentina, Peru, Uruguay) and the lack of success of the only country that resisted reform in the 1990s – Venezuela under the presidency of Caldera. Other countries, such as Brazil, have made gradual shifts to more neoliberal policies (Cammack, 1997) and have recorded improved economic performances.

However, as has already been pointed out, the continental shift away from populist

TABLE 4.8 Record of neoliberal economic reform in the 1990s

Country	Average annual GDP Growth, 1990–97	Annual inflation rate, mid-1998
Chile	7.3	5.4
Argentina	5.9	1.2
Peru	5.2	8.4
Uruguay	4.3	12.3
Colombia	4.1	20.7
Bolivia	3.9	9.4
Ecuador	3.3	33.6
Brazil	3.0	4.1
Paraguay	2.8	6.6
Venezuela	2.4	39.6
Mexico	1.9	15.1

Source: World Bank (1997); Business Monitor International *(various issues)* and Economist *(various issues)*.

and inward-orientated economic strategies and to those more based on the Washington consensus was partly linked to the wish to confront the problems of high and hyper inflation (Table 4.4). It is interesting to note that the four countries that recorded four-digit inflation in the late 1980s or early 1990s (Table 4.4) made vast strides; indeed, all four countries had only single-digit inflation by mid-1998.

Trade liberalization

The freeing up of trade has advantages and disadvantages, but at the end of the twentieth century the advantages are being stressed by a wide variety of governments and political parties – from center–left to right in terms of ideological orientation. These governments see trade liberalization as beneficial for their country's economies in a period of rapid globalization. They seem to go back to the tenets of classic economic thought in which free trade was seen to benefit resource allocation and productivity through the operation of comparative advantage at the global scale (Preston, P.W., 1996). National economies are deemed to become both more efficient and more specialized in terms of production for world markets (Box 4.2). Increasing foreign trade (exports and imports) is seen to offer the engine for national economic growth, with the concomitant advantages of increasing inward investment and improving national technological capability through the import and absorption of new innovations.

Trade liberalization does bring benefits to countries. However, it also brings problems that are sometimes overlooked by the many international economists and advisers who actively promote it. Trade liberalization has brought at least three problems to those formerly inward-orientated countries that have enthusiastically adopted it. First, it has brought severe short-term problems of restructuring in terms of production and employment. The reduction of tariffs and nontariff barriers had a very negative impact on production and employment in the formerly protected sectors (such as consumer-good manufacturing) in national economies, particularly smaller ones. Production and employment expanded only slowly in export-orientated sectors.

Second, in the smaller countries of Latin America, the growth of diversified or non-traditional exports has tended to be concentrated in primary products. Increased reliance on primary product exports brings problems in terms not only of the conditions of international trade (poor terms of trade with manufactured imports, high price volatility) but also of the long-term sustainability of export growth, because of environmental constraints on increased resource production. In Chile, this is already evident in terms of fish product exports and agricultural exports in semi-arid areas (Barton, J., 1997; Gwynne, 1993b).

Last, it has increased the importance of transnational actors in Latin American economies (Clark, 1997) not only in terms of multinational corporations but also in terms of aid agencies and multilateral organizations. Multinational companies are becoming increasingly prominent in the marketing of primary product and manufacturing exports, partly because they have the expertise in international marketing, access to capital and relevant technological innovation. This indicates a decline in national control over production in the increasingly crucial export sector.

Trade liberalization is certainly not a panacea for the economic problems of Latin America. It has brought greater advantages to the larger, more industrialized middle-income countries of Latin America than to the smaller, less industrialized and lower-income countries of Latin America (Chapter 5).

BOX 4.2 Chilean trade liberalization

The Chilean case demonstrates four advantages of trade liberalization. First, it exhibited long-term (rather than short-term) growth in exports. In Fig. 4.3, the record of export growth in Chile after trade liberalization was started in 1975 is shown. Exports doubled between 1975 and 1980 but then declined and stagnated (owing to policy reversal). From 1985 to 1995 (either side of the return to democratic governance in 1990), exports expanded fourfold – from around US$4 billion to US$16 billion. Thus rapid trade growth occurred only after 10 years of the initiation of liberalization policies.

Second, it showed a more diversified range of exports and less dependence on one or two commodities. Under inward orientation, Chile had relied heavily on copper for its export trade (copper constituted about 80 per cent of Chile's exports in 1974). Figure 4.3 reveals that, although metals exports (mainly copper) were still about half of total exports in the mid-1990s, three other specialized areas of export production had developed – agriculture (mainly fruit), fish products and forestry (particularly cellulose).

Third, it demonstrated increasing and more diversified imports. There is a significant rule to follow here: the import of capital goods (for investment) should be much higher than that of consumer goods (for consumption). The Chilean crisis of 1982/83 was compounded by imports of consumer goods far outstripping those of capital goods (Figure 4.4). The sustained economic growth of the late 1980s and 1990s was based on import growth being dominated by capital goods.

Last, it showed a more diversified range of countries traded with. When Chile relied on copper exports, its major trading partners were the United States and the main industrial countries of Western Europe. Export trade with East Asia was minimal. Trade liberalization has been associated with a mushrooming of trade with the newly industrializing countries of East Asia (particularly Japan, Taiwan, South Korea, China and Hong Kong). Exports increased from negligible proportions in 1975 to reach over US$5 billion in 1996 (Fig. 4.5); forestry and fish products have been significant as Chile has become a Pacific as well as an Atlantic trader (it is a member of APEC, the scheme of Asian Pacific Economic Cooperation).

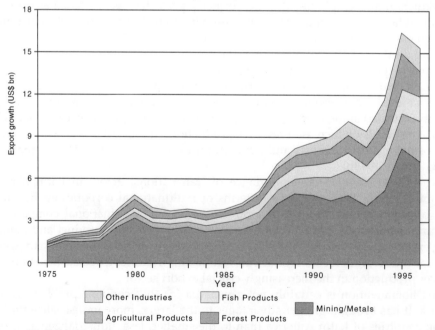

Figure 4.3 Trade liberalization and export growth (in US$ billions) in Chile, by sector.

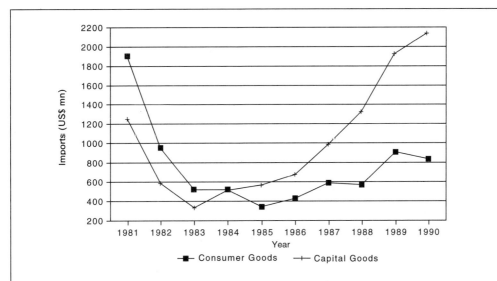

Figure 4.4 Capital and consumer good imports (in US$ millions) in Chile, 1981–90.

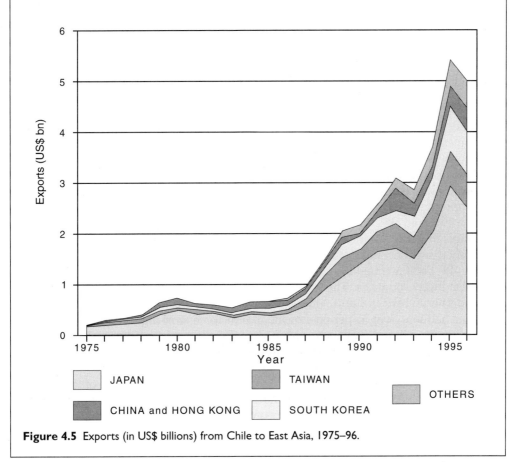

Figure 4.5 Exports (in US$ billions) from Chile to East Asia, 1975–96.

Employment change

Neoliberal reform has had major impacts on labor markets in Latin America. According to Thomas (1996: 86), neoliberal policies have targeted labor-market distortions by introducing legislation to reduce the power of trade unions and to reduce the level of legislated minimum wages. This would affect the relationship between the formal and informal sectors of Latin American labor markets (Chapters 10 and 12). It was argued that such changes would lead to a higher demand for labor in the formal sector and would thereby reduce the supply of labor in the informal sector. Such simple analysis ignores the role of rural–urban migration in urban labor markets. Indeed, Thomas (1996) argues that the lowering of wages in the formal sector has been associated with increasing unemployment in the sector and hence an increase in employment in the informal sector, with a concomitant lowering of wages in the sector.

One observable theme that emerges from the impact of market-orientated neoliberal reform on labor markets is that there is a severe short-term problem of increased unemployment resulting from the massive restructuring of the economy that the policies create. First of all the private formal sector is considerably restructured as the economy opens up to overseas markets and foreign competition. Those companies used to producing for protected national markets normally have to reduce both production and employment dramatically. Furthermore, after decades of inward orientation, national entrepreneurs are not used to the demands of highly competitive world markets, are wary of governments' long-term commitment to outward orientation and thus become involved in the export sector cautiously and slowly. Thus whilst employment declines substantially in formerly protected sectors it rises only slowly in more export-orientated firms. Thomas (1996: 89) notes that, as a result, the main increases in employment between 1980 and 1992 were in small firms and own-account workers; employment in large firms stayed broadly similar.

Second, neoliberal reform has invariably been associated with a significant decline in government employment – through the privatization of state firms and the dramatic reductions in the number of public employees both in the state bureaucracy and in social services. This changed the historic pattern of the state becoming more and more significant as direct employer. During inward orientation, the state had often been the most important generator of formal jobs in the country (Roberts, 1995: 115). However, between 1985 and 1992, the workforce in nonagricultural employment engaged in public services in Latin America fell from 16.6 per cent to 14.9 per cent (Thomas, 1996: 88). It is likely that this trend continued during the rest of the 1990s.

Table 4.9 demonstrates the broad trend of labor-market changes since 1960 for all Latin American countries. The proportion of the labor force involved in agriculture declined everywhere in Latin America between 1960 and 1990; by 1990, 13 out of 21 countries had less than 30 per cent of their labor force working in the sector. The proportion involved in industry increased in virtually all countries between 1960 and 1980, when import substituting industrialization (ISI) was still a key policy. However, between 1980 and 1990, the percentage of the labor force involved in industry declined in the seven highest-income countries of Latin America, reflecting some of the dramatic restructuring that was taking place as industrial firms lost their protected position in domestic markets and were being forced to become more outward orientated. Nevertheless, in many smaller and lower-income countries, industrial employment continued to increase.

TABLE 4.9 Employment (percentage of labor force) in Latin America: sectoral shifts, 1960–90

	Agriculture			Industry			Services		
	1960	1980	1990	1960	1980	1990	1960	1980	1990
Haiti	80	71	68	6	8	9	14	21	23
Nicaragua	63	39	28	15	24	26	21	37	46
Honduras	72	57	41	9	15	20	18	28	39
Cuba	36		18	24	–	30	41	–	51
Bolivia	55	53	47	24	18	18	21	29	36
Guatemala	66	54	52	13	19	17	21	27	30
Ecuador	59	40	33	18	20	19	23	40	48
Dominican Republic	64	32	25	13	24	29	24	44	46
El Salvador	62	43	36	17	19	21	21	38	43
Paraguay	54	45	39	18	20	22	27	35	39
Colombia	52	40	27	19	21	23	29	39	50
Peru	52	40	36	20	18	18	28	42	46
Costa Rica	51	35	26	18	23	27	30	42	47
Panama	51	29	26	14	19	16	35	52	58
Venezuela	–	15[a]	12	–	28	27	–	57	61
Brazil	55	37	23	17	24	23	28	39	54
Chile	30	21	19	30	25	25	39	54	56
Mexico	55	36	28	19	29	23	29	35	50
Uruguay	24	17	14	29	28	27	50	55	59
Argentina	21	13	12	34	34	32	45	53	55

Source: UNDP (1997, Table 16 for 1960 and 1990 figures); World Bank (1996, Table 4 for 1980 figures)
Countries listed in ascending order of gross national product per capita in 1994.
– = no data.
[a] The 1980 figure for Venezuela is taken from World Bank (1996: 195).

Services have now become the major employer of labor in Latin America – only Haiti, Honduras, Bolivia and Guatemala are exceptions at the country level. This marks huge changes in labor-market structures since 1960, when agriculture was still the main employer in all but three countries (Table 4.9). In 1990, the seven highest-income countries all had more than half their labor force engaged in service activities.

Income distribution and poverty

Poverty and inequality have long been distinctive features of Latin American economies. Decades of government intervention, inward orientation and protected markets did little to reduce inequality. Latin America was the only region in the world where the share of income going to the poorest 20 per cent of the population consistently declined between 1950 and the late 1970s (Sheahan, 1987). In the late 1970s, the percentage of income (2.9 per cent) received by the poorest 20 per cent was lower in Latin America than in any other part of the developing world and much lower than that in East Asia (6.2 per cent). Edwards (1995) argues that liberalization programs and their effects on poverty and income distribution must be placed in this context. Nevertheless, he emphasizes that 'only to the extent that poverty is reduced and living conditions of the poor are improved will the structural reforms implemented during the last decade be sustained' (1995: 252).

Liberalization programs are not usually aimed at social concerns; their priorities lie in the macro-economic sphere. However, the proponents of neoliberal reform expect that greater efficiency and success in avoiding inflation should favor economic growth and thereby reduce poverty in the long term – given that the immediate effects can be negative. Sheahan (1997) argues that the key factors influencing poverty and inequality within market-orientated liberalization programs are:

- the balance between the demand for labor and the growth of the labor force – it is important for governments to follow a labour-intensive pathway;
- the distribution of assets;
- the distribution of access to education, skills and opportunities.

Sheahan (1997: 9) maintains that liberalization programs 'do not in principle rule out redistributing assets for the sake of equalization, but their spirit certainly goes against it'. As a result, the growth of employment opportunities and the role of social programs affecting human capital are the main variables.

The growth of employment opportunities

Although liberalization programs have become associated with a short-term decline in employment, they have also become associated with subsequent long-term increases in employment. This can be demonstrated from Table 4.10, which examines the official figures for employment trends in Chile since 1973. Chile suffered two periods of austerity and employment decline as a result of economic restructuring. Between 1973 and 1976, employment (excluding make-work schemes) fell by 12.5 per cent from 2.89 million to 2.53 million. Between 1981 and 1983, during the debt crisis, employment declined by an even more severe 16.2 per cent from 3.09 million to 2.59 million. Both these declines took place as the labor force was increasing by nearly 2 per cent a year. Marcel and Solimano's (1994) econometric analysis emphasizes the significant and dominant role played by changes in unemployment for determining income shares of the poorest four deciles in Chile. However, after the further economic restructuring linked to the debt crisis, employment in Chile grew impressively – more than doubling in 13 years (from 2.59 million in 1983 to 5.22 million in 1996); as a result, adjusted unemployment (excluding make-work schemes) declined from crisis levels (31.8 per cent in 1983) to under 6 per cent during the mid-1990s.

What are the effects of sustained economic and employment growth on income distribution and poverty? As Sheahan (1997: 7) notes, programs of economic liberalization have many common features all over the world but they do not necessarily have the same consequences. Thus, the Chilean model cannot necessarily be extrapolated to other countries. Nevertheless, it is interesting to point to the work of Larranaga and Sanhueza (1994) who try to investigate the transmission mechanisms between economic growth and poverty incidence. They defined 15 sectoral subgroups and used data from national household surveys in order to decompose the change in the poverty headcount ratio between 1987 and 1992 into a growth and distributive component. The subgroups were classified by the level of educational attainment and by the economic activity of the household member with the largest income. According to their analysis,

TABLE 4.10 Employment and unemployment in Chile, 1973–96

Year	Labour force (thousands)	Employment (thousands)	EEP[a] (thousands)	Unemployment (%)	Adjusted unemployment (%)[b]
1973	3039.0	2893.1		4.8	
1974	3066.8	2784.7		9.2	
1975	3152.9	2727.3	71.5	13.5	15.8
1976	3216.4	2705.3	172.5	15.9	21.3
1977	3259.7	2796.8	187.5	14.2	20.0
1979	3480.7	3000.4	133.9	13.8	17.6
1981	3669.3	3269.3	175.6	10.9	15.7
1982	3729.5	2971.5	226.8	20.3	26.4
1983	3797.1	3091.2	502.7	18.6	31.8
1984	3937.1	3185.1	336.3	19.1	27.6
1986	4160.3	3582.0	233.5	13.9	19.5
1988	4455.0	3911.5	46.2	12.2	13.2
1990	4671.8	4391.7		6.0	
1992	4843.9	4606.4		4.9	
1994	5213.0	4904.0		5.9	
1996	5543.0	5217.0		5.9	

Source: Garcia (1993) and Banco Central de Chile (1997).
[a] Engaged on government emergency employment programs.
[b] Includes EEP figures.

about 80 per cent of the reduction in poverty from 1987 to 1992 was accounted for by the growth effect. Scott (1996: 175) thus suggests that 'trickle-down' was the major source of poverty alleviation over the period, with a tightened labor market acting as the most likely transmission mechanism.

Decline in poverty was greatest in those subgroups associated with tradeable goods production; agricultural and import-competing production, whatever their level of human capital, recorded the largest reductions in poverty. In contrast, in nontraded groups, poverty either increased or the distributive effect dominated the growth effect. The record of the agricultural sector, in general, and fruit-exporting regions, in particular, is interesting. They are characterized by low levels of industrial concentration (ownership is dispersed) and labor-intensity (although much is temporary). In contrast, Chile's other export sectors (mining, forestry and fishing) are characterized by high levels of industrial concentration and by relatively low labor intensity. This is one reason why reduction in rural poverty has been more impressive than that of urban poverty in Chile since the mid-1980s (Scott, 1996: 171).

The role of social programs affecting human capital

Sheahan (1997: 11) outlines three possible relationships between liberalization programs (LPs) and government social policies.

- Standard model: LPs are designed to allow private capital flows to determine the external balance on current account and comparative advantage to determine the structures of production and trade; state intervention is cut back to minimal levels, including social provision.
- Competitive model: LPs are designed to promote exports, investment and savings in order to raise sustainable rates of growth, increase the demand for labor and lessen dependence on external capital with the aim of building up the capacity of a country's modern sectors to compete in open international markets; social programs emphasize emergency help to sustain nutrition or health standards for the extremely poor.
- Competitive-plus-social model: an economic program similar to that of the competitive model. Social programs aim at reducing the inequality of opportunities by improving the quality of education for the poor, redistributing educational expenditures to favor primary and secondary education, providing training to increase job flexibility and promoting community projects and leadership.

Sheahan (1997: 12–20) points out that all three models have been used in Chile: the standard model (1975–83) was associated with increasing inequality; the competitive model (1984–89) brought significant improvements in poverty alleviation as a result of rising employment and targeted social programs; the third model has characterized the democratic governments since 1990 and was associated with reducing inequality, at least in the early 1990s. The lesson for other countries seems to be that sustained employment (and wage) growth is essential for poverty alleviation and that governments must not only raise the human and productive potential of the whole society but also help the poor by attacking the obstacles that hold them back. If this happens, the extreme inequalities of Latin American society may be reduced after a period of increased poverty in the immediate aftermath of liberalization programs.

THE FUTURE OF POLICY: HEMISPHERIC INTEGRATION IN THE AMERICAS?

More market-based and outward-orientated economic policies have thus become the norm in Latin America during the 1990s. In many cases, they have provided a period of economic growth after the 'lost decade' of the 1980s. Arguably, there is some evidence of more dynamic labor markets, increased employment and reductions in poverty, although the pattern remains sketchy. However, one result of the increasing openness of Latin American economies is to make countries more enthusiastic and interested in hemispheric integration.

Hemispheric integration is on the agenda for the early part of the twenty-first century, year 2005 to be precise. Two 'summits of the Americas' (Miami, December 1994; Santiago, Chile, April 1998) have approved negotiations intended to lead to free trade between North, Central and South America and the Caribbean. The agenda is presently being pushed by at least five major schemes of regional integration:

- the North American Free Trade Agreement (NAFTA) countries – Canada, USA, Mexico;
- the Central American Common Market (CACM) – Guatemala, El Salvador, Honduras, Nicaragua, Costa Rica (Chapter 5);

- Caricom (Caribbean Community) – members include Barbados, Guyana, Jamaica, Surinam, Trinidad and Tobago (Chapter 5);
- the Andean Community (formerly Andean Group) – Venezuela, Colombia, Ecuador, Peru and Bolivia;
- Mercosur – full members are Argentina, Brazil, Paraguay and Uruguay; associate members include Chile and Bolivia.

As a result, on the mainland of the Americas, only Panama and French Guiana are not members of a scheme of economic integration that is promoting free regional trade and/or the development of common markets (Fig. 4.6). The CACM and Caricom are discussed in Chapter 5; here we will focus on NAFTA, Mercosur and the Andean Community.

NAFTA is the only example so far of a scheme of economic integration involving two advanced economies and one emerging or developing economy. The differences in income and standard of living are substantial; in 1995, income per capita in the United States was over eight times that in Mexico (World Bank, 1997). Such differences have kept the focus of NAFTA very much on trade rather than wider integration; for example, there is no provision for labor mobility within NAFTA (particularly from Mexico to the United States and Canada). Prospects for trade growth have been good for all three countries – labor-intensive exports from Mexico and capital-intensive exports and plant machinery from the United States and Canada. However, considerable volatility has characterized trade. When the Mexican economy was recording growth in 1994, the United States had a US$5 billion trade surplus with Mexico. However, after Mexico's severe economic crisis that started on 24 December 1994, the next two years recorded substantial Mexican trade surpluses with the United States, about US$13 billion a year. These trade surpluses have reduced support for NAFTA amongst US trade unions (who see the migration of jobs south of the border) and the US electorate (who in addition identify Mexico as the major source of migrants and drugs). Hence when Clinton tried to expand NAFTA to include Chile by putting forward a fast-track bill to the US Congress in 1997, the bill had to be withdrawn because of the certainty of its defeat in a Congress where senators were worried by the reactions of their electorates.

The US presidents of the 1990s (Bush and Clinton) have each preferred to emphasize the wider geopolitical context. Both presidents have seen NAFTA and a wider 'free trade area of the Americas' (FTAA) as a strategic response to global trends in regional bloc formation, particularly in Europe (Gibb and Michalak, 1994) and to the dynamic manufacturing growth of East Asia. It was thought that the move to an FTAA would create greater opportunities for manufacturing producers in Latin America (compared with East Asia) and lock in shifts to trade liberalization in Latin America. At the time of the Miami Summit (December 1994), the Clinton administration thought that the best way of achieving an FTAA would be by expanding bilateral agreements with the most liberalized and free trading of the Latin American economies (starting with Chile). However, problems with the US Congress meant that by the Santiago meeting in April 1998, the FTAA would be achieved through multilateral arrangements, in which Mercosur and the Andean Community would be prominent.

Before discussing these other two schemes, a brief note about Mexico in NAFTA is necessary. First, joining NAFTA in 1992 provided a continuation of its policy of trade liberalization that had started after the debt crisis. The main advantages of NAFTA membership have been seen as export growth (the United States already accounted for

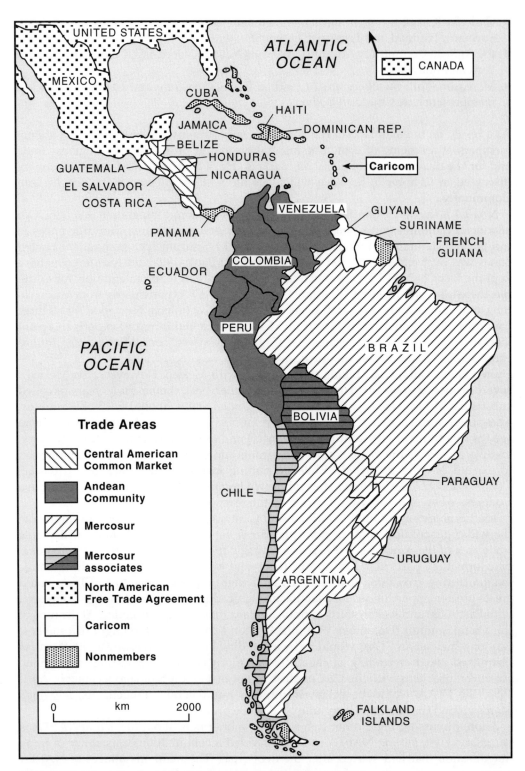

FIGURE 4.6 Schemes of economic integration in the Americas.

75 per cent of Mexico's exports in 1991) and increased inward investment (from companies wishing to have the twin benefits of a low-labor-cost location and direct access to the largest single market in the world). However, export-orientated assembly plants on the North Mexican border tend to depend heavily on cheap labor for their cost advantages, which has neither assisted the attainment of competitive advantage in Mexico nor advanced the nation's technological capability (Gereffi, 1996; Sklair, 1988).

NAFTA membership was also crucial in the aftermath of the severe economic crisis of early 1995 when the US government intervened with up to US$50 billion to support the Mexican economy. Furthermore, membership of NAFTA has also imposed global standards on Mexico in at least two areas – environmental provisions for pollution control (particularly on industry, to be fully enforced by 2002) and in terms of intellectual property rights. At the same time, Mexico is having its influence in Central America and the Caribbean reduced (Chapter 5), particularly as the smaller countries of these regions are not able to join NAFTA.

Mexico is experiencing a severe regional problem as the economic growth linked to NAFTA is being geographically concentrated in the northern region and avoiding the densely populated but poor south. Political protest against trade liberalization and NAFTA is particularly strong amongst the rural, indigenous groups of this area (Otero, 1996b). The south's 'backwardness' is becoming even more pronounced compared with that of the north – an example of how regions are being reconfigured in the context of global capitalism. Nevertheless, it could be argued that Mexico is also providing the geopolitical bridgehead between a United States increasingly wary of further integration with the rest of the Americas and Latin American countries still ambivalent about completely liberalizing trade with the most powerful global economy.

In South America, two regional trade groupings have developed, Mercosur and the Andean Community. Mercosur, based around an agreement between the two most powerful economies of Argentina and Brazil, has an agenda in which both trade liberalization and geopolitics are combined. Mercosur has made rapid progress towards freer regional trade since it began in 1991 with a tariff reduction program for reciprocal trade that finished in 1994. This was followed by the inauguration of a Common External Tariff on 1 January 1995. By 1997, trade within Mercosur had reached US$20 billion; in 1990 it had been only US$4 billion (*The Economist*, 1998a: 50).

Mercosur will find it more difficult to go beyond the objective of trade liberalization because, in particular, of its reliance on the officials of its member states. Here the officials of Brazil and Argentina can be said to dominate the agenda. The governments of these two countries seem to have pursued closer integration for different reasons. In Argentina, the radical free-market views of Economy Minister Domingo Cavallo were influential in the first half of the 1990s, and Argentina's membership of Mercosur is coordinated through the Economy Ministry. Cavallo hoped to use trade liberalization at the regional scale to discipline private firms and subject them to the 'reasonable' competition that they had avoided during the inward-orientated phase. In contrast, the Brazilian government has tended to see Mercosur as having more of a geopolitical role, and Brazil's membership of Mercosur is coordinated through its Foreign Ministry. Brazil sees itself (with a 1996 population of over 160 million) as the large, strategic country of South America and a global player to be reckoned with. Mercosur gives it the chance of leading a South American scheme of economic integration into the hemispheric plans for an FTAA in 2005. In these Brazil envisages its leadership of South America to be equivalent to that of the United States in North America; indeed, in the last two years of negotiations for the FTAA (years 2003 and 2004), Brazil and the United States will provide the joint chairs for the talks.

As a result, Brazil sees the necessity to expand Mercosur membership – and not necessarily restrict it to those countries that have liberalized their trade. Although by 1998 associate membership of Mercosur had been granted to Chile and Bolivia (both countries with liberalized trade), the next country being considered for this status was Venezuela (with a limited record of trade liberalization). This may pave the way for the incorporation of other economies of the Andean Community (Colombia, Ecuador, Peru). Although the Andean Community was created in 1969 (as the Andean Group), it was not until February 1995 that a customs union took effect – and with many exceptions. However, its common external tariff of 20 per cent for manufactured goods is similar to the highest tier of Mercosur. After 30 years of lacklustre performance, the Andean Community could well be integrated into Mercosur before the FTAA deadline of 2005.

However, the important point to emphasize is that trade liberalization at the national level has been reinforced by these regional attempts to create freer trade. Such a process has been labelled 'open regionalism', the objective of which is to strengthen the potential of the regional market while at the same time creating a basis for competing in global markets. Such regionalism has emerged as a potent new force in Latin America during the 1990s and can be seen to strengthen neoliberal policies at the national level. Substantially closer trading and investment links are now occurring between the member countries of at least two trading blocs (NAFTA and Mercosur). However, each member country is also pursuing the liberalization of trade with as many countries as possible in the wider global economy.

Acknowledgements

The author would like to thank Sylvia Chant for providing Table 4.9.

Further reading

Auty, R. M. 1993 *Sustaining development in mineral economies: the resource curse thesis*. Routledge, London. Includes a detailed analysis of the problems faced by the many mineral-rich but small Latin American countries in achieving pathways to sustained economic growth within the framework of inward orientation. It is aware of both the political and the economic factors operating.

Bulmer-Thomas, V. (ed.) 1996 *The new economic model in Latin America and its impact on income distribution and poverty*. Macmillan, London. Very useful collection of thematic studies on the new economic model in Latin America and its relationship with trade regimes, fiscal management, labor markets and financial liberalization but with a particular focus on income distribution and the reproduction of poverty. A series of case studies reinforces these themes with a particularly good and complex analysis of evidence about income distribution and poverty from the Chilean model.

Cardoso, E., Helwege, A. 1992 *Latin America's economy: diversity, trends, and conflicts*. MIT Press, Cambridge, MA. A well-organized and clearly-expressed textbook for students wishing to explore the problems of Latin American economies. Uses an historical context before examining key themes of the 1980s – debt, inflation, stabilization, economic populism and poverty. Lacks analysis of future directions and economic integration.

Edwards, S. 1995 *Crisis and reform in Latin America: from despair to hope.* Oxford University Press, Oxford. A World Bank insider (indeed former chief economist for Latin America) analyses economic transformations in Latin America in an optimistic but nevertheless critical way. Edwards clearly expresses the reasoning behind policy changes and decisions and, unlike many economists, is always aware of the political backdrop to economic crises and attempts to resolve them. However, he sees the World Bank's role with rose-tinted spectacles.

Gwynne, R. N. 1990: *New horizons? Third World industrialization in an international framework.* Longman, Harlow, Essex. Explores the nature of industrialization under inward orientation in Latin America and compares this process with the experience of industrial growth in East Asia. Addresses these issues within the context of an asymmetric world economy and focuses on issues of state intervention, trade policy, the behavior of multinational corporations and technological change.

5

GLOBALIZATION, NEOLIBERALISM AND ECONOMIC CHANGE IN CENTRAL AMERICA AND THE CARIBBEAN

THOMAS KLAK

INTRODUCTION

Central America and the Caribbean is a region of small, economically vulnerable and trade-dependent countries surrounded by larger and more industrialized countries that are moving more aggressively towards economic integration. In the hemisphere's transition towards neoliberalism, Central American and Caribbean countries tend to lag behind. This chapter surveys the ramifications of neoliberalism's outward-orientated strategies that contrast to the approaches common in the decades prior to the 1980s debt crisis that were relatively inward-looking. 'Inward' in this case refers to the geographical scales of both the nation-state and the Central American and Caribbean region, whereas 'outward' refers to the world beyond the region.

This chapter uses the labels 'Central America and the Caribbean' and 'Middle America' interchangeably to denote the countries situated between Mexico, the United States and South America (Fig. 5.1); note that Guyana, Suriname and French Guiana are traditionally considered part of the Caribbean rather than South America. 'Central America and the Caribbean' embodies more history but is a rather unwieldy term. 'Middle America', on the other hand, is a term used sparingly in the region. Central America and the Caribbean each have their own scholarly literatures, and it is uncommon to unite and compare them as in this chapter (for an historical comparison, see Grugel, 1995). Analysing Central America and the Caribbean as a single region is increasingly apropos, however, as events unfold around it. The North American Free Trade Agreement's (NAFTA's) designation of Mexico as part of North America, South America's organization into prodigious Mercosur and the weaker Andean Group (Chapter 4) and Reagan's Caribbean Basin Initiative (a trade policy for designated

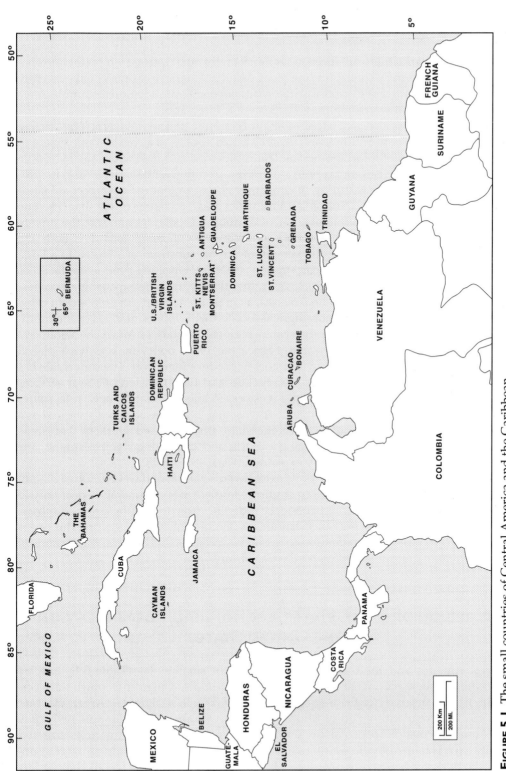

FIGURE 5.1 The small countries of Central America and the Caribbean.

'friendly' Central American and Caribbean countries) have all heightened the significance of Middle America as a meaningful region. Unfortunately, Middle America denotes a region anxious about, and reacting somewhat defensively to, hemispheric movements towards trade alliances to its north and south. Spurred on by the international climate favoring regional trading blocs, Middle American countries have recently formed the Association of Caribbean States, but beyond several regional summits have taken no firm steps towards region-wide economic integration.

Much of what was said in Chapter 4 about economic policies and problems during the twentieth century in South America and Mexico applies to a considerable extent to the smaller countries of Middle America. The decades prior to 1980 witnessed in Middle America the emergence of an array of state economic interventions including its promotion of domestic manufacturing of consumer goods. In that era of import substitution industrialization (ISI), a notion gaining widespread acceptance was that the state should control the 'commanding heights' of Caribbean economies and that the private sector should slot in under the state's guidelines for development. For both the Caribbean and Central America, the version of import substitution adopted to a large extent involved US multinational corporations (MNCs) relocating production facilities within the region to serve customers there, rather than a dramatic expansion of domestically-owned industries. In Middle America, as in South America, the 1980s debt crisis ushered in a new development paradigm associated with representatives of the multilateral aid agencies, who frequently visited regional capitals to move the neoliberal transition slowly but irreversibly along. Since the 1980s policy-makers have placed great emphasis on attracting foreign investors to produce for export. The state's role has shifted away from direct ownership, production and the provision of social services toward subsidizing export-orientated investors. While the state's new role under neoliberalism is often portrayed as one of downsizing, it is more accurately viewed as a qualitatively different relation between the state, investors and workers. For some neoliberal activities, such as promoting exports and competing for investment, the state's role has actually considerably expanded (Klak, 1996, 1998).

This chapter relies on Chapter 4 for its extended discussion of the evolution of macroeconomic policies in Latin America (including Middle America), and aims primarily to note ways that Middle America's experience is distinct from that of the larger countries. Because it can build on the previous chapter's economic policy groundwork, this one is able to devote considerable attention to the social, political and sectoral dimensions of development in Central America and the Caribbean.

A REGION OF SMALL STATES, ECONOMICALLY DEPENDENT ON THE UNITED STATES

Central America and the Caribbean is a region of small states (Table 5.1), but, by international standards, what is meant by small? Middle American countries are certainly tiny compared to the regional giants of Brazil, Colombia, Argentina and Mexico. The population of all of Middle America was about 72 million in 1997, only three-quarters that of Mexico alone. But our perception of Middle America's size and significance is tainted by the emphasis in the media and in scholarship on huge countries. Note that when size is measured by population, the mean for all the world's

TABLE 5.1 Basic indicators for Middle American countries and territories

	Area (square miles)	Population 1996 (thousands)	Infant mortality rate, 1996	Per capita GDP (US$)[a]	Human development Index, 1993[b]
Central America					
Costa Rica	51 100	3 534	13.3	5 500 (1996)	31
El Salvador	21 040	5 561	30.3	2 080 (1996)	115
Guatemala	108 890	11 685	49.2	3 460 (1996)	112
Honduras	112 090	5 751	40.2	2 000 (1996)	114
Nicaragua	129 494	4 386	44.1	1 800 (1996)	117
Panama	78 200	2 693	24.6	5 300 (1996)	43
The Caribbean					
Anguilla	35	10	23.0	7 600 (1994)	na
Antigua and Barbuda	171	66	17.2	6 600 (1994)	40
Aruba	75	68	8.2	18 000 (1994)	na
Bahamas	5 382	259	23.3	18 700 (1994)	26
Barbados	166	257	18.7	9 800 (1995)	25
Belize	8 867	219	33.9	2 750 (1994)	67
British Virgin Islands	59	13	19.1	10 600 (1991)	na
Cayman Islands	102	35	8.4	22 500 (1994)	na
Cuba	42 804	10 951	9.0	1 300 (1995)	79
Dominica	290	83	9.6	2 450 (1995)	65
Dominican Republic	18 704	8 089	47.7	3 400 (1995)	87
French Guiana	33 399	151	14.6	6 000 (1993)	na
Grenada	133	95	11.9	3 000 (1995)	77
Guadeloupe	687	408	8.3	9 200 (1995)	na
Guyana	83 000	712	51.4	2 200 (1995)	103
Haiti	10 597	6 732	103.8	1 000 (1995)	145
Jamaica	4 244	2 595	15.6	3 200 (1995)	86
Martinique	421	399	7.1	10 000 (1995)	na
Montserrat	40	13	11.8	4 500 (1994)	na
Netherlands Antilles	308	209	8.9	10 400 (1994)	na
Puerto Rico	3 515	3 819	12.4	7 800 (1995)	na
St Kitts and Nevis	104	41	18.9	5 380 (1995)	45
St Lucia	238	158	20.0	4 080 (1995)	76
St Vincent and Grenadines	150	118	16.8	2 060 (1995)	73
Suriname	63 251	436	29.3	2 950 (1995)	75
Trinidad and Tobago	1 978	1 272	18.2	12 100 (1995)	38
Turks and Caicos Islands	193	14	12.6	6 000 (1992)	na
US Virgin Islands	136	97	12.5	12 500 (1987)	na

Sources: The World Factbook (1997) and UN (1996).
na = not available.
[a] The year for which data were collected is shown in parentheses.
[b] Rank out of 174 countries.

countries (33.6 million people) is much larger than the median (6.7 million people). This is because the mean population is positively skewed by a few huge countries such as the United States, Brazil and Mexico. In fact, one out of every four countries in the world is smaller than Jamaica. By world population standards, El Salvador (6.2 million people) and Haiti (7.5 million people) are actually average sized. Similar conclusions would be reached if size were measured by area rather than population. Comparisons of this sort suggest the importance of examining unexceptional countries such as those of Middle America if we are to grasp the possibilities for and obstacles to development throughout the Third World (Table 5.1).

Middle America is also a region of economically vulnerable countries. The first column of data in Table 5.2 expresses trade dependence as a ratio of exports and imports to gross domestic product (GDP). The lower the value, the greater the share of a country's economic activity that involves domestic suppliers, producers and consumers. In other words, lower values mean more economic autonomy. The data indicate that, by and large, Middle American countries are considerably more trade dependent than are the larger and more industrialized countries represented in the table. Values of well over 100 for three of the Middle American countries indicate extreme trade dependence.

Note the exceptions to the above generalization. Guatemala is the least trade dependent of the Middle American countries, with a value similar to that of Chile, the most trade dependent of the South American countries shown. Indeed, Chile's pronounced primary-product-based export orientation has been the subject of considerable attention and debate in scholarly and policy circles (Chapter 4). But Chile distinguishes itself from Guatemala regarding the diversity of trading partners. Chile is not only considerably less dependent on imports from and exports to the United States, but also its exports there are now surpassed by those going to the European Union (EU) (and to Japan).

The patterns just described for trade dependency in Middle America are extended in Table 5.2 by the data for imports to and exports from the United States. For virtually every country shown, the United States is the largest importer and exporter. At the same time, note that the United States is the least trade-dependent country in the table (it has a value of 13 in the left-hand column). This is testimony to the profound economic power of the United States in the American hemisphere. The only exception to the rule of US trade dependency among the Middle American countries is St Lucia, whose banana shipments to a guaranteed UK market dominate its exports. Dominica and St Vincent display similar trade patterns. Unfortunately for these eastern Caribbean islands, their lack of trade dependency on the United States has only created a bigger problem. The United States has successfully argued to the World Trade Organization that the guaranteed market in the United Kingdom is illegal and must be terminated (this issue is discussed further below).

Besides Chile and St Lucia, the other exception to the rule of US trade dependency is Brazil, which also stands out for its exceptionally low value in the left-hand column for overall trade dependency. Brazil, together with Argentina, stand out in the table as the most economically autonomous of the Latin American countries. Note that the United States dominates the international trade of Mexico, but its value of 22 in the left-hand column indicates more economic autonomy than most other countries, especially those of Middle America.

Trade dependence is a key economic characteristic for all of Middle America, but countries differ in terms of level (Table 5.2). The smaller islands of the Eastern

TABLE 5.2 Trade dependency: selected Middle American countries in comparative perspective

	Ratio of exports plus imports to GDP (%)	Exports to United States (%)	Imports from United States (%)
Middle America			
Bahamas	130	51	55
Barbados	117	13	36
Dominican Republic	48	48	60
Guatemala	35	30	44
Honduras	59	53	50
Jamaica	82	47	54
Panama	77	39	40
St Lucia	166	22*	34
Trinidad and Tobago	96	48	48
South America and Mexico			
Argentina	20	9	21
Brazil	15	17*	23
Chile	38	15*	25
Colombia	28	39	36
Mexico	22	85	69
Industrialized Countries			
United States	13	na	na
United Kingdom	48	13*	12*

Sources: data on the ratio of exports plus imports to gross domestic product (GDP) are from Baumol and Wolff (1996: 877); these data are for 1950–90 with slight differences for Bahamas, Barbados and Jamaica. Data on exports to and imports from the United States are from World Factbook; these data are for 1993, 1994 or 1995, depending on the country.
Note: The United States is the largest export outlet and import source for all countries listed except those marked with asterisk. For St Lucia, 56 per cent of exports go to the United Kingdom; for Brazil, 28 per cent of exports go to the European Union; for Chile, 25 per cent of exports go to the European Union; for the United Kingdom, 13 per cent of exports go to Germany and 15 per cent of imports come from Germany.

Caribbean are more trade dependent than are the larger islands and mainland countries. Regional diversity is also suggested by the ranks of Middle American countries on the Human Development Index (HDI), a combined measure of life expectancy, literacy, access to education and per capita purchasing power (UN, 1996). As Table 5.1 shows, Middle American countries are distributed among the top two-thirds of the world's countries on the HDI index (that is, ranking 117 or lower out of 174 countries). Only Haiti falls in the bottom third of the distribution along with most of Sub-Saharan Africa.

Despite the variations noted above, there is much regional commonality. On average, Middle America is more dependent on trade, and is less industrialized, than are South America and Mexico. Middle America's economic dependence on the United States is also an important element of regional commonality, as distinct from the situation in South America. Middle America's relatively high levels of trade dependency distinguishes how neoliberalism is applied there compared with most of South America.

Neoliberalism puts pressure on already highly-trade-dependent Middle America to export more. Middle America's relatively low levels of output and industrialization also contribute to its peripheral status in the global economy.

The economic power of the United States in the region also allows it to express its geopolitical power. Within the past few decades, direct or covert intervention by the United States into many Middle American countries (for example Guatemala, Cuba, Guyana, the Dominican Republic, Nicaragua, El Salvador, Panama, Grenada, Haiti) attests to the importance the superpower places on pursuing and maintaining its interests in the region (Blum, 1986; Walker, 1997; Dupuy, 1997). The US economic embargo against Cuba is now well over three decades old and was only strengthened with the passage of the Helms–Burton Amendment in 1996 (LeoGrande, 1997).

Middle America's economic and geopolitical dependency on the United States is a constant factor weighing on the region's development policies, in contrast with the relative decision-making autonomy enjoyed by South American countries, especially the larger ones. Middle American countries must always take account of how the United States will react to their policy proposals. For Cuba's socialist regime, US animosity has been a constant preoccupation since the 1959 revolution.

DEVELOPMENT POLICIES PRIOR TO NEOLIBERALISM

The Caribbean was, from the outset of European colonialism, wholly outwardly orientated and economically dependent on a few primary product exports. In Central America, on the other hand, the haciendas which historically dominated were never as thoroughly focused on serving export markets as were the plantations of the Caribbean (West and Augelli, 1989). The legacy of this historical difference is somewhat more agricultural diversity in Central America than in the Caribbean, laid on top of the greater amount of arable land in the much larger mainland territories (Table 5.1). However, both regions have inherited highly unequal distributions of agricultural land. Throughout Middle America, rural peasants and small farmers have been unable significantly to expand production and income levels because the colonial or post-independence state has been unable or unwilling to implement serious land reform accompanied by appropriate infrastructure and extension services (de Janvry, 1981; Barry, 1987; Mandle, 1996).

Since World War Two, Caribbean states and large landowners have tried to reinvigorate the profitability of the monocrop plantation systems and associated export sectors by replacing an outmoded crop with another that appeared to have more promise. For example, various Caribbean countries substituted bananas for sugar, arrowroot or cocoa, coffee for cocoa, or citrus for sugar. However, such limited adjustments to the countries' agrarian production did not bring back the profitability of the colonial economies (Conway, 1998). Compared with the islands, there has been somewhat more diversity to Central America's agrarian exports, of which the five most important are coffee, bananas, cotton, sugar and meat. Still, these commodities have suffered since the early 1970s from deteriorating terms of trade. Further, they do little to support the basic needs of the vast majority of Central Americans who are not large landowners. Local protein consumption has actually fallen with the increase in land devoted to the production of meat for export (Barry, 1987).

The development policy experiments in Middle America in the 1950s to 1970s divide geopolitically into two camps – socialist and dependent capitalist – each with distinctive risks (Richardson, 1992: 102). The first set of policies that pursues development through socialistic routes has emphasized redistribution, equity, social welfare and prioritization of working-class needs. This route has purposefully attempted to turn away from dependence on the United States, which emerged as the region's political and economic hegemon around 1900 with the taking of Cuba and Puerto Rico from Spain and the Panama Canal Zone from Colombia. As the regional hegemon, the United States has interpreted the socialistic experiments as threats and has confronted them. Thanks in large part to pressure by the United States and its allies, the list of Middle American countries that have undergone failed experiments with alternatives to mainstream capitalist development reads like an epitaph to Third World socialism: Guatemala 1945–54 under Juan Jose Arevalo and Jacobo Arbenz, Guyana 1961–64 under Cheddi Jagan, Jamaica 1972–80 under Michael Manley, Grenada 1979–83 under Maurice Bishop, and Nicaragua 1979–90 under Daniel Ortega. Enemies from within and without (mainly the United States) worked to capitalize on any of the inevitable errors in leadership and to sabotage these experiments before their developmental capabilities could be ascertained (Blum, 1986; Sunshine, 1988; Booth and Walker, 1993). Only Cuba retains elements of socialism, but even Cuba is following the rest of the Caribbean region by relying on international tourism as its economic mainstay (Pattullo 1997), and by courting foreign capital with special labor codes and incentives, as discussed further below.

The essential challenge for the socialist regimes of Middle America has been to harness an underindustrialized and highly unequal peripheral economy and to expand rapidly the productive forces and to redistribute resources more fairly. This is in itself a tall order, made more difficult because it runs against the tide of global capitalism and US hegemony. Such departures raise broad questions such as 'How can a country in Middle America, in the US "backyard", pursue a noncapitalist path without the trade, capital, and blessing of the United States and the rest of the largely capitalist world?' 'How can internal resources be progressively redistributed when there are few to go around, and when societies are poor and unequal and therefore prone to internal divisions, unrealizable pent-up popular expectations, and patronage systems?' The challenges are many and the successes, not surprisingly, are few.

A second, contrary and far more common development policy tack has reinforced and extended ties with the North Atlantic region. It is represented by Puerto Rico's Operation Bootstrap industrialization commencing in the 1950s, by Europe's Banana Protocols dating back to 1957 with the granting of special market access to 12 former colonies, including eight in the Caribbean, by Reagan's Caribbean Basin Initiative (CBI) of the 1980s and by the present neoliberal transformation. The positive and negative impacts of the first two of these capitalist policies are briefly reviewed below.

As an economic development policy, CBI has two main components (it was initially a disguised US military aid policy against leftists in Central America). CBI and related US tax provisions encouraged a further opening of Middle American markets to US products and a relocation of US garment factories to Middle America to take advantage of cheap labor and subsidized factory space (Deere et al., 1990). The United States now has a large and growing export surplus with Middle America, which, along with the rest of Latin America, is the only world region with which the United States has a trade surplus. CBI should therefore be seen as a vanguard policy for a reconstituted US regional hegemony under neoliberalism, to which we turn for the remainder of the chapter. In general, Middle America's dependent capitalist paths to development avoid US hostility and instead aim to exploit opportunities availed by strong trade and policy ties to advanced capitalist countries. The risk is that the results will replicate history whereby Middle America has gained relatively little in subordinate relationships with core countries.

Puerto Rico's Operation Bootstrap continues to be viewed by political leaders in the region as an industrial development model, although upon closer inspection it can be seen as nonreplicable and domestically exclusive. Puerto Rican industrialization derives from its unique Commonwealth relationship with the United States. Operation Bootstrap began in 1947 when US tax law amendments gave corporations multiyear tax exemptions on any net income deposited in banks on the island. For their part, Puerto Rican officials promised industrial infrastructure and low-wage, nonunion and, by Caribbean standards, skilled labor (Thomas, 1988; Cordero-Guzman, 1993). First to relocate to Puerto Rico were light industries such as clothing, shoes and glassware, then came the heavier and more environmentally damaging industries of petroleum refining, petrochemicals and pharmaceuticals. They contributed to impressive annual economic growth rates of 6 per cent in the 1950s, 5 per cent in the 1960s and 4 per cent in the 1970s. All totalled, Puerto Rico's real GDP growth rate for 1950–90 was the eighth highest of the 74 countries worldwide for which data are available, and the highest in the hemisphere (Baumol and Wolff, 1996). In 1979 Puerto Rico had the highest per capita level of US imports, modern transportation facilities and 34 per cent of all US foreign direct investment. The 22 US drug companies manufacturing in Puerto Rico during the 1980s saved US$8.5 billion in income tax exemptions (Freudenheim, 1992). The subsidies from the various sources described above from the US mainland to Puerto Rico amount to around US$9 billion annually, almost as much as all of the US aid to the rest of the world combined (de Blij and Muller, 1998).

The geopolitical price of Operation Bootstrap is great dependency on US subsidies, capital and trade. Few local manufacturers have emerged or have networked with those from the mainland. In 1996 US Congress began a 10-year phase-out of Puerto Rico's corporate tax exemption, leaving the island entrenched in a bankrupt model of dependent industrialization (Pantojas-Garcia, 1990). Puerto Rican people have made the most of their special political status by migrating to and circulating back from the US mainland in great numbers and by relying on the federally mandated social welfare net (Conway, 1998). The human face of Puerto Rico's industrial expansion includes the fact that real per capita income has not grown relative to that of the mainland United States during Operation Bootstrap and that unemployment has often been in excess of 20 per cent since the 1970s (Cordero-Guzman, 1993; Grugel, 1995).

Europe's Banana Protocols began in 1957 and were followed in 1975 by a succession of four Lomé Conventions. They have provided special market access to more than 70 Third World nation-states that were once European colonies and which are now designated as the ACP (Africa–Caribbean–Pacific) countries. The special market access applies mainly to primary sector commodities such as coffee, sugar and bananas. Only 12 of the ACP countries export bananas to Europe through Lomé, and eight of these are members of Caricom (Caribbean Community and Common Market) (Fig. 5.2). Measured by volume traded, bananas are the world's most important fruit or vegetable, the EU is the banana's largest market and the EU's banana subsidies are estimated to be worth US$2 billion annually (Wiley, 1996, 1998; *The Economist*, 1997b). Like Puerto Rico's Operation Bootstrap, banana farming in the Eastern Caribbean provides an example of the double-edged sword of growth/prosperity and dependency/vulnerability associated with relying on special trade preferences from the core capitalist countries.

In the 1950s British public authorities and private shipping firms encouraged its

FIGURE 5.2 The weekly banana shipment from Dominica to Britain. The World Trade Organization has ruled that the Eastern Caribbean's preferential market access is illegal and must be eliminated. Photograph by James Wiley.

Caribbean colonies to shift from sugar cane to bananas. Britain offered a guaranteed market and good prices for all the bananas that they could grow. Farmers in St Lucia, Dominica, St Vincent, and, to a lesser extent, Grenada and elsewhere in the Caribbean responded wholeheartedly, digging up or abandoning many other crops to specialize in bananas (Welch, 1994; Wiley, 1998). In St Lucia, for example, the growing reliance on bananas continued through the late 1980s as thousands of additional acres were converted or deforested to grow them (Barrow, 1992). As a result, Dominica, St Lucia and St Vincent have each earned more than half of all their foreign exchange from bananas, which has placed them among the countries of the world most dependent on the export of a single cash crop.

Dependence on a guaranteed overseas market for a single cash crop has had its positive side. It fueled a growth of prosperity on the islands and especially for their many small family farmers, whose prominence and vitality are unusual in a region dominated by large landholdings (Barrow, 1992). Banana exports helped Dominica, St Vincent and St Lucia climb to the top half of countries ranked by the Human Development Index (Table 5.1). Additionally, only the smaller Caribbean territories with special access to European markets, primarily for bananas, were able to avoid the fluctuations and decline in Middle America's traditional agricultural exports during the 1980s (Schoepfle and Pérez-López, 1992).

EU rulings in recent years, however, began a gradual phase-out of the ACP countries' special market access (de Cordoba, 1993; Wiley, 1996). A 1997 World Trade Organization (WTO) decision completed the phase-out for bananas by ruling that the preferential access to the British market is illegal. The United States brought the case to court after extensive lobbying by Chiquita's chief executive officer Carl Linder who exports bananas from Central America and would like a larger share of the EU market (Wiley, 1996; Herbert, 1996; *The Economist*, 1997a). The EU's preferential banana market will soon close (barring the Caribbean islands' successful appeal against the WTO decision) and banana farmers will be forced to find new sources of income.

Bananas grown on small family farms in the Eastern Caribbean cannot compete with the volume, size and price of bananas from Ecuador and Central America grown on mechanized plantations where wages are 50–75 per cent lower and unions are repressed (Barry, 1987). However, the fruit has become too central to, and entrenched in, the rural societies of the Eastern Caribbean for farmers to abandon it quickly. In St Lucia for example, more than two-thirds of agricultural land was still devoted to bananas in 1997. Public policy devoted to the promotion of nontraditional exports has therefore needed to concentrate on the industrial and tourism sectors, while continuing to work towards diversifying agriculture (Pattullo, 1996; Klak, 1998). The monumental task for the Eastern Caribbean, and for Middle America more broadly, is to replace traditional sources of income with revenue from sources suitable to the present era of more open trade relationships and to avoid the vulnerability associated with relying on a single product and North Atlantic market (Table 5.2).

THE DEBT CRISIS AND THE NEOLIBERAL REMEDY

> Latin America [had no] desire to flaunt free trade, but . . . was forced to do so. [In this period,] most of Latin America followed the liberal doctrine of free trade. Advocates of protection in Latin America found little favor in government circles which were beholden both literally and intellectually to creditors in the developed world (Skidmore, 1995: 228).

This profile could well apply to the 1980s and 1990s when in fact it was written to characterize the period 1880–1914. The point is to stress that now is not the first time that Latin America has been lectured on the benefits of free trade. Within Latin America, Middle American countries have over time remained especially vulnerable owing to their distorted, underdeveloped and peripheral economies. As the example of bananas illustrated, these countries have overemphasized a few export-orientated primary products, particularly agriculture and in a few cases mining. They have also relied on imported industrial and commercial goods and on foreign capital and expertise to feed people, to service industry and to finance internal capital expansion (Grugel, 1995). Even Trinidad, the region's wealthiest country owing to its petroleum deposits, has made only modest progress towards indigenous industrial diversification while at the same time it has accumulated US$2 billion of foreign debt (Mandle, 1996; Klak, 1998).

The current development policies in Middle America focus on attracting foreign investors to generate new exports, and date back to the debt crisis of the 1980s. The foreign debt situation for Middle American countries parallels that of the hemisphere's larger countries, as told in Chapter 4. In the 1970s international banks were awash in deposits and therefore eager to make loans to public and private interests in Latin America and the Caribbean (Corbridge, 1993). When interest rates rose in 1980–81, the world economy sunk into recession and the vulnerable primary product and tourism sectors of Middle America lost many of their customers. Economic decline was precipitous region-wide. By the end of 1983, the economies of Barbados, Dominican Republic, Guyana, Haiti, Jamaica and of Trinidad and Tobago shrank an average of 17 per cent in real terms compared with 1980 (Conway, 1998). The interest rate hike also contributed to making the foreign debts of most Middle American countries unserviceable. In sum, an array of related factors, from poor agricultural yields and prices, to wasted windfall loans and political thinking that seeks industrialization through MNC investment, led to governmental insolvency by the 1980s (Mandle, 1996).

The many interrelated problems outlined above have provided strong motivation for governments across Middle America to approach the international development agencies seeking debt rescheduling, additional loans and, in the words of its leaders, the International Monetary Fund's (IMF's) 'blessing' (McBain, 1990; Killick and Malik, 1992). Middle American governments one after another, conservative and socialist alike, have had to sign onto a series of austerity agreements with the IMF and World Bank in exchange for financial bailout. The international agencies' demands of fiscal conservatism, privatization, economic opening and aggressive promotion of MNC investment have the effect of homogenizing government policies across Middle America (and the Third World) so that distinctive development strategies are history. They are likely to continue along the homogeneous neoliberal track in the future, in part because of the ongoing constraints of foreign debt. In 1996 the total foreign debt of Latin America and the Caribbean stood at US$622 billion, or US$189 billion more than in 1987 when bankers and First World politicians considered that the crisis was at its worst (IADB, 1997: 256; Rosen, 1997). Measured against population size or various economic indicators, the foreign debts of Jamaica, Guyana and Nicaragua were by the late 1990s the hemisphere's worst and among the most burdensome in the world (Box 5.2 below).

THE MEANING AND IMPACTS OF GLOBALIZATION IN MIDDLE AMERICA

Over recent years 'globalization' has become one of the most commonly used terms in corporate, political and academic discussions. Despite its ubiquity, it is unusual to find a clear definition of globalization and even rarer to see it interrogated against evidence. Some writers on the political left disparagingly dismiss globalization as a corporate takeover scheme (Rieff, 1993), but many other influential commentators are more optimistic about its impacts (Reich, 1991; Qureshi, 1996). Of particular concern here are some of the positive spins given to globalization trends suggesting that there are many new opportunities for peripheral countries such as those of Middle America. Many (but not all) of these positive interpretations of globalization in the academic literature fall under the umbrella notion of a 'global village'. Global village ideas suggest that in various ways the world is coming together and balancing out. Such global village ideas are misleading, however, when considered in light of the history and contemporary situation in Middle America; the regional situation is unfortunately less balanced and promising than global village ideas suggest. Further, current political and economic transformations must be specified at the local level rather than subsumed under overly generalized notions of globalization. There are six components of the global village concept that are worth addressing with respect to contemporary trends in Middle America.

First, it is claimed that the world is 'shrinking' as a result of greater international connectivity (Giddens 1990: 64; World Bank 1995a; Watts 1996: 64–65). Middle America, however, has long been heavily integrated into the global economy, as suggested by the earlier discussion of its trade dependency. The Caribbean, in particular, is historically perhaps the most globalized of world regions. Since the 1500s it has been controlled by outside powers, based economically on imported labor, cleared to create monocrop landscapes of sugarcane or bananas and reliant on the import of virtually everything else needed to sustain local populations (Richardson, 1992; Mandle, 1996). In Central America, export-orientated production of sugar, bananas, beef and other agricultural commodities has been dominant since before the twentieth century. So for Middle America current 'globalizing' trends represent another round of powerful external influences for a region historically shaped by exogenous decisions and events.

Second, traditional distinctions between core and peripheral regions of the world are said to be blurring (McMichael, 1996). Writers such as Kearney (1995: 548) claim that 'globalization implies a decay in [the] distinction' between core and periphery. This structure of inequality is still highly relevant to an understanding of US–Middle American relations, however (Klak, 1998). The fact that production (of certain items) is more geographically dispersed across nation-states does not indicate an accompanying fundamental redistribution of control over and benefits from production. The constraints imposed on Central American and Caribbean workers by their own capitalist classes and states, and the constraints imposed on Caribbean countries by core states, especially the regional hegemon, the United States, produce vast international gaps in income and living standards.

Third, it is said that foreign investment, trade and opportunities for development are more widely distributed across world regions and nation-states (Qureshi, 1996). However, global economic integration is highly selective in favor of developed countries and a few developing countries, none of which are in Middle America. Fully 80 per cent of all world trade is within the core triad, although it is less than 20 per cent of world population (Dicken, 1992; Hirst and Thompson, 1996). Of the world stock of direct foreign investment, 81 per cent was located in the EU, North America and Japan as of 1991, up from 69 per cent in 1967 (Koechlin, 1995: 98). There has been an increase in the share of direct foreign investment flows to certain developing countries, but none of the major recipients (headed by China, Mexico and Argentina) is in Middle America. There is indeed a global integration apace with regard to the basic ideas about development and neoliberal policy, and there are pressures and incentives to increase trade, but the material rewards are enormously unequal.

Fourth, it is claimed that there is now a convergence into one world economy, to which all places and people can, and generally do, find export market niches (McMichael, 1996). But for peripheral regions such as Middle America, market niches are narrow, highly competitive and fraught with obstacles. They are largely outlets for selling nontraditional goods and services, (that is, fruits, vegetables, flowers, garments, electronic products, processed data and tourism) to wealthy Western consumers (Fig. 5.3). However, Middle America as a whole has thus far had very limited success with filling new export market niches (Klak, 1998).

Fifth, media and cultural influences are supposedly more widespread and multilateral (Patterson, 1994; Kearney, 1995). It is a stretch, however, to compare the northward influence of such artifacts as reggae music, Mayan handicrafts or merengue music with the multitude of cultural impacts in the opposite direction. US affluence and opportu-

FIGURE 5.3 Electronics assembly in Costa Rica. Owing to its international reputation for educational attainment, political stability and environmental quality, Costa Rica has been the most successful Middle American country for attracting foreign investment in an array of export sectors paying reasonable wages. Photograph by Thomas Klak.

nity, often romanticized, is especially well-known, deeply ingrained and alluring to people of Middle America. They are prone to set their living standard goals in accordance with what the US media ascribe to the United States. And the imbalance in media flows is increasing with Middle America's economic crisis and neoliberalism, as local media have been slashed. And the imbalance in the flows of influence between the United States and Middle America is greater for economic and geopolitical issues than for cultural issues.

Last, people themselves are said to be more integrated through immigration and communications. However, the increasing international flows of people and information should not obscure the region-specific and class-specific motives for and access to emigration and communication. In pursuit of opportunities for economic development and also for social and cultural expression, Caribbean people have engaged in more twentieth century emigration than those of any other region. Central Americans have also emigrated in vast numbers in recent decades, especially to flee civil war, oppression and the associated lack of opportunities. The societal porousness and citizens' coping strategies are not a global phenomena but are particular to the historical and geographical context of the region and its inhabitants (Simmons and Guengant, 1992). The massive scale of contemporary emigration and remittances is the result of growing marginalization and pessimism about the economic prospects back home. Middle America is economically stagnant, and current globalization trends present Middle Americans with few promising options.

The main point of these critical comments on globalization is to suggest that current economic and political trends are not really globalized, but rather are highly uneven geographically, in terms of impacts and control (Dicken et al., 1997). Peripheral regions such as Middle America are certainly shaped now, in the era of globalization, as they have been under previous phases of capitalism, by the ideas and actions of outside investors and political leaders. Currently, Middle American governments are under pressure to create incentives that will attract investors in order to export a greater quantity and range of products (Klak, 1998). The following sections look more closely at these exogenously originating political and economic transformations in Middle America.

THE CURRENT NEOLIBERAL DEVELOPMENT MODEL IN MIDDLE AMERICA

Because Middle American economies are smaller, less diverse and less industrialized than their South American counterparts, the reform period since the 1980s has been especially painful to its working people, and the new outward orientation has much less to build on. While Latin America has long held the unfortunate distinction as the world's most socially unequal region, things have only worsened since the debt crisis and the neoliberal bailout. For Latin America and the Caribbean as a whole the share of population living in poverty increased from a historical low of around 22 per cent in 1982 to over 30 per cent in the 1990s. With the exception of Costa Rica, Central America is more unequal than the average Latin American country (IADB, 1997: 41). This helps to explain why Central America's poverty rate is now almost 60 per cent (Latinamerica Press, 1998).

With the most notable exception of Haiti (Box 5.1), Caribbean countries are more socially equal than both Central American and Latin American averages. This helps to account for the Caribbean's lower poverty rates and relatively good standing on social welfare indicators (Table 5.1). However, the Caribbean's relative equity is canceled out by its relative economic weakness (Table 5.2). Small size, limited industry and an historical lack of international competitiveness leave the Caribbean in a precarious position during the neoliberal transition.

The international development agencies (chief among them, the World Bank, the IMF and USAID) are encouraging Middle American governments to promote production for export rather than for domestic markets and to reduce social welfare spending. In most cases domestic production and social welfare spending were already inadequate prior to the cuts (Box 5.1). A recent comparative study of the public sector reforms under neoliberalism in three Middle American countries (Costa Rica, Nicaragua and Jamaica) echoed these observations (Evans et al., 1995: 44–5). The study identified two important commonalities in the new role of the state in these countries. Internationally, the relationships of these Middle American countries are increasingly neocolonial in nature. As the authors put it, 'the countries themselves are able to exercise less control over their economic policy than was the case in many former colonial regimes'. Domestically, policies have turned noticeably more hostile to the interests of the working class. This has occurred directly through policies that have dramatically cut public services, employment and wages, and indirectly through increased unemployment. In addition, governments have weakened trade unions through repression and division. In particular, the abject dependency of emasculated Nicaragua in the late 1990s stands in acute contrast to the widespread hope in the early 1980s among most of its people for a socially just society (Box 5.2).

BOX 5.1 World Bank Recommendations for Haiti

Haiti has for many years been the poorest country in Latin America (Table 5.1). For the vast majority of the Haitian population, conditions have been characterized by a high infant mortality rate, low life expectancy, low birth weights, high maternal deaths, inadequate caloric intake, squalid housing conditions and a lack of basic services such as potable water and sewer lines. The country is also economically dependent, importing around 70 per cent of all that it consumes. It is in this context that the World Bank in 1985 offered advice to the Haitian government in a report intended to have restricted distribution, but which became public. The report, entitled 'Haiti: policy proposals for growth', suggests the following reorientation of national priorities:

> The development strategy must be export-oriented ... [Domestic] consumption ... will have to be markedly restrained in order to shift the required share of output increases to exports ... More emphasis will have to be put on development projects that support the expansion of private enterprises in agriculture, industry and services. Private projects with high economic returns should be strongly supported with accordingly less relative emphasis on public expenditures in the social services (quoted in Wilentz, 1989: 272–3).

Given how reliant Haiti is on foreign assistance and investment, World Bank recommendations such as these carry a lot of weight in government circles.

BOX 5.2 Nicaragua's Sandinistas and their successors (source: Walker, 1997)

In the 1970s, mounting society-wide disenchantment with the 'mafiacracy' of Anastacio Somoza helped the Sandinista rebel army depose him by the end of the decade. Over the next several years the ruling leftist junta was able to demonstrate effective leadership. Accomplishments included effective social policies such as in a literacy campaign and agrarian reform, a mixed economy which grew at 7 per cent annually per capita and a clear electoral victory in 1984. For the remainder of the 1980s, however, Ronald Reagan sought to make the Sandinistas 'cry uncle' (as he put it) by orchestrating a cut-off of its international loans and the Contra war. The latter's many costs included the more than half of Nicaragua's public resources that were needed for defense, devastation of rural schools and clinics that the Contras viewed as targets, and 30 000 Nicaraguan lives. By 1990 Nicaragua's war-weary voters rejected the Sandinistas in favor of the United States' preferred candidate, Violeta Chamorro. On her watch the country reversed policy direction and began what William Robinson characterized as 'a close U.S. tutelage in the process of reinsertion into the global system.' Nicaragua's USAID program leapt to become the world's largest, and included funds to replace Sandinista textbooks with ones beginning with the Ten Commandments that USAID's director believed would help 'reestablish the civics and morals lacking in the last eleven years.' By the late 1990s, more than half of all Nicaraguan workers were unemployed or underemployed. And the US$500 million dollars worth of international aid keeping the Nicaraguan economy afloat has made current conservative–populist President Arnoldo Alemán highly subordinate to the wishes of core capitalist countries. Despite Nicaragua's 'economic straitjacket,' however, it has managed to retain some meaningful vestiges of the Sandinista period. These include vibrant grass-roots activism, politically-astute citizens, gender equality laws, and autonomy laws protecting its culturally distinct Atlantic Coast people.

Middle American governments are aggressively promoting and subsidizing nontraditional exports (Klak, 1996, 1998). These efforts can be likened to a 'shatter-shot' approach to searching out export market niches. Even in the smallest countries, policy-makers are actively promoting investment in a host of nontraditional activities. In tiny Dominica, for example, these range from tourism, assembly operations and data processing, to vegetables, fruits, seafood and cut flowers (Fig. 5.4; see also Wiley, 1998). Such experimentation raises the essential question of whether Middle American countries can identify product niches with considerable promise or whether they are replacing monocrop and single-market dependence with a new form of vulnerability. In other words, are Middle American countries trying to do many things at once while not doing any of them well?

The nontraditional export activity that employs the most people and is most widespread in Middle America concerns the assembly factories. Virtually every Middle American state has created export processing zones and has taken loans to build factory shells within them. Garments are by far the most common product, although electronics, plastic goods and shoes are also represented. Tens of thousands of mainly young females are employed in such factories in Jamaica, Haiti and each Central American country. Most of the Eastern Caribbean countries have more than a thousand workers each. The Dominican Republic has by far attracted the most assembly activity and has over 160 000 factory workers. But even there, the assembly plants are low-paying economic enclaves with minimal positive impact on the local economy (Willmore, 1994; Kaplinsky, 1995). And although it is widely claimed that the assembly plants provide

Figure 5.4 Dominica's nontraditional industrial exports, on display at the Dominica Export Import Agency (DEXIA). Is Middle America successfully filling a neoliberal export niche or undergoing a new round of dependency? Photograph by James Wiley.

net economic gains for the host country that assertion would need to be validated through a thorough cost accounting which, to my knowledge, has never been undertaken. The gross income from low factory wages and subsidized rents and utilities would need to be weighed against the public costs of promoting the country as an investment site, of building, operating and maintaining the factories and related infrastructure and of training workers (Klak, 1996, 1998). The main beneficiaries of the low-cost products made in Middle American assembly plants are the holders of the corporate brand names they carry, the department stores where they are sold and US consumers who buy the low-priced products (more information on these issues is available through the Campaign for Labor Rights website at www.compugraph.com/clr).

Before examining the prospects for nontraditional agriculture, it is worth noting the continued significance of traditional primary product exports. While the region's traditional exports such as sugar, bananas, coffee, cotton and bauxite have lost value in recent years, it is premature to deem them irrelevant to the role of Middle American countries in the global economy. In 1990, primary products (mostly traditional ones but including a steadily growing share of nontraditional exports) were still 46.5 per cent of the exports sent to the United States by CBI countries. By then, manufactured products had surpassed primary products as a share of total, but this reversal is attributable more to primary export decline than to manufacturing growth (Deere and Melendez, 1992). As noted above, new exports are mainly associated with the assembly plants.

Export goods such as fruits, vegetables and flowers, labeled NTAEs (nontraditional agricultural exports), by definition have high value by volume and area under cultivation. Central America has been more successful than the Caribbean in meeting the demand for NTAEs in the United States and other northern markets. By 1991 Central America's NTAE earnings exceeded US$175 million, while those of the Caribbean were under US$90 million (Thrupp, 1995). The Caribbean's limited success is especially notable given its more desperate need for new sources of foreign exchange. One indicator of this need is that Caribbean earnings from traditional agricultural exports have fallen faster than those of Central America.

A more general problem with NTAEs throughout Middle America is that the neoliberal image of a small-scale farmer/entrepreneur rising up to meet new export market niches – perhaps substituting snow peas or strawberries for bananas – does not match the reality. Although the experience varies across contexts and products (and therefore leaves hope for greater small-scale sustainable production), the following general characteristics of NTAEs in Middle America restrict benefits to smaller-scale farmers:

- markets are highly competitive, require connections to complex international commodity chains (Gereffi and Korzeniewicz, 1994), continually draw new entries from countries worldwide and risk saturation;
- products are highly perishable, require expensive transportation, are subject to wide-ranging price fluctuations and entail risk for producers;
- the vast majority of NTAEs are native to temperate rather than tropical climates; this disadvantages Middle American farmers whose knowledge is of local crops;
- crops are usually planted continuously and intensively in monocultures, and buyers demand perfect-looking produce;
- the above features often lead to the problem of a 'pesticide treadmill', to related problems of human and ecological health and unsustainability and to residues on crops entering the United States (Conroy et al., 1996);

- the above features, combined with the need for large capital investments, contribute to dominance by large foreign firms, for which local people mainly serve as low-wage labor.

These factors have meant that small-scale farmers have generally not been competitive (Thrupp, 1995). The NTAE sector is characterized by dominance by firms from the United States and elsewhere in the core triad, inadequate state support to develop the sector, shaky performance and low to no growth for small-scale local producers and poor working conditions for the employees. In essence, core–periphery relations are maintained despite shifts in economic sectors.

Neoliberalism's market niche strategy may have its conceptual appeal, but Middle America's experience with it shows that it is fraught with difficulties, as the data-processing sector further illustrates. The development prospect of this sector entails incorporating local labor and nurturing local firms to take advantage of expanding opportunities in the global data-processing industry that earns US$1 trillion yearly, of which information-processing services is a component. Mullings (1995, 1998) draws on a detailed analysis of the rise and fall of information services in Jamaica to explain why this industry, which has real potential (however narrow) for growth in employment, wages, managerial expertise and backward linkages into the local economy, has stagnated in terms of all these criteria. She identifies the problem in terms of inadequate state support for local firms; continued policy steering and dampening by traditional, nondynamic private elite; investment fear on the part of foreigners; and an extremely narrow role allotted to Jamaican firms and workers by US outsourcing firms. Rather than propelling Jamaica to a heightened position in the international division of labor as the neoliberal model predicts, the information services sector has slumped and entrenched the gender, class and international inequalities that have long characterized this peripheral capitalist country.

Another relatively new and globally orientated economic activity that Caribbean political leaders have eagerly pursued by offering investment incentives is offshore finance. The islands that are winners and losers in the casino capitalism associated with the quest for hypermobile international finance capital have their own set of pre-existing attributes and consequent problems. Islands that have been most successful at attracting offshore banking and related activities are relatively small territories, even by Caribbean standards, that continue to fly the Union Jack: The Cayman Islands, the British Virgin Islands, the Turks and Caicos Islands and, until volcanoes made two-thirds of the island uninhabitable in 1997, Montserrat. The British dependencies confer the highest level of political stability and investor confidence. Besides the British dependencies, the Bahamas (which is closely tied to the US economy; Table 5.2) and the Netherlands Antilles are also notable for their ability to attract financial holdings from thousands of foreign firms. In some places, such as the Caymans and the British Virgin Islands, offshore finance has become such a major economic component as to alter dramatically domestic politics and social conditions (Roberts, 1994; Evans, 1993). But most islands have seen little of the sought-after mobile capital or, in fact, capital of any kind. In fact an island's ability to attract furtive capital into an offshore finance sector is negatively correlated with its need for new sources of foreign exchange earnings. Most islands are now independent countries with many features viewed as unattractive to finance capital, including weak and unstable economies, poverty, high unemployment and social tensions. These and other colonial and neocolonial legacies leave most of the

Caribbean isolated and uninteresting with regard to the offshore financial component of the current round of reordering of the global economy.

THE SPECIAL CASE OF CUBA: ISLAND SOCIALISM AMID GLOBAL CAPITALISM

The above examples of nontraditional sources of foreign exchange earnings indicate that neoliberalism is wholly based in the hegemony of global capitalism and the associated international system of states. But even socialist Cuba has needed to turn to a development policy that offers incentives to foreign investors. Cuba's crisis is not unrelated to that of the capitalist countries of the region. Since the nineteenth century Cuba has essentially had a one-crop (sugar cane) exporting economy with the concomitant vulnerabilities of output and price fluctuations and deteriorating terms of trade. In part Cuba's crisis is unique to its experience in CMEA (Council for Mutual Economic Assistance), the now defunct trade alliance among state socialist countries led by the former USSR. Rather than embarking on a major post-revolutionary economic diversification effort from the 1960s through the 1980s, Cuba was encouraged to specialize primarily in sugar exports and to import most other requirements from other CMEA members. A one-for-one trading deal with Russia, sugar in exchange for petroleum, encouraged Cuba's specialization and yielded a US$5 billion annual subsidy compared with open-market prices (note, however, that little sugar actually trades on the open market). The USSR also availed to Cuba a generous line of credit that in 1995 left an outstanding debt of US$10 billion in addition to the island's US$9 billion of debt to Western creditors (*The World Factbook*, 1997).

Cuba's CMEA trade relationship delivered a reasonable assortment of industrial inputs and consumer goods until the unexpected events of the late 1980s. The collapse of the Soviet bloc over the period 1989–91 meant that Cuba lost 75 per cent of its trade flows and imports as well as aid worth 22 per cent of national income (Marshall, 1998). Cuban imports fell from US$8.1 billion dollars in 1989 to only US$2 billion in 1993. Cuba's economy went into a free fall (Fig. 5.5). Independent of a country's political–economic organization, it is hard to imagine any nation enduring such vast economic losses and weathering through the necessary transitions without a major internal upheaval.

Castro coined the euphemism 'The Special Period in the Time of Peace' to describe the post-Soviet era of massive shortages, daily hardships and new policies that encouraged private initiative and courted foreign investors. Domestic policies now include the selling of agricultural surplus at the farmer's markets directly to consumers, parallel market activity for many consumer goods, legalized use of the US dollar and the beginnings of a conversion of the entire currency to a dollar equivalent. Production policies include encouraging foreign investment by allowing (at least partial) foreign ownership of enterprises in Cuba, promoting tourism, reorientating state investment to enhance product development for global markets and reconfiguring large state farms into autonomous and productive cooperatives (Susman, 1998).

Obtaining food and other daily necessities now preoccupies many citizens, especially Habaneros who are most isolated from rural food production. Urbanites who considered curbside gardening or bicycling beneath their sophisticated lifestyles grudgingly

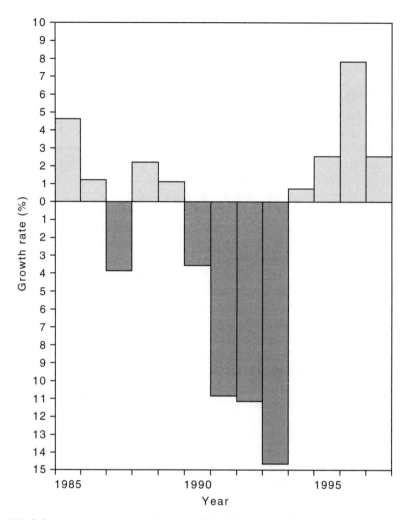

FIGURE 5.5 Cuban economic growth rates, 1985–97. Sources: data compiled by Paul Susman, Department of Geography, Bucknell University, Lewisburg, PA, USA, from UN *Anuario Estatistico de America Latina y el Caribe*, Santiago, Chile; UN (various years); other. *Note*: figures for 1985–89 are global social product, whereas those for 1990–97 are the more standard gross domestic product; for an analysis of their differences see Zimbalist and Brundenius (1989).

adapted to necessity. Faced with a crisis of shortages of basic needs, theft of state resources and black marketeering, Cuban officials have been forced to legalize a range of capitalistic practices. In a rare public opinion poll, most Cubans pointed to the farmers' markets legalized in 1994 as the most important recent reform, well ahead of self-employment and legalization of the dollar (Marshall, 1998: 287). To compensate for the agricultural bias against the capital city, farmers' market vendors in Havana are taxed at a rate of 5 per cent of the value of their goods, while everywhere else the rate is 15 per cent (Marshall, 1998). The farmers' markets have expanded to offer prepared food, and a growing number of Habaneros have made their homes into informal restaurants, called *paladares* (Segre *et al.*, 1997). Further, the foreign tourism sector has acted like a vacuum drawing Cubans of all stripes away from skilled professions and toward dollar-earning possibilities. A highly trained professional now realizes she or he can earn much more driving a taxi cab for tourists (Segre *et al.*, 1997). Besides turning to tourists as a source of income, Cubans have needed to become very imaginative in order to obtain enough income or in-kind transfers to make ends meet (Box 5.3).

The Cuban economy bottomed out in 1993 and has since haltingly recovered (Figure 5.2), bringing with it greater availability of food and basic necessities. However, the ongoing crisis of Cuba's CMEA-based economy is suggested by the fact that the 1998 sugar harvest of around 3 million tonnes was the lowest since 1959. Cuba has therefore had little alternative but to join most other islands within the membership of the

---

BOX 5.3 How people survive during economic crises: a comparison of Cuba and Jamaica

Cuba's economic difficulties have forced its citizens to resort to survival tactics, adding new meaning to the verbs 'to invent' and 'to resolve'. Cubans regularly use the phrase '*hay que inventar*,' or 'one has to invent,' to explain their alternative ways to sustain themselves. In many cases '*inventar*' involves stealing or committing a crime against the state, as in the following example. One young man in the colonial city of Trinidad in south–central Cuba describes his own creative adaptation this way:

> To resolve my economic difficulties, I've begun to collect used packages of Popular, a state brand-name for cigarettes. My friend [*socio*], who works at the tobacco farm, sells me tobacco and paper. With a handmade rolling machine, I roll cigarettes and repackage them for sale on the black market. That is what '*hay que inventar*' means to me.

Using the tobacco and paper that his friend steals from the farm, this man is able to sell his hand-rolled version of a pack of Populares, which sell for 10–11 pesos in state stores, for 5 pesos. Those purchasing the black market cigarettes are well aware of their illegal origins and accept them as part of the current survival strategies. Cuban authorities admit that nearly 30 per cent of all tobacco is stolen for illegal sale. (Source: Brandon J. Cabezas, Latin American Studies Program, Ohio University, based on field work in August, 1997; and Chauvin, 1998:5.)

In Jamaica, the great majority of adult residents of poor communities in Kingston and other cities are unemployed. They need to come up with many different strategies to earn money in order to meet their basic needs. These strategies range from legal ones, such as temporary employment, to illegal ones, such as robbery or selling drugs, and include many others in the gray area in between, as in this example:

In the gray area, residents developed innovative income-raising strategies, often centered on 'hustling' (*ad hoc* buying and selling, using begged or borrowed money). A sophisticated hustling system had grown up around the nearby prison, with women buying food for the prisoners, arranging for visitors' food parcels to get to the prisoners, buying goods from prison warders to sell outside, and begging/'negotiating' for money from prison guards on the paydays in exchange for sex (source: Moser and Holland, 1997: 24).

Caribbean Tourism Association and to rely on foreign tourism as its main economic sector. Cuba's tourism receipts reached US$1.5 billion in 1997, six times higher than in 1990 (Reuters, 1998). And while the United States continues to try to isolate Cuba, other countries such as Canada, Mexico and Spain are taking advantage of the profitable opportunities in tourism, as well as in mining, utilities and many other sectors availed by the lack of competition from US firms.

THE ROLE OF MIGRATION IN MIDDLE AMERICA

Prolonged crises and hardship have required that Central American and Caribbean people be very creative both at home and abroad. In reaction to neoliberalism, a growing number of people have tried to emigrate to North America and Europe, where about 5 million Caribbean islanders have gone since 1945 (Klak, 1998). Central American migrants have been fewer but have also increased in number in recent decades (Pastor, 1989: 6). In recent years, the enticing messages from relatives, friends, and the ubiquitous mass media from the north have conveyed that there is a better life abroad. The resulting Middle American diaspora is indeed large. To cite an extreme example, for every five people born in St Kitts and Nevis and alive today, two now reside in the United States. For Jamaica, a more typical Caribbean case, the ratio is 6 to 1. All totalled, Jamaica sent 480 600 legal immigrants to the United States in the period 1970–96, while Cuba and the Dominican Republic sent 547 300 and 668 700 each (de Souza, 1998). A recent poll in the Dominican Republic found that half the population now has relatives in the United States, and more than two-thirds would emigrate if given the chance (Klak, 1998).

Not all people leave for good, however. Many see opportunities for employment and education in North Atlantic countries as part of a multidimensional international strategy to make ends meet and advance economically. Some people from Middle America who are working abroad send remittances home to immediate family members. How much is sent home varies greatly per household and over time. Macro-economic statistics for remittances understate their significance because they principally record bank transfers, leaving unrecorded money and goods sent through the mail or with travelling relatives and acquaintances. Nonetheless, official data on remittances reveal they contribute about 5–7 per cent to the GDPs of Jamaica, the Dominican Republic, Haiti and Belize (Latinamerica Press, 1997).

Earnings from the north are used in Middle America to meet basic needs, acquire a plot of land, build a home, finance the children's education and start a small business (Chevannes and Ricketts, 1996; Portes and Guarnizo, 1991). For Mexico and Central America, the positive impacts of 'migradollars' on living standards, investments and overall economic conditions are clear (Durand *et al.*, 1996). Male garment-factory workers from the Dominican Republic have been notably successful in transferring skills acquired on the job in the New York metropolitan area back to ownership and management positions on the island (Portes and Guarnizo, 1991). Jamaicans now refer commonly to a local phenomenon called 'barrel children', the parents of whom work in North America and who largely survive on the contents of the overstuffed containers that regularly turn up at customs. The combined effect of these growing trends towards emigration, remittances and circulation is a 'deterritorialization' of Middle American societies (Olwig, 1993: 206).

REGIONAL TRADING BLOCS IN MIDDLE AMERICA

It is said that Central America and the Caribbean 'belong to two different cultural universes' (Grugel, 1995: 155). It follows that they have had negligible economic ties historically and have developed their own regional trading blocs recently. The Central American Common Market (CACM) consists of five countries with a long history of linkages: Guatemala, El Salvador, Honduras, Nicaragua and Costa Rica. For centuries Spain administered them as one colony, and for fifteen years after independence in 1823 they were a united republic. The obvious Central American outlier from this group is Panama, which was a province of Colombia until its independence in 1903 (Barton, J., 1997). Panama now participates in Central American presidential summits without commitment to implementing CACM's common external tariff, an important component of any trading bloc.

The Caribbean region is less economically united than Central America, a reflection of the islands' greater physical fragmentation (Figure 5.1) and greater historical fragmentation by European colonial powers (Richardson, 1992). Because most Caribbean countries are now English-speaking, they dominate the region's trading bloc created in 1973 and called Caricom (the Caribbean Community and Common Market). The 15 full members of Caricom as of 1998 were 12 English-speaking countries of the Caribbean region, one British colony (Montserrat), Suriname and most recently Haiti. Caricom accepted Haiti's membership in 1997 contingent on it moving more aggressively in the neoliberal direction, that is, towards privatization, public spending reduction and promotion of foreign investment. Haiti has already undergone many neoliberal reforms. For example it reduced tariffs on imported rice from 50 to 3 per cent in 10 years, causing the local market to be flooded with US rice and local production to fall by almost 50 per cent (CPT, 1998). The Caribbean trading bloc reinforces the IMF/World Bank/USAID dictates; Caricom views these neoliberal reforms as necessary prerequisites for all of its members if the bloc is be globally competitive and eventually enter NAFTA.

The Bahamas is the only other English-speaking Caribbean country choosing not to belong to the Caribbean Common Market, although it does participate in the Caribbean Community. In this regard, the Bahamas' position is similar to Panama's relative to CACM; it participates in regional summits of heads of state but not in trade or finance ministers' meetings, and is not committed to Caricom's common external tariff. In recent years the Netherlands' Antilles and several of the region's Spanish-speaking countries have attained observer status in Caricom, and some now seek full membership (Demas, 1997; Klak, 1998).

CACM and Caricom countries, together with Mexico, Colombia, Venezuela, Panama and the Dominican Republic, have recently united under the Association of Caribbean States (ACS). The first ACS summit was held in 1995. Thus far ACS has not functioned as a trading bloc but rather as a forum for discussion of common economic concerns and possible future linkages for trade, transport and tourism (Demas, 1997: 74).

The five-member CACM took effect in 1960 two years after the establishment of the original six-member European Economic Community. The parallel and precocious origins of CACM and the EEC led observers to wonder if they would develop and strengthen in tandem. CACM got off to a brisk start during its first decade when intraregional exports grew from 7 to 26 per cent of total (Table 5.3). Central America's policies that emphasized economic opening, fiscal conservatism and regional trade integration

TABLE 5.3 Central American and Caribbean trading blocs: intraregional exports as a percentage of total exports, selected years

	CACM	Caricom
1960	7.0	na
1970	26.0	na
1980	24.2	8.9
1985	15.5	13.0
1990	15.2	12.3
1992	23.2	11.8
1994	22.7	15.6
1996	21.5	na

Sources: Bulmer-Thomas (1998) and Demas (1997).
na = not applicable or available; Caricom was created in 1973.

generated world record economic growth rates of around 5 per cent during 1965–75. Since then CACM has faltered, however. While a full 70 per cent of the exports of the now 15-member European Union are within the trading bloc, intra-CACM trade accounts for only 22 per cent of total exports (Bulmer-Thomas, 1998). Why did CACM show such early promise and then stagnate?

The core of CACM's impressive growth in the 1960s was the states' successful cultivation of import substitution industrialization (ISI) firms, which to a large extent involved MNCs from the United States establishing factories in the region and exporting within the CACM (Grugel, 1995). The internal market for these industrial goods was quickly saturated, however, given the large share of the region's 11 million people at that time living in poverty. While economic growth and regional trade statistics looked impressive, the disruption of peasant economies and the inequities of state policies were fomenting guerilla wars in three of the five countries – Nicaragua, Guatemala and El Salvador (Walker, 1997). Trade also proved to be highly imbalanced across CACM countries. On the one hand, the mutual geographical accessibility of the main urban-economic regions of El Salvador and Guatemala helps to explain why their trade has dominated CACM. On the other hand, Honduras, the least developed member country, accumulated a large trade deficit in the 1960s. This, along with its infamous 'soccer war' with El Salvador in 1969, led to Honduras' withdrawal from CACM in 1970. Regional relations only worsened in the 1970s from the region's civil wars, and then in the 1980s from Reagan's efforts to isolate the Sandinista government. World recession and the debt crisis in the 1980s further dampened CACM's trading prospects. The neoliberal bailout offered by the World Bank and IMF since then has discouraged ISI and intra-regional trade, while it has emphasized extraregional exports of nontraditional agricultural and manufactured goods (Bulmer-Thomas, 1998). As a result, the five CACM countries sustained annual export growth rates of between 11 and 31 per cent from 1992 to 1995 (Colburn, 1998).

Despite the seemingly terminal doom for CACM as of the 1980s, it rebounded in the 1990s, although not to its peak trade level that occurred around 1970 (Table 5.3). Many of the early shortcomings have again surfaced. Honduras rejoined the group in 1990 but has once again accumulated large trade deficits. Trade between CACM countries continues to be predominantly stand-alone industrial products, rather than components of an integrated commodity chain or agricultural products. The lack of the former suggests the shallow development of regional trade and economic integration. The latter were legally restricted until the 1990s. Even since their inclusion, however, agricultural exports have not diversified, and constituted only 3–4 per cent of CACM trade as of 1996 (Rueda-Junquera, 1998). Throughout CACM's existence, trade restrictions have diverted more imports from third countries away from the region than they have increased production within the region. In the parlance of economists, CACM has caused more trade diversion than trade creation (Nicholls, 1998). That is not necessarily bad, however. Trade-dependent regions such as Central America benefit from greater internal production and reliance even if domestically produced goods are more expensive than imports. Last, while capital and products now move more freely across Central American borders, political leaders have been unwilling to consider facilitating labor mobility in the same way (Bulmer-Thomas, 1998).

In comparison, Caricom has never approached CACM's regional trade levels, suggesting the former's lower levels of commitment to trade integration and its limited economic complementarity. Intra-Caricom trade as share of total exports fluctuated in the 8–16 per cent range over the period 1980–94 (Table 5.3). Further, whereas intra-Caricom exports were worth US$5.9 billion in 1980, they amounted to only US$4 billion in 1994 (Demas, 1997: 124). Caricom's market is also more limited. Whereas CACM's population stood at 31.5 million in 1997, Caricom's was only 6 million, until Haiti joined the bloc and added another 7.5 million people (mostly impoverished rural peasants). Caricom has not been able to fix and adhere to a common external tariff (CET), although members pledged to implement a CET of 20 per cent by 1998 (Demas, 1997). Skilled labor has been able to move but not unskilled labor (Demas, 1997: 50). Caricom's limited commitment to regional integration is also illustrated by the fact that Caricom nationals need a passport to enter another country in the trading bloc; North Americans or Europeans can enter with only a driver's license. Further, the lack of leverage against and common position *vis-à-vis* cruise ships means that cruise lines play one island off another in an effort to reduce charges (Pattullo, 1996).

As has typically been the case historically for Middle America, exogenous factors have overwhelmed its progress toward regional economic development. Both CACM and Caricom have suffered as regional integration involving outside countries has become a higher priority. In 1994 Costa Rica entered a bilateral trade agreement with Mexico, a move that impedes further progress toward a common external tariff for CACM because it blurs the distinction between countries that are internal and external to the trading bloc. Caricom has similarly signed free-trade agreements with Colombia and Venezuela (Demas, 1997). NAFTA creates an even larger obstacle, as Bulmer-Thomas has recently suggested. His critical observation on how CACM countries' attention has been diverted toward NAFTA applies equally well to Caricom countries: 'Unrestricted access to the US market, however implausible, has been seen as more desirable than the less exciting, but more realistic, goal of regional integration' (Bulmer-Thomas, 1998: 320).

More broadly, NAFTA has decreased Middle America's economic importance in the global economy and has relocated manufacturing investment to Mexico. Mexico's foreign investment and exports have grown relative to Middle America since 1994. Middle American policy-makers now see Mexico as a major threat to their efforts to develop by attracting foreign investment into assembly operations. CACM and Caricom have reacted to NAFTA in a dependent way. They have been pressing the United States for trade concessions. The most urgent recent aim is parity with Mexico within NAFTA, so that Middle America will no longer be comparatively disadvantaged as an export platform. Caribbean and Central American countries seek NAFTA membership despite their competitiveness *vis-à-vis* the United States for little beyond low-wage labor. And they do this despite all the hardship in Mexico over NAFTA's initial years (Anderson *et al.*, 1994; Schrieberg, 1997). For example, open unemployment increased in Mexico from 3.4 to 7.3 per cent during NAFTA's first two years, while real median salaries fell by 50 per cent and interest rates and foreign debt soared. Mexico's per capita GDP was lower in 1997 than in 1982. NAFTA's benefits have also been economically and spatially unequal. Only 2.2 per cent of Mexican companies account for 80 per cent of the country's non-oil exports (Gayle, 1998). Further, benefits have been concentrated in Mexico's northern states. In southern states such as Chiapas, impacts are largely negative and they have fueled the Zapatista rebellion.

That Middle American policy-makers want to make their countries like Mexico suggests their comprador class positions. They are able to benefit personally from international ties that may not be in the interest of local residents (Cardoso and Faletto, 1979a). Indeed, investment promotion agencies have expanded throughout Middle America under neoliberalism. Middle America continues to engage in a 'race to the bottom' that characterizes the majority of the world's countries (Green, 1995). This irony symbolizes neoliberalism's hegemony and Middle America's abject dependency and limited development options.

Unfortunately, the economic marginality of Middle America described above breeds political marginality. As Demas (1997: 20) laments for Caricom, 'We have already conceded far too much, far too quickly, by way of trade liberalisation'. Middle America's preparations for and pursuit of NAFTA parity has been met by stonewalling on the part of the United States. A trade bill to extend NAFTA-like benefits to Middle America did surface recently in US Congress. However, the bill was pulled in October 1997 because its backers realized they lacked the votes. Opponents of such a bill are an odd mix of conservatives opposed to Third World 'aid' packages, liberals who want more labor and environmental protections in such a bill and nationalists who resist any favorable treatment to foreign interests such as those of Middle America. The US government has offered little more than rhetoric claiming its interest in deeper economic ties with Middle America and regional cooperation toward mutual and complementary development. It is difficult not to conclude that Washington's policy toward the Caribbean – if such scattered actions and mostly inactions deserve to be called 'policy' – is self-serving, narrow and shortsighted.

A VERY CAPITALISTIC ENTERPRISE: MIDDLE AMERICA'S GROWING DRUG ECONOMY

With so few opportunities for growth through neoliberalism materializing, the production and transhipment of drugs through Middle America has become widespread. Individuals have taken to growing and exporting marijuana and to trafficking cocaine and, increasingly, heroin from the Andean region through Middle America on the way to the world's largest market in the United States (Griffith, 1997). The growing number of tourists moving in and out of Middle America facilitates the movement of drugs to the United States by air and by boat.

A recent study, by Guatemalan researcher Edgar Celada, of Central America's increased drug trafficking characterizes it as 'geographic fatalism' – the region is sandwiched between South American producers and North American consumers. The same study found Central America expanding its drug activities beyond trafficking to production and consumption (Jeffrey, 1998). Problems are similar in the Caribbean. Puerto Rico had the highest per capita murder rate in the United States in 1995, with two-thirds of the deaths associated with drugs. A growing number of Caribbean people are being deported from the United States and the most common cause is drug activity. At the same time Caribbean jails are already overcrowded and have been judged to be unsafe and unhealthy (Griffith, 1997).

Expansion of drug activity in Middle America is also occurring with respect to social classes. The 'democratization' of drug activity now includes growing numbers of politicians, judges, businesspeople and ordinary citizens. Military personnel are involved in both drug running and enforcement (Dinges, 1988). Following the conclusion of Central America's civil wars and peace accords, there is now a remilitarization of the region through US funding for antidrug efforts. Neoliberalism's open economies and nontraditional exports are also entangled with the drug trade. Another Guatemalan researcher, Mario Maldonado, laments the growing number of local 'businessmen who earn their wealth hiding cocaine in shipments of broccoli and cut flowers' (Jeffrey, 1998: 2). Guatemala joins most other countries of Middle America where residents now assume, often quite accurately, that sudden increases of expenditure among locals on gold jewelry or a new car, home or restaurant are attributable to drug trafficking.

CONCLUSIONS

Central America and the Caribbean have traditionally been studied separately. Today, the homogenizing impacts of US regional hegemony, economic globalization and neoliberalism suggest the fruitfulness of combining and analysing them as Middle America. Over recent decades Middle Americans have learned that US dominance of, and intervention in, their region makes developing a diversified set of investors, export sectors, markets, local interpretations of 'democracy' and foreign policy relations extremely difficult. The region's hurried experiments with various democratic socialist paths to development are therefore mostly things of the past. Instead, Middle America has converged on a set of policies that pursues dependent capitalist development under the tutelage of the United States.

The current global political economy presents the small, underindustrialized and trade-dependent countries of Middle America with many risks of further vulnerability and marginality, but also some narrow options and opportunities. The greater economic openness associated with neoliberalism is exposing domestic firms to competition with larger and more competitive firms from abroad. Such competition has been fierce both for old and for new producers from Middle America. The few survivors from the region have been able to fill the market niches that neoliberalism trumpets, although too often such competitiveness is attained primarily at labor's expense.

However, international economic integration can potentially have other positive impacts in Middle America if it helps to expose and dislodge internal obstacles to development and human rights. Economic exposure puts pressure on Middle American countries to confront internal problems such as economic stagnation under the weight of local mercantilist capital and inefficient local production methods from which only a few have benefited. Global integration can also help to expose government mismanagement, graft and nepotism, that has long plagued many Middle American countries. Nontraditional export production means that there are now more external interests connected to Middle American countries that can expose (but also potentially benefit themselves from) illegal or inhumane practices. Issues of human rights can be brought to international attention through greater global connections. For example, the international campaigns in recent years against Nike, Reebok, Kathie Lee Gifford apparel and Disney (and their subcontractors) for their Third World employment practices demonstrate these trends (NLC, 1997). The campaigns are making internationally known employment and living conditions that have long been sequestered by distance, foreign culture and closed societies. Thus economic globalization is primarily a top-down process shaping the lives of Middle American people, although it can also open opportunities for confronting conditions of domination, exploitation and stagnation.

Acknowledgements

The author wishes to thank (without implicating) the following people for supplying important information included in this chapter: Dennis Conway, Alex Dupuy, Dennis Gayle, Garth Myers and Jim Wiley. Thanks are also extended to the editors for their thoughtful suggestions on an earlier draft.

Further reading

Escobar, A. 1995 *Encountering development: the making and unmaking of the Third World*. Princeton University Press, Princeton, NJ. Escobar is a major figure in the Latin American movement to find 'alternatives to development,' often termed the 'antidevelopment' school of thought.

Hirst, P., Thompson, G. 1996 *Globalization in question: the international economy and the possibilities of governance*. Polity Press, Cambridge. This is the most thorough and careful interrogation of the ideas and evidence behind economic globalization to date. It refutes many truisms about globalization and offers empirical evidence to document the extent and the precise ways in which global economic integration is actually occurring.

Klak, T. (ed.) 1998 *Globalization and neoliberalism: the Caribbean context*. Rowman and Littlefield., Lanham, MD. Through the analytical lens of political economy, the book examines the impacts, adjustments and coping strategies found in the Caribbean as it undergoes a rapid and profound transformation. Issues addressed include development policies, nontraditional exports, external relations, the environment, tourism, class and gender relations and human migration.

Sachs, W. (ed.) 1992 *The development dictionary: a guide to knowledge as power*. Zed Books, Atlantic Highlands, NJ. This collection provides a critical etymology and eco-political interrogation of the principal concepts used in the discourse of Third World development. The book aims to break open for careful historical and contemporary examination the major concepts and ideas associated with the theory and practice of Third World development. Representing a 'south-centric' perspective on development, it is essential reading for anyone who resists the intellectual blinders of mainstream development economics.

Thomas, C.Y. 1988 *The poor and the powerless: economic policy and change in the Caribbean*. Monthly Review Press, New York. The most thorough and insightful account of the various models of state-promoted noncapitalist paths toward development of Caribbean countries since World War Two.

Part 4

Globalization and environmental change

6

NATURAL RESOURCES, THE GLOBAL ECONOMY AND SUSTAINABILITY

WARWICK E. MURRAY

LATIN AMERICA – THE PARADOX OF PLENTY?

Latin America has served as a resource periphery for the global economy since colonial times. Indeed, the continent was conquered largely for that purpose. It is a region extremely rich in natural resources. Given the abundance of 'gifts of nature' one might intuitively expect Latin American countries to be among the most 'developed' in the World. This is clearly not the case – Latin America has lagged, and continues to lag, behind the industrialized countries of the world in terms of many development indicators. Since the theorizations of the structuralist school, an increasing array of commentators have arrived at the idea that resource abundance may actually operate as a 'curse' which, under certain conditions, can prejudice long-run sustainable development (Auty, 1993). Various policy initiatives based on such ideas, which have aimed at breaking the continent's dependence on resource-based development, have not been fully successful. Thus, at a general level, the countries of the region remain on the periphery of the global economy – largely dependent upon the global 'core' for consumer goods and for markets for their primary products.

Latin America's role as a global resource periphery has been profoundly damaging to development prospects since its inception. Unquantifiable damage has been wrought upon society, the economy and the environment given the premium attached to specialization in primary product exports. This continues to be the case. The neoliberal hegemony that currently exists in Latin America – fuelled by the imperatives of deepening globalization – have compounded Latin America's developmental problems. This situation seriously threatens the prospect of attaining sustainable development[1] in the continent. At the heart of this issue lies a perplexing paradox. Why should a region so blessed with natural resources and gifts of nature have failed to 'develop' and industrialize to the extent that other regions have? This is the 'paradox of plenty' (Kant, 1997) which seems to lie at the core of Latin America's social and economic woes.

This chapter attempts to describe and account for Latin America's historical and contemporary role as a resource periphery in the global economy, to analyse some of its impacts and to stress the pressing need to move beyond this situation in the coming century. In order to achieve the latter, a call is made for the abandonment of neoliberalism and for the reworking and incorporation of relevant structuralist and dependency ideas into a new framework which intends to diminish the ill effects of specialization in primary product production and trade.

THE HISTORY OF RESOURCE EXPORTS AND THE ROLE OF THEORY

This section has two interlocking aims: first, to outline briefly the history of Latin America's role as a resource periphery; second, to outline how various 'turns' in theory have informed, and have been informed by, such developments. For the sake of clarity, it is useful to conceptualize the evolution of these mutually determining elements in five overlapping time periods.[2]

Colonialism and natural resource exploitation

The roots of Latin America's role as a global resource periphery were laid during Spanish and Portuguese colonization. Initially, colonial interest lay in precious metals such as gold and silver. Later, as transport technology evolved, Latin America began to serve as a major supplier of agricultural products whilst consolidating its role as a mineral and metals supply source. This began with the shipment of tropical fruits (mainly bananas) and sugar cane. Production was largely organized through plantation systems, which required large amounts of labor, leading to the development of the African-sourced slave trade. Beginning in the late eighteenth century, a wider range of products, including wheat and beef, was brought into the system. In this case, production was based on the *latifundio–minifundio* dichotomy that was created through the *encomienda* system of land organization initiated by the colonial powers (Chapter 11). Under this system, the economic benefits of export agriculture were concentrated largely in the hands of a small landed elite.

Independence and the influence of neoclassical theory

The independence of the Latin American countries beginning in the early nineteenth century did little to alter the basic orientation of the economy within the context of the global system. Essentially, this involved the supply of primary products destined for the markets of former colonial powers and other countries (including Britain and later the United States) within the rapidly industrializing global 'core'. The range of exported products was relatively wide: cereals and cattle from the Southern Cone countries (Argentina, Chile and Uruguay); precious metals and minerals from a range of countries (gold from Venezuela, copper from Chile); and various agricultural cash crops (coffee from Central America and Brazil, sugar cane from Cuba and Brazil and bananas from Central America and the Caribbean). Characteristically, there were few backward and virtually no forward linkages into the system and capital was generally foreign-owned. This structure was later referred to as an 'enclave' economy (Cardoso and Faletto, 1969, 1979b; Bulmer-Thomas, 1994).

In general, at the time, continental specialization of the above nature was seen as advantageous in terms of the prospects it offered for economic development. was based on the principles encapsulated in the classical model of 'comparaantage' (CA), developed by Ricardo in the nineteenth century. This theory on the assumption of perfect competition, that global welfare would be maximized through the mechanism of global free trade. This, the theory argued, would allow the working of CA within the international economy, where CA is defined as relatively low opportunity cost in the production of a given good or service. For many Latin American countries the 'laws' of CA indicated that they should specialize in the production for export of agricultural and other primary products. Given the important political role of the landed class within Latin American society in the early part of the century, this theory was widely supported (Bulmer-Thomas, 1994).

The Great Depression and the structuralist response

The optimism surrounding the prospects for development based on primary product specialization was broken by the impact of the Great Depression. During the 1930s global recession led to an enormous decline in international trade. This hit Latin America in a profound way. In general, it appeared that economies specializing in the export of primary products were impacted in a particularly harsh way relative to the industrialized economies of the global core. Structuralist theory, developed in ECLA (Economic Commission for Latin America) in the late 1940s, evolved in an attempt to explain this effect (Kay, 1989). During the 1950s, and up to the beginning of the 1980s, this approach became the most influential in terms of the formation of economic policy in the continent.

Structuralist theory was explicitly critical of neoclassical and other 'eurocentric' theoretical derivatives regarding development. Crucially, theoretical ideas regarding 'comparative advantage' and 'gains from free trade' were heavily challenged by structuralists, who argued that development based on the export of primary products was not sustainable in the long run (Sunkel, 1993b). The centre-pin of structuralist analysis was laid by Raul Prebisch in his theory of 'secular decline' (Prebisch, 1949). Prebisch offered empirical evidence, and theoretical explanation, pointing towards the tendency of long-term decline in the 'terms of trade' (the ratio of the index of export prices to the index of import prices) within countries specializing in the export of relatively unprocessed natural resources. Central to the argument was the observation of, and explanation for, the trend of decreasing primary product prices over time. This trend was argued to be the case for reasons of both demand and supply. On the demand side, it was noted that primary products face income-inelastic demand, that is, as income rises demand for such goods rises less than proportionately. In contrast, manufactured goods were seen to be characterized by income-elastic demand. The combination of these two factors implied that, as the global economy grew, the relative price of primary products would decline. On the supply side, it was argued that organized labor and oligopolistic marketing structures in the industrialized countries meant that such countries could catch the gains of economic upswings, resist losses during downswings and capture productivity increases (most notably in terms of increased real wage rates). Furthermore, the deterioration of primary product prices during downswings, or following productivity increases, was partly explained by relatively high levels of competition between less developed countries (LDCs) and the existence of surplus labor in such countries. Based on the preceding, the general policy conclusion offered by Prebisch and the ECLA school was industrialization (Hunt, 1989).

In order to achieve their policy objectives, the structuralist school developed, and was instrumental in, the application of import substitution industrialization (ISI). This policy proposed the substitution of imports at increasingly complex levels over time – from light manufactures (consumer nondurables followed by consumer durables), intermediate goods, to basic capital goods. Despite the distinctly 'protectionist' and 'inward-orientated' nature of structuralist ISI, it is important to mention the structuralists were neither anti-trade nor anti-export *per se*. To the contrary, it was hoped that development in the industrial sector would provide the basis for exports in the future (Kay, 1989). Indeed, the ECLA structuralists argued for the creation of Latin American export cartels. However, the immediate concern remained the structural transformation of the economy away from primary product exports.

By the end of the 1970s it was clear to many that the 'structuralist' approach needed reappraisal. The Latin American economies remained, on the whole, dogged by foreign exchange problems and debt, industrial sectors had not expanded at the rates that ISI had promised, rural labor had not been absorbed effectively, social differentiation and poverty were endemic and the continent's economy remained highly specialized in primary production and resource exports. Thus, even before the crisis of the early 1980s alternative frameworks were being explored.

Dependency analysis and the global periphery

The structuralists themselves were among the first to cast doubt upon the possibility of long-term development as proposed by early structuralist analysis (Kay, 1989). To a certain extent this led to the development of dependency analysis – a collection of perspectives which both propelled and was propelled by a number of successful and attempted socialist revolutions in the region beginning in the late 1950s. This group of analysts rapidly split into two 'schools' – the neo-Marxists and the reformists. The former utilized nonorthodox Marxist class-based analysis to explain the perpetuation of inequalities in 'peripheral', resource-exporting countries (Frank, 1967; Emmanuel, 1972). This represented a radical departure from structuralist thinking (Kay, 1989). On the whole, for this school of thought development of the 'periphery' would not be possible unless withdrawal from the global capitalist system was effected – in many cases it was argued that this would necessitate social and political revolution. The reformists on the other hand represented a more explicit extension of structuralist thought – with greater emphasis placed on the need for substantial reform of constraining 'structures' and the attainment of 'self-sustaining' development based on restructuring away from primary product exports (Cardoso and Faletto, 1969; Sunkel, 1973). The contribution of the dependency analysts to development thinking has been enduring. However, in the case of Latin America, the school has not proven highly influential in development policy-making except in a few isolated cases (notably Cuba).

Neoliberalism and the revived rationale for resource exports

During the 1980s a combination of external factors, vested interests and real economic hardship led to the eclipsing of a number of interesting theoretical advances, such as dependency analysis, by the revival of interest in free-market economic development policies. The debt crisis beginning in the early 1980s was the watershed which suggested to many that a new perspective was sorely required. Neoliberal theory placed the blame for the lack of development in Latin American countries (and other LDCs) squarely at the feet of the 'interventionist' state central to ISI. This body of theory essentially represented a reversion to neoclassical thinking based on the supposed virtues of free-market resource allocation. During the 1980s neoliberal development economists and international lending institutions stressed, and in the case of the latter insisted, upon market reforms – commonly referred to as structural adjustment policies. In general, such reform emphasized the role of market liberalization and deregulation in order to maximize economic growth. This liberalization would necessitate a significant 'rolling back' of the state. In terms of trade policy, one of the cornerstones of the neoliberal argument was the importance attached to the potential global welfare 'gains from trade'. This provided the rationale for 'outward orientation' which was recommended for the Latin American economies (Gwynne, 1990) and has underpinned the increasing role of exports in total income over the recent past.[3] Given the region's comparative advantage in primary products, theoretical rationale supporting specialization in such exports once again informed economic policy decisions. This has had a profound restructuring impact in many countries as production has retrenched in the primary export sector (Green, 1995). This process has been facilitated by the 'globalization' of the world economy, involving increased flows of trade and investment initiated by industrialized countries searching for cheap products from global resource peripheries (Le Heron, 1993; Whatmore, 1995). For reasons which will be made clear in the remainder of this chapter, this latest 'turn' is precipitating widespread damage to society, the economy and the environment and promises to be non-sustainable in the long run. In many ways, the paradox of plenty continues to haunt Latin American development.

LATIN AMERICA AS A CONTEMPORARY GLOBAL RESOURCE PERIPHERY

It is now timely to present some empirical evidence which illustrates Latin America's role as a resource periphery in the contemporary global economy. This is achieved through a brief analysis of continental primary product exports and key resource export sectors.[4]

The importance of primary product exports

Aggregate figures suggest that primary product exports have become of less proportional significance over time. In 1970, such exports accounted for 89.2 per cent of the value of total regional merchandise exports.[5] By 1994, this had fallen to 50 per cent (Fig. 6.1; Table 6.1). In terms of absolute nominal values, however, primary product exports continue to rise. In 1970, such items earned US$12 748 million, rising to US$87 423 million by 1994. The rise was especially marked between 1970 and 1980, which is partly explained by rapidly rising oil prices at that time.

Industrial exports have risen substantially in a number of Latin American countries – most notably Mexico and Brazil. These countries can be seen as part of an evolving global 'semi-periphery' characterized by the inflow of foreign investment in industrial activities attracted by low-cost factors of production (especially labor). These large countries account for an overwhelming proportion of all Latin American exports. Owing to this, aggregate data are somewhat misleading as they underestimate the

TABLE 6.1 Proportional importance of primary product exports in total merchandise exports (%), 1970–94

	1970	1980	1990	1994
Ecuador	98.2	97.0	97.7	92.6
Nicaragua	82.2	81.9	91.8	86.9
Venezuela	99.0	98.5	89.1	86.2
Peru	98.2	83.1	81.6	86.1
Honduras	91.8	87.2	90.5	85.0
Chile	95.2	88.7	89.1	83.6
Belize	–	82.4	84.6	83.0
Panama	96.5[a]	91.1	83.0	82.3
Paraguay	91.0	88.2	90.1	78.7
Bolivia	96.8	97.1	95.3	77.8
Guatemala	71.9	75.6	75.5	68.7
Argentina	86.1	76.9	70.9	67.2
Colombia	91.0	80.3	74.9	63.1
Trinidad and Tobago	87.2	95.0	73.3	57.5
Uruguay	82.4	61.8	61.5	57.1
El Salvador	71.3	64.6	64.5	55.3
Brazil	86.6	62.9	48.1	45.2
Mexico	66.7	87.9	56.7[c]	22.6
Barbados	74.7	47.5	56.7	–
Costa Rica	81.3	70.2	72.6	–
Total	89.2[b]	82.6	67.2	50.0[d]

Source: CEPAL (1995).
– = no data.
[a] Does not include canal zone. [b] Excluding Belize [c] Preliminary figure [d] Does not include Barbados and Costa Rica.

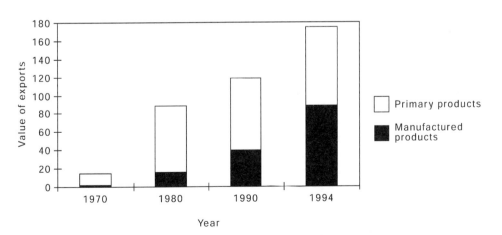

FIGURE 6.1 Value of manufactured and primary product exports (in US$ billions), 1970–94. Source: CEPAL (1995).

continued importance of primary product exports in the remaining countries. As Table 6.1 shows, in 1994, in 16 of the 20 countries listed, primary products accounted for over 50 per cent of the value of total merchandise exports. In Ecuador, the proportional role of primary product exports stood at over 90 per cent. In seven other countries (Nicaragua, Venezuela, Peru, Honduras, Chile, Belize and Panama), such products accounted for over 80 per cent of the value of export goods. In most cases, reliance has fallen over the past two decades – but certainly not to the extent that aggregate figures might suggest.

Further evidence for the overwhelming proportional role of primary products in most countries is revealed when the major export goods for each country are considered. Table 6.2 shows that in 19 of the 20 selected countries, the major export earner in 1994 was a primary product (except for Bolivia). In 6 of these cases (Belize, Ecuador, El Salvador, Honduras, Panama and Venezuela) the major export contributed over 30 per cent to total merchandise export earnings. In 14 cases, the next two most important exports were also primary products. In general, countries are highly reliant on a nondiversified range of primary product exports. In 7 of the 20 cases, export earnings from the top 3 contributors accounted for over 50 per cent of merchandise export earnings. Furthermore, it is important to note that trade data underestimate the importance of primary product exports as, in some countries, illegal products (such as coca in Bolivia, Colombia and Peru) also make important contributions (Weeks, 1995).

Major primary product export sectors

Energy

Latin America is a significant net exporter of energy, which is the region's major foreign exchange earner. A range of countries export either oil, gas, electricity, oil products and/or coal both to countries in the region and beyond. Total net energy exports equalled 106 million tonnes of oil equivalent (mtoe) in 1995, having recovered from a low of approximately 40 mtoe during the mid-1980s (CEPAL, 1995). The overwhelming component in energy exports is crude petroleum (101 mtoe in 1995), coming from the oil-producing countries: Colombia, Ecuador, Mexico, Trinidad and Tobago and Venezuela. Of these, Venezuela is by far the largest exporter (see later case study, pages 143–46), followed by Mexico. Both of these countries are increasingly exporting to energy-scarce Pacific Asian countries such as Japan and South Korea. Furthermore, as the major Latin American economies, Brazil and Mexico, continue to expand, interregional oil exports are rising. Oil product exports are also important, although they have declined in value over recent decades. Countries such as the Netherlands Antilles and Venezuela are the major sources of such items. Electricity exports do not go beyond the region and are almost exclusively accounted for by hydroelectric power (HEP) exported from Paraguay to Brazil. Coal exports are less important – the only significant net exporter being Colombia, which exports mainly to its Latin American neighbors. Energy resource exploration is now high in the continent as extraregional and interregional energy demands rise. In the future, it is likely that a number of further countries will become net exporters. Recent approval of plans for a partly British Gas backed natural gas pipeline from Bolivia to Brazil, and progress in a large HEP scheme between Paraguay and Brazil, are indicative of likely future trends.

TABLE 6.2 Major export products in Latin America, 1994

Country	Major export product	%ª	Next two major exports products	%	Top 3 exports
Argentina	vegetable oils	8.2	crude petroleum, soya bean oil	19.6	27.8
Barbados	raw sugar	17.5	circuit breakers, casks for metal	16.9	34.4
Belize	raw sugar	33.8	fruit juice, men's clothing	28.4	62.2
Bolivia	jewellery	14.6	gold, tin	20.4	25.0
Brazil	iron ore	5.3	coffee, vegetable oil	9.8	15.1
Chile	refined copper	23.2	copper ore, wood pulp	15.1	38.3
Colombia	coffee	22.4	crude petroleum, coal	16.6	39.0
Costa Rica	bananas	24.9	coffee, meat	15.9	40.8
Ecuador	bananas	31.2	coffee, cocoa beans	32.9	64.1
El Salvador	coffee	32.5	medicines, paper bags	7.6	40.1
Guatemala	coffee	21.2	raw sugar, bananas	18.3	39.5
Honduras	coffee	30.7	bananas, seafood	29.9	60.6
Mexico	crude petroleum	10.9	cars and motor parts, insulated wire	13.1	24.0
Nicaragua	coffee	21.7	meat, seafood	20.3	42.0
Panama	bananas	38.3	seafood, fish	18.5	56.8
Paraguay	soya beans	27.2	raw cotton, soya bean oil	24.8	52.0
Peru	meat meal	16.9	refined copper, gold	19.5	36.4
Trinidad and Tobago	petroleum products	29.4	crude petroleum, ammonia	29.9	59.3
Uruguay	meat	10.7	cars and motor parts, rice	17.9	28.6
Venezuela	crude petroleum	62.3	petroleum products, iron pyrites	30.7	93.0
Latin America	crude petroleum	10.4	petroleum products, coffee	6.7	17.1

Source: CEPAL (1997).
ª Refers to percentage of total merchandise exports.

Metals and minerals

Of all the primary product export sectors, metals and minerals has the longest history in the continent. It remains extremely important in the region's economy and is currently the third most important primary sector in terms of export earnings (CEPAL, 1995). After sharp declines in the 1980s, due to depressed international prices, the sector partly recovered, earning US$24.3 billion in 1994. This nominal figure, however, is still nearly US$10 billion lower than it was in 1982. Major mineral and metal products include copper from Chile and Peru, iron ore from Brazil, aluminium from Venezuela, and tin from Bolivia. The region is a net exporter of such goods – although production is concentrated in selected countries. Currently, mineral exploration is high in the continent (Gilbert, 1997), and production in all major sectors is rising (CEPAL, 1995). The markets for mined products are wide, including the United States and a number of countries in Europe. As with all primary product exports, Pacific Asia (especially Japan) is becoming a market of increasing importance.

Agriculture

Currently, agriculture is the second major primary product export sector in value terms. In 1994, 20.6 per cent of total export earnings were accounted for by the sector, falling from a figure of over 40 per cent in 1970. However, many of the Latin American countries are heavily reliant on agricultural exports (Table 6.3). Although the proportional importance of such exports has fallen, the region remains a significant net exporter. This figure recently rose to reach a net of US$24.5 billion in 1995 (see Table 6.4). The largest agricultural exporter in terms of absolute value is Brazil, followed by Argentina, Mexico, Colombia and Chile (CEPAL, 1995).

The most important traditional agricultural exports include coffee, bananas and sugar, largely from Central America and northern South America, and meat, cotton and vegetable oils, largely from South America. Recently, the importance of agricultural exports has been compounded by neoliberal policy in a number of countries seeking to promote the export of counterseasonal nontraditional agricultural items including, for example, fresh fruits from Chile (see later case study, page 146), winter vegetables from Guatemala and cut flowers from Colombia. Foreign investment has played an impor-

TABLE 6.3 The proportional role of agricultural exports in total merchandise exports (%), Latin America, 1970–94

	1970	1980	1990	1994
Paraguay	66.2	75.3	90.1	72.2
Belize	–	74.8	77.7	68.5
Honduras	72.9	73.6	69.8	68.1
Nicaragua	73.8	76.3	73.7	67.2
Guatemala	70.5	70.1	71.3	63.6
Panama	65.5[a]	48.4	58.9	58.9
Argentina	84.5	68.8	56.5	49.4
El Salvador	70.9	77.1	55.5	48.2
Uruguay	71.7	48.0	46.8	41.8
Colombia	81.2	77.2	35.4	38.3
Ecuador	79.5	25.1	29.2	36.5
Brazil	71.1	46.3	27.9	28.9
Bolivia	6.4	9.9	20.8	22.0
Chile	3.2	8.5	15.2	15.8
Peru	16.9	8.2	7.4	11.2
Trinidad and Tobago	8.4	2.0	5.4	7.6
Mexico	54.2	12.0	10.3[c]	6.5[d]
Venezuela	1.4	0.4	1.8	2.4
Barbados	53.7	36.9	27.1	–
Costa Rica	78.9	65.8	58.8	–
Total	44.1[b]	29.0	24.0	20.6[e]

Source: CEPAL (1995).
[a] Does not include canal zone.
[b] Preliminary figure.
[c] Includes goods for processing (*maquila*).
[d] Excluding Belize.
[e] Does not include Barbados and Costa Rica.

TABLE 6.4 Net value of agriculture, fish and forestry exports from Latin America, 1990 and 1994 (US$ billions)

	Agriculture		Fisheries		Forestry	
	1990	1995	1990	1995	1990	1995
Central America and the Caribbean	2.91	2.44	0.47	0.63	−1.27	−2.03
South America	17.27	22.27	2.26	3.68	−1.44	3.22
Latin America	20.17	24.51	2.74	4.31	0.17	1.20

Source: FAO (1996).

tant role in the rise of nontraditional agricultural exports (NTAEs) as multinational companies have set up packing and, in some cases, production facilities in order to supply 'exotic' products for the global market (Barham et al., 1992; Murray, 1998). The process of the globalization of agriculture has facilitated this growing trend (Le Heron, 1993; Friedland, 1994). Traditionally, markets such as the United States and Europe have played the major roles. However, NTAEs are increasingly being marketed in Pacific Asia and the Middle East. Furthermore, agricultural trade between Latin American countries is rising (Weeks, 1995).

Fisheries

One of the fastest growing primary product export sectors in Latin America is fish and fish products. Between 1990 and 1995 the value of net exports rose from US$2.74 billion to US$4.31 billion (Table 6.4). As with agricultural products, much of the recent increase in trade is being marketed in Pacific Asia where dietary tradition makes the demand for such items especially high. High levels of foreign investment from East Asian countries, such as Japan and South Korea, in fishing operations in places such as Chile, Ecuador and Peru have underpinned this trend. Chile is the largest fish and fish product exporter, followed by Peru, Ecuador and Argentina. Increasingly, in-shore fisheries are playing an important role, especially in the case of salmon exports from Chile (Barton, 1997b).

Forestry

Another recent growth area is forestry exports. In this case, a clear regional demarcation exists. In Central America and the Caribbean all countries (excluding Nicaragua) are net importers (FAO, 1996). Indeed, this part of the region imported products valued at over US$2 billion in 1995 (Table 4). In contrast, a number of South American countries are net exporters. The largest of these at present is Brazil, followed by Chile and Panama. Despite the fact that the majority of Latin American countries are net importers, the large value of products exported from key countries means that the region as a whole is a net exporter. This has been underpinned by neoliberal policy in a number of countries, especially Chile, seeking to raise exports to forest-scarce countries – especially in Pacific Asia (Gwynne, 1993a).

SPECIALIZATION IN RESOURCE EXPORTS AND THE QUESTION OF SUSTAINABILITY

As the above has shown, in general, Latin American countries continue to rely heavily on the direct exploitation of their natural resource wealth in order to generate export income. In the future, if outward-orientated policies persist, it is probable that this specialization, based on comparative advantage, will further increase. The increasing globalization of the world economy will provide added impetus for this trend. What are the prospects for sustainable development based on resource export specialization? What are the major problems associated with reliance on primary product exports?

The 'terms of trade' constraint

There is considerable evidence that the original structuralists' concern with the 'secular decline' of primary product prices is of contemporary relevance (Grilli and Yang, 1988; Barham *et al.*, 1992; Ocampo, 1993; Weeks, 1995). This general trend has had a negative impact on the evolution of the terms of trade for most Latin American countries during the present century and over recent decades (see Table 6.5).

Naturally, the trend will be affected by the exact mix of exports and imports. In the case of countries exporting coffee and cocoa, the effect has not been so pronounced.

TABLE 6.5 Evolution of the terms of trade (percentage change) in Latin America, 1972–80 and 1980–92

Country	1972–80	1980–92
Argentina	−8.3	−2.4
Bolivia	+7.1	−6.7
Brazil	7.1	−6.7
Chile	−8.3	0.1
Colombia	4.1	−1.4
Costa Rica	0.3	−0.2
Ecuador	7.1	−5.0
El Salvador	7.1	−5.0
Guatemala	−1.5	−2.2
Honduras	0.0	−1.3
Mexico	6.0	−4.6
Nicaragua	−3.2	−2.8
Panama	−0.1	2.6
Paraguay	−2.9	0.7
Peru	−6.8	−1.8
Uruguay	−8.1	1.0
Venezuela	12.5	−8.1
Latin America (weighted mean)	−1.2	−2.3

Source: adapted from Weeks (1995: 97).

However, in countries showing a relative specialization in metal production, cereals or the production of agricultural inputs (for example nitrates) the decline over the twentieth century has been especially marked (for example in Chile, Argentina, Peru) (Ocampo, 1993). Oil-exporting countries, which initially experienced a rise in their terms of trade in the 1970s, suffered as prices 'bottomed out' towards the late 1980s (Weeks, 1995). Given the continued relevance of the terms-of-trade problem, the original structuralist calls for the organization of export cartels, commodity price support systems and regional coordinated response in Latin America are as relevant today as they have ever been (Ocampo, 1993; Sunkel, 1993b).

Price volatility

Within the broad decline of primary product prices it is often the case that world market prices fluctuate considerably. This is particularly the case for energy, metal and mineral products – the demand for which are intimately tied to general economic conditions. The post-war global economy has been characterized by volatility (Gwynne, 1990). Consequently, a number of the energy, mineral and metal exporting economies of Latin America have been subject to rapid changes in economic fortunes. In particular, such volatility has precipitated difficulties associated with managing 'windfalls' and riding out recessions. It is widely argued that windfalls, such as those conferred on the oil exporting economies after the first oil hike, have not been utilized in ways which engender sustainable development. Authors such as Auty (1993) point towards the process of 'rent seeking', whereby powerful interest groups have been able to lobby for a disproportionate fraction of rents. Furthermore, funds have also been used for large-scale projects and 'showpieces', as was the case in Venezuela after the first oil hike (Gilbert, 1997). In contrast, during global recessions such economies have often been hit very hard, given that past policy errors have significantly depleted funds. There have been some efforts to offset this 'boom–bust' process. For example, the Chilean copper-stabilization fund founded in 1987 saves 'windfall' receipts during times of rising copper prices to be used to augment government expenditure in downswings (Gwynne, 1996b). In general, however, short-term political and institutional interests are such that it is often difficult to escape the negative implications of 'windfall' rents and price fluctuations.

Dutch disease and 'unbalanced' development

For commentators such as Auty (1993) the phenomenon of 'Dutch disease' explains why energy, metal and mineral exporting economies have performed so badly despite their significant resource endowments. Indeed, it lies at the core of his 'resource curse' thesis. Dutch disease results from the appreciation of the exchange rate brought about by a rapid rise in inflows of mineral rents. When commodity price booms occur, or production rapidly increases, such rents will rise sharply. Appreciation will have the effect of making exports to the global economy more expensive. During booms this will not impact the mineral exporting sector as international prices remain high. However, it has the effect of making manufacturing and agricultural exports uncompetitive. The major problem arises during periods of recession, when mineral and metal prices are low. During such times, it may prove difficult to diversify into nonmineral sectors and boost exports in those sectors. This situation was observed in the Latin American mineral exporting economies during the global recession of the mid-1980s. Dutch disease may also arise in the case of economies specializing in agricultural exports. There was evidence pointing towards this type of process following the Chilean NTAE 'boom' of the 1990s (Hojman, 1995; Gwynne, 1996b).

Protectionism and regionalism in the global core

A factor which has aggravated the secular decline in the long-run price trajectory of resource exports through the restriction of demand is the high level of protectionism existent in many developed countries and the trading blocs of which they are part. This is the case in minerals and metals markets and, most notoriously, in the agricultural sector (McMichael, 1993). Guoymer *et al.* (1993: 231) claim: 'nearly all of the industrial countries hold resources in agriculture behind a panoply of protectionist barriers that insulate the primary sector from competition'. This has had the effect of distorting patterns of world trade enormously leading to a situation where 'production, specialization and trade in agriculture are determined by the comparative strength of policies not by comparative advantage' (Hitiris, 1989: 67). Thus, 'most of the world's food exports are grown in industrial countries where the costs of food production are high, and consumed in developing ones where costs are lower' (World Bank, 1986: 154). Of the advanced economies the United States, the European Union (EU) and Japan are renowned for their highly protective stance with regard to domestic agriculture. In the United States for example in 1986, 36 per cent of total output value was accounted for by subsidies (Apey, 1995). In the recent Uruguay round of the General Agreement on Tariffs and Trade (GATT) negotiations on agriculture took on a high profile. Although certain concessions were achieved, no great breakthroughs with regard to the prospects for LDCs were made. A further contemporary development in the global economy which has led to protectionism is the increased formation of regional trading blocs. Individual developing countries can, potentially, gain if they are able to join a suitable group. However, in practical terms it is the case that more developing countries are left out than included (Grant, 1993).

The Latin American countries are attempting to overcome the potential ill effects of global regionalization in two main ways. First, a number of countries have attained, or are vying for, membership of the growing number of free-trade organizations. For example, Mexico is a member of the North American Free Trade Agreement (NAFTA). Further to this, Chile and Mexico are members of the Asian Pacific Economic Cooperation (APEC). Second, there is growing impetus towards regional integration within Latin America. Most notably, the Mercosur, which comprises Uruguay, Argentina, Brazil and Paraguay as full members, accounts for a growing proportion of regional exports. This type of process may also have the impact of offsetting negative price trends in global markets such as those discussed previously as, in theory at least, a 'common-front' approach grants the region greater bargaining power in global markets (Gwynne, 1995).

'Enclave' economies

The problem of 'enclave' economies, which tend to characterize primary product export sectors, has received attention in the Latin American development literature since structuralist writings and took on extra importance in dependency analysis. This problem is most associated with energy, minerals and metals mining but is also relevant to certain types of agricultural, forestry and fishing operations (especially large-scale operations). Mining and energy operations may perform as enclaves in two main ways. First, activities are generally capital-intensive and large-scale. This means that they may produce large revenues, but generate little in terms of employment and other linkages into local, regional and national economies. In the past, given high levels of foreign ownership in the Latin American mining sector, a good deal of surplus was repatriated. More recently, higher levels of domestic ownership have partly reduced this problem, although the trend towards privatization is reviving some concerns (Gilbert, 1997). Second, mining and energy operations are often geographically isolated, being located some distance from major cities. As such, these areas often form distinctive geographical economic zones which can lead to the exacerbation of social and economic inequalities at the regional scale.

Foreign ownership and control

The issue of foreign ownership has caused enormous controversy in Latin America. High levels of foreign ownership can raise a number of problems. First, it can lead to technological and financial dependence as multinational companies (MNCs) act as the major, and sometimes only, importers and diffusers of the relevant capital and technology. Second, it can lead to an outflow of profits that could be captured if national potential were developed. Third, flexible sourcing means that MNCs are relatively 'footloose' *vis-à-vis* national firms and may withdraw their operation in one particular country without much notice. Finally, in general there is little incentive for the companies to behave in a way which is explicitly beneficial for the sustainable development of a particular country. For example, firms involved in NTAE production or distribution will actively search for a variety of locations with low labor costs and natural resource abundance. This may compound the 'locking-in' of countries into low-skill, low-productivity activities. In contrast, if the foreign 'element' is controlled more effectively, the benefits in terms of technology transfer and adaptation, tax revenues, access to foreign markets and foreign exchange, availability of capital for smaller farmers and employment can be large.

Foreign ownership in Latin American resource sectors has traditionally been high. This is especially the case in energy and other mineral sectors given the high levels of capital and technology required to set up production. This has led to confrontations between governments and foreign companies and inspired a number of expropriations and nationalization of sectors where foreign control was high, including Mexican oil (1938), Peruvian oil (1968), Venezuelan oil (1976), Bolivian tin (1952) and Chilean copper (1973). In the neoliberal 1990s the benefits of foreign investment have been given greater weight, which has led to number of privatizations, including Bolivian tin and silver mines and Mexican copper (Gilbert, 1997). Currently, high levels of foreign ownership are evident in NTAE sectors which are most explicitly geared towards export for 'luxury' markets in the developed world (Barham *et al.*, 1992; Sunkel, 1993b). According to Barham *et al.* (1992) this is notably the case in Chile, Guatemala and Costa Rica. Foreign ownership is also high in forestry and fisheries activities as evidenced by MNC dominance in Chilean sectors (Gwynne, 1993a; Barton, 1997b).

Food security

Recent restructuring towards agro-exports has reduced food security as traditional staples have been replaced by export cash crops. Furthermore, protectionist policies and the 'dumping' of surpluses practised by industrialized countries, along with internal controls on food prices and overvalued domestic currencies in Latin America, have contributed to the rising import of food into the continent (Kay, 1995). Figueroa (1993: 296) claims 'the per capita availability of food products has declined in the region ... ECLAC/FAO estimates show that this drop was around 10% between 1981 and 1986'. There is some disagreement over the root cause of this insecurity. Some (such as Teubal, 1992) blame the impact of MNCs, with their increasing emphasis on production geared towards the needs and wants of high-income countries. Others (such as Scott, 1985) are more critical of internal government policies – in particular controlled prices and low investment in rural areas. Whatever the relative contribution of these factors, there can be no doubt that the shift to NTAEs has aggravated this problem. This calls into question the wisdom of basing development efforts along these lines – not least as the problem of food insecurity is likely to impact most forcibly on the poorer sections of society (Figueroa, 1993; Kay, 1995).

Impacts on environmental sustainability

The environmental impact of primary product exploitation in Latin America has received increasing attention in recent years (Figueroa, 1993; Furley, 1996). Studies have tended to emphasize threats to ecological sustainability created by environmentally insensitive activities. Among other things, relevant concerns have included the pollution of river systems and air by large-scale mining activities, the impact of pollution created by the petroleum and petrochemical sectors, the ecological impacts of large HEP schemes in a number of countries (Furley, 1996), the overexploitation and depletion of natural resource stocks [for example, the depletion of the fishery stock in Chile (Barton, 1997b) and the deforestation of Brazil within the imperative of 'expansion of the agricultural frontier' (Gligo, 1993)]; the intensive use of inputs and the effects on ecosystems and human health (Murray and Hoppin, 1990), issues of water supply and the effects of irrigation-intensive production (Gwynne and Meneses, 1994) and issues of increased run-off and flooding resulting from clear-cutting and soil erosion (Leonard, 1987).

The restructuring towards primary sector export activities precipitated by neoliberal policies, and the noninterventionist ideology which characterizes this approach, is increasing pressures on the physical environment. Increasingly the region's natural resource wealth is being consumed as current income, rather than being managed as a basis for sustainable development. A range of commentators agrees that this issue should be of central concern (Barham et al., 1992). However, the explanations for (and therefore potential solutions to) environmental degradation differ. For example, in a recent World Bank study, Edwards (1995) argues that the problem is that environmental impacts have been undervalued within cost–benefit analysis. This argument is placed firmly within the 'Western' environmental economics paradigm and assumes that the market can be altered at the margins in order to ensure the efficient and sustainable utilization of the environment. In his assessment of the current environmental problems faced in Latin America, Gligo (1993) is more radical, arguing that perpetrators of environmental damage do not have to bear the full cost of their actions – which, instead, is diffused across the wider society. Often, argues Gligo, wider society does not have a say in deciding whether activities should take place. Rather, such decisions are largely concentrated in the hands of those with political and economic power – a group which includes resource company owners and managers. In other words, the latter group of theorists argues that the problem is 'structural' rather than 'marginal' in its nature.

Inequality and social conflict

Often, growth in primary product sectors has had the effect of exacerbating social, economic and geographical inequalities. In the case of mining, the 'enclave' nature of production, low employment and other multiplier effects and the repatriation of profits, means that benefits have often not 'trickled down' to poorer members of society. In the case of the agricultural sector, there is considerable evidence to suggest that restructuring towards NTAEs and the associated 'modernization' of agriculture is distributionally regressive (Barham *et al.*, 1992; Figueroa, 1993; Kay, 1995). In spite of the 'scale-neutral' characteristics of many of the technologies used in the intensive production of agricultural goods for export, access to such technologies has most certainly not proven 'scale-neutral' (Cornia, 1987). Capital constraints faced by peasant and small farmers mean that, in general, they have not been able to participate fully in the system. Furthermore, the relative rise in the price of tradables precipitated by neoliberal restructuring has conferred greater benefits upon larger-scale farmers as production on such farms involves a greater proportion of goods for export (Figueroa, 1993). The negative impact for most small farmers during the structural adjustment period was aggravated by the withdrawal of social expenditure which accompanied austere fiscal/monetary stabilization (Sunkel, 1993b; Vergara, 1994; Kay, 1995). Tensions created by growing social and economic differentiation are creating political conflict, as witnessed in the recent Chiapas uprising and continuing unrest in Peru and elsewhere. For some commentators, these processes raise the potential for revolution (Vergara, 1994).

THE DEVIL'S EXCREMENT? CASE STUDY OF VENEZUELAN OIL EXPORTS

> OPEC's (Venezuelan) founder, Juan Pablo Perez Alfonso, observed, 'we are drowning in the devil's excrement' (quoted in Kant, 1997: 4).

One of the most intriguing puzzles when it comes to issues of resource-based development in Latin America is the failure of Venezuela to live up to its developmental promise given its status as a major oil producer and exporter. Oil production began in the 1920s, and by the end of that decade Venezuela became the second largest producer and largest oil exporter in the world (Gilbert, 1997). At this time, its reserves were among the largest in the world. Under the Gómez dictatorship (1910–35), production was controlled by large multinational oil companies, drawn in by attractive government incentives. Gross domestic product (GDP) grew at unprecedented rates during this time. However, as Gilbert (1997: 184) notes, 'Venezuela became rich whilst much of its people remained poor'. Furthermore, the productive potential of the economy was not increased, the economy became enormously reliant on oil revenues and dependence on foreign capital and manufactured imports rose substantially (Kant, 1997). The situation improved slightly with the fall of the dictatorship – as plans to diversify the economy and protect Venezuelan labor were introduced. In particular, a new oil law was signed in 1943 in order to reduce dependence on foreign capital. The Jiménez dictatorship, which took power in 1945, largely reversed these favorable trends. Among other scandals, a large fraction of oil revenue was used to promote 'showpiece' developments in Caracas. Under democratic rule after 1958, deals were struck which were less favorable to foreign capital. Eventually, in 1976 the industry was nationalized. However, Gilbert (1997: 185) argues that 'the nature of relations between the foreign oil companies, the dependent middle class, the government bureaucracy, and the poor remained unaltered'.

From boom to bust

By the time of nationalization, oil had brought major macro-economic benefits to the Venezuelan economy. The rate of economic growth between 1920 and 1976 averaged 2.6 per cent per annum – which was very high compared with the other Latin American countries (1.7 per cent) and even the industrialized countries (2.1 per cent). In 1976 the level of per capita income (US$1344) was the highest in the continent. According to Kant (1997), these conditions underwrote the political stability of democracy, as organized groups saw their demands on government expenditure satisfied at comparatively low tax rates. Investment and salaries grew, life expectancy rose and access to education was increased.

In the 1980s, however, this relative prosperity was rudely interrupted. On the back of oil rents, Venezuela had accumulated huge foreign debt in the 1970s. The debt crisis of the early 1980s augmented this burden, and the country slid into deep recession. Between 1981 and 1989, per capita income fell by 25 per cent, labor's share in total income fell dramatically, productivity declined (especially in non-oil sectors) to levels more than half the average for the region and inflation rocketed to 80 per cent (Kant, 1997). At the same time, imports remained high, which augmented balance-of-payments problems. Between 1980 and 1988, given the decline in global oil prices, the value of oil exports in domestic currency fell by over 50 per cent. Economic problems were particularly harsh in the non-oil exporting sector of the economy as previous rents had not been used to diversify the economy. For example, the average growth in non-oil GDP was only 0.5 per cent between 1986 and 1990. The most regressive impacts were saved for the poorer sections of society who suffered a sharp decrease in relative welfare during the 1980s. In 1989, the level of poverty rose to 53 per cent; up from 22 per cent in 1982. By the same year, 22 per cent of households did not have sufficient income to meet daily food requirements. In the 1990s, improvements have been slow and currently real per capita income remains at levels below those recorded in 1960. As a consequence, the country has struggled to hold onto the democratic political system (Kant, 1997).

Explaining Venezuelan failure

Why did this happen? How could what was once Latin America's great development hope descend into a social and economic disaster? A range of arguments, which to a certain extent complement each other, have been put forward. Gilbert (1997) argues that the inherent nature of the oil sector has a lot to do with the problems eventually experienced. In particular, he points to more traditional 'enclave'-type arguments in attempting to explain distributionally-regressive impacts. First, he points to the low employment generation of the Venezuelan oil industry, leading to low trickle-down of export benefits. Second, he argues that oil exports produced little manufacturing growth, which led to low economic multipliers. Third, he points to the unbalanced and limiting geographical characteristics of the oil industry in Venezuela. In particular, he notes how the major oil regions (Zulia, Anzoategui and Monagas) are isolated from the major economic regions of the country. Furthermore, he argued that oil rents were used by the government in ways which promoted 'urban bias'. In particular, he argues, 'Caracas has gained most from the oil revenues; its bureaucracy, industrial sector, and middle-class population have flourished' (1997: 187). In Auty's (1993) explanation, 'inherent' problems are also given importance. Central to Auty's 'resource curse' thesis is the problem of Dutch disease unleashed in the Venezuelan economy during the various oil booms it has enjoyed. This made diversification during the recession of the 1980s almost impossible as the non-oil sectors had become uncompetitive internationally.

In a more recent evaluation of Venezuela's economic woes, Kant (1997: 5) adopts an eclectic political-economy/political-science approach which attempts to 'capture the underlying political and institutional processes that set off economic laws and market forces in the first place'. Kant extends ideas regarding rent-seeking behavior earlier proposed by Auty and, in doing so, seeks to explain the institutional roots of 'Dutch disease'. She does not subscribe to the idea that poor development performance is a necessary outcome of reliance on commodity exports, arguing that 'it is not inevitable. Paradoxes can be resolved and development trajectories altered ... There is nothing inescapable about the future repercussions of petroleum or any other commodity' (1997: 242). Instead, it is the nature of the political institutions which evolve in response to commodity booms which partially shape the eventual developmental outcomes. In the case of Venezuela, rent-seeking behavior, patronage and self-interest permeated political institutions governing the economy. Thus,

> rather than creating a self-sustaining development model potentially independent of petroleum, Venezuela was living on a false economy, borrowed money, and borrowed time. Because state structures were formed precisely to perpetuate this model, Venezuela's readjustment to a productive nonrentier economy promised to be among the most difficult in Latin America (Kant, 1997: 235).

Kant argues that as Venezuelan political institutions evolved they took on an agenda of their own and became more interested in survival (and in some cases enrichment) than in national economic management. Thus, the Venezuelan state failed to develop the 'diversified tax structures, professionalized civil services, and more representative and equitable institutions' (1997: 242) necessary for sustainable social and economic development.

On balance, a mixture of characteristics 'inherent' to the oil industry and ineffective government management most probably explain Venezuelan problems. If effective and long-term-visioned political institutions are put into place, the potentially negative impact of commodity-export development can be at least partially offset. This begs a role for both the state and 'civil society' (including, for example, nongovernmental 'bottom-up' pressure groups and unions) in creating such an environment.

Unfortunately, the hegemonic neoliberalism which currently pervades Latin America frowns upon state intervention almost *per se*. Although this approach can help to reduce the impact of rent-seeking behavior and bureaucratic failure it prejudices the establishment of an effective regulatory framework. Furthermore, the scope for the generation of an effective civil society within a neoliberal framework is uncertain as pressure groups are often viewed with suspicion. Overall, adherence to neoliberalism prejudices the attainment of sustainable development in Venezuela and in regional economies where energy and other mineral exports play an important role.

FRUITFUL EXPORTS? CASE STUDY OF CHILEAN NONTRADITIONAL FRUIT EXPORTS

As discussed previously an important example of contemporary Latin American links into the global economy is that of NTAEs. The example of Chilean nontraditional fruit exports (NTFEs) is considered the most notable and, for some, most 'successful' example of such restructuring. However, as the subsequent discussion argues, the Chilean case study also highlights a number of problems associated with attaining sustainable social, political and economic development based on NTAEs.

The neoliberal model was implemented in Chile some years before it was in Chile's neighbors, subsequent to the military coup of 1973. The General Pinochet regime, under the advice of Milton-Friedman-inspired 'Chicago Boys', implemented a range of free-market reforms which intended to reverse the structuralist-informed inward-orientated model of development prescribed by previous governments. One of the central objectives of such reform was to stimulate exports. This was achieved with considerable success and had the effect of stimulating an enormous increase in nontraditional fruit exports, as the various comparative advantages which Chile possessed in this activity were allowed to operate in global space.

Between 1974 and 1995 the nominal value of Chilean fruit exports rose from US$30 million to US$1146 million (Fig. 6.2). Growth was especially high in the 1980s as a number of peso devaluations and export incentives took effect (Murray, 1997). Volume increases were also remarkable, rising from just over 50 000 metric tonnes in 1977 to over 1 300 000 metric tonnes in 1995 (ODEPA, 1996). Grapes and apples have been the most important species and have represented over 60 per cent of NTFE earnings since 1974. Traditionally, the major markets for Chilean NTFEs have been Europe and the United States; recently, a greater proportion of exports have been sent to East Asia and Latin America. The fruit boom initiated deep social and economic change in certain localities, such as in the Norte Chico and the Northern Central Valley, where the economic base has been transformed rapidly from relatively diversified production for national markets to specialized (in some localities monocultural) export-orientated production. In the country as a whole, land planted under fruit rose from approximately 65 000 ha in 1977 to just under 190 000 ha in 1996 (ODEPA, 1996).

The heady rise in Chilean NTFEs can be explained partly in terms of the large 'comparative advantage' such products enjoy on global markets. These advantages are both 'natural' and 'institutionally induced'. 'Natural' advantages include: the 'counter-seasonality' of production with respect to the major markets in the northern hemisphere, especially the timing of harvests for Christmas markets in the United States

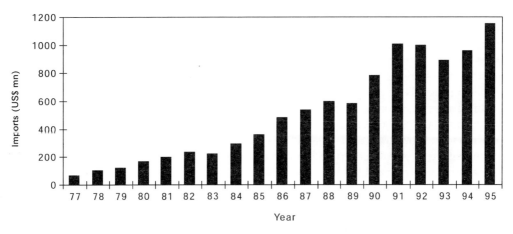

FIGURE 6.2 Value of Chilean nontraditional fruit exports (in US$ millions), 1977–95. Sources: Asociación de Exportadores (various years) and ODEPA (various years).

and Europe; the existence of ample supplies of fertile land and water in certain areas; and one of the most favorable range of climates in the world (from subtropical to temperate), ideal for the production of a wide range of fruits. Institutionally induced comparative advantages include: state-led investment in the fruit sector during the presidency of Eduardo Frei (sen.) (1964–70); various episodes of land reform under Frei (sen.), Allende and Pinochet which brought an end to the inefficient *latifundio–minifundio* land holding system; and the significant lowering of costs through post-coup labor reforms (Murray, 1997). Further to these advantages external conditions were conducive to NTFE growth. In particular, the 1970s and 1980s witnessed the acceleration of the 'globalization of agriculture' – especially in the fruit sector. This involved sharp increases in the consumption of 'exotic' fruits in industrialized markets, sourced largely by investments in a range of Latin American and other peripheral countries by multinational companies (Whatmore, 1995; Murray, 1998).

Foreign investment and the role of export companies

The role of multinational investment in the Chilean NTFE sector has been central. Although the earliest exporters were domestic companies (such as David Del Curto and Copefrut), during the 1980s foreign (especially US) capital rose to dominate the sector. Important examples include Unifrutti, Dole, United Trading Company and Chiquita-Frupac. By 1994, three of the top four companies (accounting for over 40 per cent of exports) were foreign owned and only four in the top 10 were Chilean-owned. Foreign firms were critical in the introduction of the organizational system which dominates the industry and links domestic producers to the global market. This system has three basic components. First, companies are responsible for marketing, providing facilities for the packing and storage of fruit and gathering produce in sufficient quantities to justify large-scale investments and obtain bargaining power. Second, companies undertake research and development in the adoption and adaptation of fruit varieties and technologies and are largely responsible for transferring these to the growers. Third, and crucially, firms provide finance for growers, acting effectively as banks. The vast majority of growers are linked into the system through contracts with such companies. These contracts are extremely exacting and have become increasingly tight in the 1990s – leading to high levels of economic and technological dependence upon export companies among growers. For some commentators, however, one of the main advantages of this organizational system is that it has permitted the participation of many small-scale growers who would otherwise have been unable to gain access into the sector. Others argue that the system has evolved into one which exploits such growers and has led to increasing levels of indebtedness, landlessness and marginalization (Murray, 1997).

Macro-economic success – at what cost?

At the macro-economic scale of analysis, the restructuring towards NTFEs has proved a resounding success. It has helped lay a firm foundation for the high average levels of economic growth and trade surpluses generally recorded since the debt crisis of the early 1980s. Perhaps most importantly, it has helped diversify the Chilean economy away from reliance on mining exports. Between 1971 and 1994 the proportional value of mining exports in total exports fell from 85 to 44 per cent. However, despite this success, when one reduces the geographical scale of analysis, it becomes apparent that the distributional impacts of NTFE growth have been highly regressive (Murray, 1998). Spatial and social inequalities have both been exacerbated by the fruit boom. In terms of the former, many rural localities and a number of regions have been unable to participate in the system. In particular, in rural regions where environmental and economic conditions are not conducive to fruit cultivation (generally north of Region III and south of Region VIII), farmers continue to rely upon traditional, low-margin products for the national market. Such farmers have faced a range of deep problems over recent years (Kay, 1997). These spatial inequalities are causing political tension, recently exemplified by protest marches by 'traditional' farmers in the Central region. Free market ideology, which has continued under the democratic *Concertación* governments[6] of Aylwin (1990–94) and Frei (jun.) (1994–2000), has been characterized by the absence of an explicit regional policy aimed at reducing the growing spatial imbalances.

Differentiation between socio-economic groups has also been exacerbated through the workings of the boom. In particular, nonlandowners, temporary workers (especially women) and small farmers have seen a decline in their relative socio-economic position in rural society. One of the major impacts of the boom has been to raise the demand for labor to pick and pack fruit. A significant proportion of the increased demand is temporary in nature, leading to economic insecurity in the workforce and to a range of problems associated with the flow of migrants to fruit-producing areas during the harvest season (Gwynne and Ortiz, 1997). Female labor has formed a central part of the labor force employed in the packing houses (Fig. 6.3). Some argue that this process is positive in that it has provided many women with their first opportunity to engage in paid employment outside of the home (Bee and Vogel, 1997). However, women in the packing plants generally receive less than the men 'in the fields' for a day's work, often having to return home to work a 'double day' (Barrientos, 1997).

NTFE growth has also had a differential impact on farmers of different scales of operation. Medium-scale (20–50 ha) and large-scale (50 ha or more) farmers and an increas-

FIGURE 6.3 Seasonal female labor in one of Chile's fruit-packing plants. The boom in fruit exports in Chile has created a large market for temporary labor, employed mainly during the harvest season (March to May in apple-growing regions). The gender division of labor is marked: men pick the produce in the fields; women work in the packing plants. Photograph by W. Murray.

ing number of urban suitcase-farmers, have prospered enormously from the boom. In contrast, very-small-scale farmers (*minifundistas*) have been precluded from participation because of the high costs of setting up an orchard (up to US$35 000 per hectare). Small-scale farmers (5–20 ha) initially entered the system in large quantities. However, they have found it very difficult to survive in the market – particularly in the 1990s. Sales of land are rising as larger-scale farmers and export firms move in to take over the *parcelas* of heavily indebted small-scale farmers (Murray, 1997). Part of the problem in this context are financially demanding contracts with export firms which allow a significant passing on of costs to politically disorganized growers. Failed growers often fall back on low-paid temporary labor or informal activities. Thus, marginalization, landlessness, and proletarianization are becoming increasing realities in export-orientated regions. Again, given the neoliberal model favored by government, little more than marginal efforts have been made to address the above problems.

The question of sustainability

Doubts are increasingly being raised concerning the sustainability of the Chilean fruit export sector. Given the importance of the sector within the national economy and almost complete reliance on the sector in certain regions and localities, this should be an issue of great concern. First, sustainability is being threatened in an environmental sense. The rapid expansion of fruit cultivation, especially in marginal environments, is placing stress on ecosystems. In particular, water shortages, water contamination due to pesticide and fertilizer leakage and soil salinity due to excessive irrigation are rising. Problems of decline in soil fertility are accentuated on small-scale fruit farms where monocultural practice has become dominant. Second, economic sustainability is threatened. As Figure 6.2 shows, during the early 1990s, the value of Chilean NTFE had 'plateaued' somewhat (if figures for the real value of exports were shown this plateau would be more significant). This was due to a range of interacting internal and external factors. Externally, rising global competition from southern hemisphere producers, protectionism, technological change allowing other fruit producers to increasingly impinge upon the Chilean counterseasonal market and shifting consumption patterns towards higher quality and new types of 'exotic' fruit in the industrialized countries have caused a decline in the real price of Chilean fruit on world markets (Murray, 1998). Internally, problems in the sector have been accentuated by the Chilean state's reluctance to regulate in order to offset mounting economic threats. Thus, there have been only limited efforts to invest in quality, diversify the export range, diversify markets, develop value-added production and invest in infrastructure. Finally, increased tension in the countryside is bringing troubling political implications in a country which is still very much in 'transition' to democracy. Given growing disillusionment with the *Concertación* government's ability to solve pressing agrarian questions and distributional issues in general it is possible that rural voters may abstain from voting or choose more hard-line alternatives on the left or the right. The shift to the right (along with the low turnout) which took place during the senate elections of December 1997 points towards this trend (Agüero, 1998). Any further move to the right is likely to damage the prospects of building effective regulatory institutions and worsen income distribution in the sector.

Although the Chilean NTFE sector is far from collapse, the ability to sustain its central role as a provider of diversified export earnings and a livelihood for hundreds of thousands of people is contingent upon state action (including the promotion of civil organizations, such as small grower unions) designed to regulate and diminish the negative impact of powerful global forces on regional and local economies and social groups. Adherence to neoliberal principles seriously prejudices this possibility in Chile and a number of other NTAE exporting countries where similar trends are being observed (Carter *et al.*, 1996).

CONCLUSIONS – THE RELEVANCE OF OLD IDEAS?

In many ways, at the dawn of the new millennium, the countries of Latin America face the same problem which has existed since colonial times. For sure, there have been advances in economic diversification and, crucially, industrialization in a number of places. Countries such as Brazil and Mexico are, to a certain extent, breaking away from resource export dependency. However, the majority of countries remain dangerously reliant on primary products for export income generation. In this way, they remain highly vulnerable to external conditions and the whims of the global market place and face considerable internal problems in managing rentier, enclave and, sometimes, foreign-dominated export sectors. Furthermore, ecosystems and local environments are placed under stress given the continued premium on resource exploitation. Although this situation is not a consequence of globalization and neoliberalism, the latter's emphasis on the short-term logic of comparative advantage has compounded the problem. Furthermore, noninterventionist logic has augmented the distributionally regressive impacts and negative environmental implications associated with resource export development. As this chapter has shown, neoliberalism has largely failed as a remedy for primary-product-dependent economies as it clearly prejudices sustainable development.

To return to the original question; why the paradox of plenty? This chapter has argued that this can be explained, to a varying extent in different countries, by a combination of the 'inherent' problems associated with primary-product-led development, the impact of asymmetric power relations which exist in the global economy and the general failure of Latin American governments to create effective institutions to regulate regressive impacts and promote economic diversification. Government failure was recognized by neoliberals and caused a counter-revolution against state intervention which is now hegemonic. It would seem that neoliberal theory has widely missed the mark. The state is not the *root* of the problem, although it can clearly become *part* of the problem. At the core, significant structural impediments to development – including, most notably, the region's continued role as a global resource periphery – continue to retard meaningful progress. Free-market economics is incapable of resolving this condition – simply 'rolling back' the state can make matters far worse. In this sense, it is crucial that the model, in its pure form, is abandoned. In order to transcend the paradox of plenty, the time is ripe for the reincorporation of a number of the ideas of dependency and structuralist thinking which, as this chapter has endeavored to show, remain relevant. Such ideas must be altered to reflect changing global realities and past weaknesses (Dietz, 1995; Kay, 1989). However, at the center of such an endeavor should be an attempt to foster the creation of an effective, sustainability-minded, environmentally-conscious and politically-inclusive state which can regulate the less fortunate impacts of resource-export-based development and implement long-term plans to move beyond it.

Endnotes

1 In this chapter, 'sustainable development' is defined in a broad way. To be sustainable, development must be viable in the long run in a number of closely interlinked ways. Thus, sustainability refers to more than purely ecological or environmental sustainability. In particular political, economic and social sustainability are considered equally important.
2 The dates given for the various periods should be seen as rough indicators only. There is a good deal of overlap in the timing of the influence of the various theories and events.
3 Overall, in 1994 17 per cent of Latin American gross domestic product (GDP) was accounted for by exports (UNDP, 1997). This figure has risen over the past 15 years as countries in the region have adopted outward-orientated policies. This aggregate figure underestimates the important role of total exports in most countries as in two large economies, Argentina and Brazil, exports account for less than 10 per cent of GDP (UNDP, 1997). In most economies exports account for over 20 per cent of GDP and this figure is rising.
4 It is widely held that trade data for a number of the Latin American economies are not fully reliable. In some cases data collection methods are considered imperfect. In others the role of illegal exports (such as drug-based items), which are not counted, are of crucial importance. Thus, the analysis in this section should be seen as indicative of general trends only.

5 That is, exports of goods and not services. Service exports from Latin America have risen over the recent past (for example, tourism) (Gilbert, 1997). However, merchandise exports dominate export portfolios in the continent as a whole.
6 Democracy was restored in Chile in 1990. The *Concertación de Partidos por la Democracia*, a coalition of center-left parties made up predominantly by Christian Democrats and led by two successive presidents from that party, has governed since that time.

FURTHER READING

Barham, B., Clark, M., Katz, E., Schurman, R. 1992 Nontraditional agricultural exports in Latin America. *Latin American Research Review*, **27**(2), 43–82. Gives a good introduction to most of the issues surrounding the growth of nontraditional agricultural exports in Latin America since the shift to outward orientation. Provides useful definitions, a comprehensive reference list and a number of country case studies.

Bulmer-Thomas, V. 1994 *The economic history of Latin America since independence*. Cambridge University Press, Cambridge. Provides a detailed and in-depth discussion of the evolution of the role of resources and resource exports in development over the past two centuries.

Kant, T.L. 1997 *The paradox of plenty: oil booms and petro-states*. University of California Press, Los Angeles, CA. Delivers the most contemporary and, arguably, most sophisticated analysis of the development of resource-rich Venezuela. Of general relevance to understanding the political economy of countries specializing in the export of primary commodities.

Weeks, J. (ed.) 1995 *Structural adjustment and the agricultural sector in Latin America*. Macmillan, London. Gives a valuable entry point into the literature on Latin American agriculture through a collection of contemporary essays on a wide range of thematic and country studies.

7

THE POLITICAL ECONOMY OF SUSTAINABLE DEVELOPMENT

ROBERT N. GWYNNE, EDUARDO SILVA

Sustainable development has become an increasingly popular paradigm, both in academic and policy circles, in the last two decades of the twentieth century, as it has been realized that 'environment and development are not separate challenges: they are inexorably linked' (Brundtland, 1987: 37). Sustainable development is the product of combining environmental and developmental concerns and although it is subject to disciplinary biases it usually refers to decisions affecting the sustainability of consumption, production, the resource base and to livelihoods obtained from the resource base (Redclift, 1987). It has been defined as 'development that meets the needs of the present without compromising the ability of future generations to meet their own needs' (Brundtland, 1987: 43). Meanwhile contemporary development studies have envisaged a political economy framework as one which studies the relationships between market-based economics and democratically-based politics. The study of political economy attempts to push enquiry past the purely economic to uncover the root causes of the political and social characteristics of capitalist production:

> Part of the limitations of the sustainable development thinking and the reformist technical guidelines ... is their failure to address political economy. Without a theory of how the world economy works, and without theories about the relations between people, capital and state power, sustainable development thinking is locked within a limited compass (Goldin and Winters, 1995: 200).

It is certainly true that the globalization of Latin American economies and the rapid population growth within Latin American countries are radically affecting Latin American environments. The globalization of Latin American economies has been closely linked to the political and economic objectives for higher economic growth and the policies of economic reform that have been introduced since the mid-1980s in order to achieve it (Chapters 4 and 5). As a result, trade and investment have expanded, particularly in terms of nontraditional exports. In Latin America (as opposed to East Asia), nontraditional exports have tended to be concentrated more in natural resource areas than in manufacturing (Chapter 6).

Thus, recent export growth has been particularly evident in the agricultural, fishing, forestry and, to a lesser extent, mining sectors. The emphasis on trade has meant clearing more forest for timber exports and to make more land available for export agriculture and ranching. Such was the case with soy production in Bolivia in the 1990s and cattle ranching in Costa Rica in the 1980s. The emphasis on trade also produces pressure for more prospecting for and development of mining (oil in Ecuador, gold and gems in Venezuela, iron in Brazil) as well as heightened industrial use of water and more intensive fishing (World Resources Institute, 1994).

International capital markets reinforce the pattern. Latin American nations are saddled with the heavy burden of a debt overhang – the legacy of the debt crisis of the 1980s (Chapter 4). Servicing those debts is a key condition of continued creditworthiness in international capital markets. Doing so requires generating even more exports, or else too much precious foreign exchange will go to pay the creditors and not enough will remain for development needs (Miller, 1991).

This increased resource extraction has brought into focus the question of the long-term environmental sustainability of such economic policies. Thus, environmental factors cannot be omitted from the question of contemporary development in Latin America. If Latin American countries achieve further economic growth in future years, their use of nonrenewable resources and their contribution to greenhouse gas emissions will increase, particularly if the level of heavy industry increases. Thus, provisions must be made to promote the efficient use of resources and the minimization of wastes. At the same time, Latin American countries have been experiencing rapid population growth, particularly in urban areas (Chapters 10 and 12). It is in these urban areas that the huge inequalities present in Latin American society become most evident. Policies must be formulated to ensure that the development needs of the poor are met without imposing unsustainable levels of resource use (WHO, 1992).

It is generally agreed that future policies must integrate both ecological and human needs. If a reasonable balance is not found, then sustainable development will be difficult to achieve. It is therefore essential to consider the ways in which Latin American countries might set about achieving the goal of sustainability. The issue of geographical scale is important here. At continental and national scales of analysis, resource use and misuse can be seen as major environmental questions. However, at the scale of the city, the issue of basic environmental requirements becomes more fundamental.

The interdependence between the environment and society ensures that ecological possibilities are inextricably linked to social and economic policies. Thus the aim of Latin American societies, within the scope of sustainable development, should be to achieve social, economic, political and ecological targets whilst simultaneously minimizing the level of local, regional and global environmental damage (Figure 7.1). Sustainable development must concern the rational use of resources – minimizing the use of nonrenewable resources and ensuring that renewable resources can be used for the long term rather than the short term. However, the essential point about sustainable development (and one that can be ignored by environmentalists) is that it must aim to meet basic human needs – access to an adequate livelihood, access to adequate shelter and a healthy environment and some form of participation in decision-making that affects those basic human needs (Figure 7.1).

In this, there is a marked contrast between the agendas of environmentalists in the

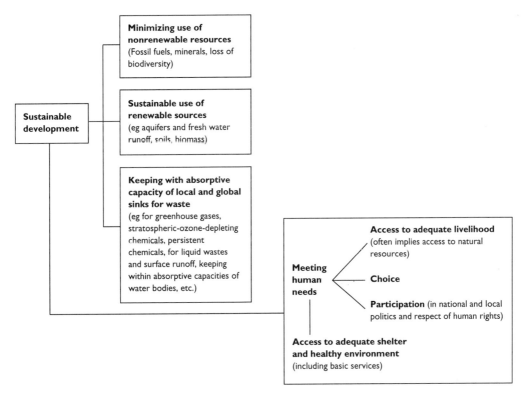

FIGURE 7.1 Components of sustainable development. Source: Mitlin (1992: 3).

more advanced economies and those that focus on the developing world. Satterthwaite et al. (1996: 2) point out that in the advanced economies most of the concern for the environment is related to chemical pollution in the air and water, damage to the natural environment and the scale of resource use and waste. However, Satterthwaite et al. (1996: 2) forcefully argue that on a global scale there are 'two, more serious, environmental crises that are forgotten or their importance is downplayed', particularly by politicians from advanced economies. They are more serious because of their immediate impact on human health, and especially on the health of children. The first is the ill health and premature death caused by pathogens in the human environment – in water, food, air and soil. Each year these contribute to the premature death of millions of people. The second environmental crisis is the hundreds of millions of people who lack access to natural resources on which their health and/or livelihood depend. The issues provoked by this analysis will be focused on in the latter half of the chapter.

The question of environmental sustainability must be firmly set within the framework of social relations. Virtually all national governments in Latin America have declared aims of achieving high economic growth, greater equity and environmental sustainability. However, at present, there are significant trade-offs between these declared aims. This chapter will examine these issues in two parts. First, in a section on the political economy of natural resource use, we will examine the conflicts in resource use between the aims of economic growth and environmental sustainability with particular reference to forestry. Second, the focus in the section on the environmental questions in the city will be on facing up to the environmental crises being encountered in Latin American cities and the trade-offs between the aims of equity, economic growth and environmental sustainability. Is the Latin American city sustainable? Can it keep up with the absorptive capacity of local waste production and can it adequately meet basic human needs?

THE POLITICAL ECONOMY OF NATURAL RESOURCE USE

Economies use natural resources in ever-increasing quantities to produce the goods and services we consume. Until recently, it was widely assumed that natural resources were 'free', in the sense that their extraction need not consider environmental side-effects. This view has changed. It is now generally recognized that the unrestrained extraction of natural resources all too often results in their rapid depletion and in environmental degradation. The consequences include the health hazards inherent in water and air pollution and the poisoning of land with agrochemicals and pesticides. Local and global climate change threaten. Species extinction contributes to the loss of biodiversity, sapping the vitality of the gene pool life relies on for its creation; it also robs us of economically useful species, thus jeopardizing food sources and depriving us of potential medicinal compounds and so on. Unrestrained mining of minerals raises the spectre of scarcity of the basic inputs our economies depend on (Pearce and Turner, 1990; Daly and Townsend, 1993).

There are two categories of natural resources: renewable and nonrenewable. Renewable natural resources, as the name suggests, are those replaced by nature at rates in proportion to human lifecycles, such as forests, water and animals (including fisheries). Nonrenewable natural resources largely refer to minerals. Once extracted, natural regeneration occurs in geological time, which is far too slow a process to be useful to humans; some resources can be reused through the use of scrap (such as copper and aluminium). Land suitable for agriculture and ranching falls somewhere in between.

Growing recognition exists that these resources are interconnected (Pearce and Turner, 1990; Commoner, 1990). This is especially true for renewable natural resources. For example, forests are crucial for watersheds. They maintain rainfall patterns that feed them. Furthermore, they fix soil at the banks and on slopes, thus reducing erosion, siltation and maintaining oxygenation of the water for river life. The watersheds feed rivers and lakes that agriculture, energy creation and other economic activities rely upon. Clear-cutting the forests that watersheds depend upon affects the entire chain (Myers, 1992).

The extraction of nonrenewable resources, however, also has effects on surrounding ecosystems and the interconnections inherent in them. Mining operations frequently have devastating impacts on local ecosystems. Oil exploration in tropical forests can pollute land and water to the detriment of local human and animal populations; in Ecuador's Amazon, an environmental crisis has occurred from the poisoning caused by oil mining by-products not being reinjected back into the wells but instead being left in open pits. Placer gold mining in Amazonia has poisoned water supplies and fisheries with mercury, posing potentially daunting health risks, particularly for children (Schmink and Wood, 1992). In northern Chile, the open-cast mining of porphyry copper ores has released large amounts of arsenic into the atmosphere; the mining towns of

Chuquicamata and Calama both have high rates of arsenic poisoning and cancer among their inhabitants (Comisión Nacional de Medio Ambiente, 1992; World Resources Institute, 1994).

As our awareness and knowledge grow, all nations face the problem of how to balance natural resource use with sensitivity to related environmental issues. The question, however, is particularly pressing for developing nations. On average, they rely on unprocessed natural resources for their economic growth to a much greater extent than do advanced economies. Exports of mineral and agricultural commodities supply a greater proportion of the savings and investment necessary for their economic development. Latin American countries depend on the hard currency generated by commodity exports to trade for food, technology and research and development key to their economic growth (Cardoso and Faletto, 1979; Furtado, 1976). Thus, a push for economic expansion, or a drive for a development leap, usually requires an increase in the rate of natural resource extraction to sell more on international markets.

Sustainable development and natural resource use

In this economic context, the leaders of Latin American nations (and of most developing countries) frequently protest that growing environmental sensitivity in developed nations forces unacceptable trade-offs upon them. For example, they bridle at demands from developed nations for the preservation of large tracts of land (usually forests) in the interest of biodiversity. Latin American leaders argue such demands are unjust. Developed nations became great by exploiting their natural resources not by preserving them; now they seek to prevent developing nations from enjoying the same benefits (White, 1993).

Policy-makers in developed countries, and their allies in developing countries, respond that this is a false characterization of the problem. Preservation (nonuse of nature) does not stand at the core of policy demands; conservation – commercial use with (as much as possible) the maintenance of the natural setting – is the goal. The issue, then, is not whether to use natural resources, it is how to use them while minimizing environmental damage (Pearce and Turner, 1990; Rosenberg, 1994).

According to economists, the problem is one of formulating discrete choices over how to use resources and minimize the environmental impact (Pearce and Turner, 1990; Carley and Christie, 1993; Goldin and Winters, 1995). For renewable resources the choice is over competing sustainable uses, by which economists usually mean sustained-yield harvest: the point at which the rate of renewal is at least equal to the rate of extraction. Thus, the resource is never exhausted. For nonrenewable resources the choice is over acceptable rates of depletion, allowing time for technological advances and substitution to reduce the need for the mineral. Choices may also include the means by which nonrenewable resources are extracted to minimize their impact on surrounding ecosystems.

Of course, the formulation is deceptively simple. First, there are many difficulties in calculating those points. Second, and more significantly, decisions over minimizing the environmental consequences of natural resource use involve more than discrete calculations of public choice; they also encompass political and social processes (Goldsworthy, 1988; Hurrell, 1991; Hurrell and Kingsbury, 1992). These are most acute in the distributional issues inherent in natural resource extraction: who will benefit from its use? In short, how natural resource extraction and sensitivity to the environment are combined also entails a struggle between social groups for control of the resource and the benefits of its exploitation.

In the drive for economic growth, dominant domestic and foreign socio-economic groups linked to agricultural, industrial and service corporations benefit the most from the extraction and sale of resources. When it comes to settling frontiers, as occurred in Brazilian Amazonia, they may demand or profit from road-building and other large-scale projects (such as mining and urbanization). Moreover, colonization programs may accompany such ventures to provide (in the absence of land reform) land to landless peasants or new settlements and jobs for the unemployed in more established urban areas (Bunker, 1985; Cockburn and Hecht, 1988).

In the planning stages, these schema may include environmental protection in resource extraction, as well as extension services to colonists for successful farming. But these good intentions may be abandoned during policy implementation. Such was the case in Brazil. The government institutions charged with administration and oversight of the plans lacked the capacity to carry out their mandates (Bunker, 1985). Centralized bureaucracies and fiscal crises prevented effective follow-through. Moreover, in the Amazonian states of Brazil, those institutions were 'captured' by large-scale ranchers and mine owners. They siphoned off meagre credit appropriations from the colonists. As a result, colonists cleared land and, lacking access to credit and know-how to farm their fragile tropical parcels successfully, then sold to ranch owners. Ranch owners bought land as a hedge against inflation and to capture state subsidies for beef produced in Amazonia. Colonists cleared more land for subsistence and the cycle was repeated (Hecht, 1985; Schmink and Wood, 1992).

In addition to global and national-level development strategies, the plight of the colonists on the Brazilian frontier highlights a second, oft-cited, source of environmental degradation in natural resource use: poverty. Extreme necessity forces people to abuse natural resources, especially renewable ones. Dense populations of poor people are forced continually to clear forests for fuel and land in slash-and-burn cycles. They also overgraze pasture, are unaware of the fragile character of most tropical soils and use primitive, polluting technologies in small-scale mining. Poor economic performance in any given country exacerbates the problem. It increases the number of desperate, impoverished people (Ascher and Healy, 1990; IADB, 1991; Annis, 1992).

Awareness of these problems led analysts to formulate the concept of sustainable development. The Brundtland Commission's *Our common future* (1987) first popularized the idea, recognizing the interrelation of economic development, overcoming poverty and safeguarding the environment. However, formulating policies for natural resource use that effectively link these three overlapping systems is no easy matter. In addition to technical difficulties, funding for programs is scarce. As a result, policy-makers are faced with significant trade-offs between different policy alternatives (Redclift, 1987, 1992).

Large-scale sustainable development

One policy response favors the utilization of large-scale private enterprise to foster economic growth, a posture well-anchored in the free-market tenets underlying globalization, although public enterprise may also participate in such ventures (Box 7.1). These companies have the financial wherewithal to invest in both development and environment, thus relieving responsibility from fiscally-strapped governments. With respect to natural resource extraction and the environment, the task is to persuade such firms to engage in sustained-yield practices, to adopt pollution-abatement technology, to avoid disfiguring and degrading surrounding ecosystems and to respect autochthonous cultures when present. It is assumed that project requirements (when funding is largely by multilateral development banks) and environmental regulations will accomplish those goals. Moreover, private sector projects and their wider economic effects provide employment for the local population. This addresses the link between poverty and environmental degradation. A focus on creating national parks and protected area systems to ensure biodiversity preservation (supported with international funding) complements the approach (World Bank, 1992).

Latin American nations' drive for economic advancement requires some large-scale development of natural resource extraction. The fiscal weakness of Latin American states and the decline in development funding from multilateral development banks certainly open space for more private sector involvement in the process. However, excessive reliance on large-scale development, private corporations and national parks (viewed as set-asides) presents serious obstacles for achieving sustainable development. There are at least three difficulties with the approach.

First, no matter what the rhetoric, the environment is not high on the hierarchy of issues confronting Latin American states; economic growth is. The fiscal debility of the state itself compounds the problem. As a result, the ministries and agencies charged with regulating, overseeing, coordinating and enforcing environmental policy are weak

---

BOX 7.1 Industrial-scale sustainable development

The Greater Carajás Program The Greater Carajás Program spearheaded the Brazilian government's Amazonian policy in the 1980s. It focused on export-orientated mineral projects, which allowed this regional development program to address major national economic problems. The program covers a wide area in three northeastern Amazonian states: Pará, Maranhao and Amapá. This vast development project involves state-owned mining companies, local capital and European and Japanese transnational corporations. After the completion of environmental impact reports, it has also received funding from the World Bank. Extraction and smelting of iron ore has been developed in Carajás; two aluminum complexes have been established, one near Belém and another on the Atlantic coast near Sao Luis; a hydroelectric plant at Tocurui was also part of the plan. Port facilities near these industrial projects have also been expanded (Neto, 1990).

Tree Plantations in Chile Chile boasts of a timber industry that is the envy of Latin America (Silva, 1997a). In response to the military government's free-market economic policies (1973–89), and with the aid of substantial government subsidies, a number of powerful Chilean conglomerates invested heavily in mainly-for-export timber plantations (Gwynne, 1993a). Joint ventures with international corporations or wholly-owned subsidiaries of foreign companies also entered the market (Gwynne, 1996a). The bulk of the wood was from radiata pine (Clapp, 1998). In 1994, forest sector exports topped US$1.5 billion, making it one of the leading export industries. It also employed about 95 000 people, roughly 2 per cent of the economically active population.

and lack the capacity to carry out their mandates. Indeed, sometimes the mandates themselves are deliberately narrowly circumscribed, as was the case in Chile (Silva, 1996–97). Supporters argue that market incentives help circumvent these shortcomings. Markets for pollution vouchers and opportunities for companies to pay for forest preservation and plantations to offset greenhouse gas emissions – joint implementation ventures – are often cited. This amounts to the 'browning' of the environment (Nielson and Stern, 1997). Such approaches rest on the uncertain assumption that we can calculate 'acceptable' levels of pollution. They also ignore the fact that we do not know the cumulative effects of many chemicals acting together. What may be an 'acceptable' level for one pollutant may turn out to be quite hazardous to public health when combined with the 'acceptable' levels of many others. In the final analysis, irrespective of ideological posturing, government action has been the most effective source of corporate sensitivity to the environment.

Second, large-scale, corporate-orientated resource extraction does not adequately address the livelihood needs of impoverished rural populations. Nor does it sufficiently protect native peoples and their cultures from the ravages of modernization. Large-scale agribusiness and mining do provide some employment, albeit at very low wages and frequently in substandard working conditions. Moreover, the capital-intensive ventures throw many more peasants off the land than are employed, driving them further onto the frontier to clear more forest for land or into urban shantytowns, putting more pressure on already woefully inadequate services (Painter and Durham, 1995; Montbiot, 1993). The pattern is aggravated by the influx of desperately poor people from other regions of the country into the area where the new concerns are being set up. These migrants compete with local communities, often disrupting and displacing them. The resulting social tension frequently erupts in rural violence (Schmink and Wood, 1992).

Third, the land hunger of these poor and displaced people places great pressure on a country's protected areas, making unworkable the dream of nature preservation. Land invasions into national parks are common both for the purpose of subsistence agriculture and for placer (small-scale) mining. Again, social tension among migrants and native peoples frequently erupts; it is often exacerbated by the influx of entrepreneurs that follow them to buy cleared land or to wrestle it forcibly from them (Barraclough and Ghimire, 1995). This pattern has repeated itself several times in Brazilian Amazonia (Schmink and Wood, 1992; Ozório de Almeida and Campari, 1995).

As the Brazilian case exemplifies, the problem is aggravated by the weakness of the responsible state institutions: the extension services of ministries of agriculture, environmental ministries and of parks and forest services. Bureaucratic rivalry may also intervene. Agriculture and mining ministries may (at least implicitly) support invasions. Since they are higher on the hierarchy of ministries (and better organized) than agencies of the environment they often nullify the latters' mandate.

The grass-roots development approach

Given these problems, dealing with the livelihood, or social justice, component of sustainable development calls for complementary efforts; or, as some argue, a completely different approach to natural resource use. We call this the grass-roots development approach, which focuses on strengthening local communities and fomenting small-scale economic activity rooted in sustained-yield practices (Friedmann and Rangan,

1993; Ghai, 1994; Ghai and Vivian, 1992; Utting, 1993; Schumacher, 1973). It privileges the values of local autonomy, solidarity, self-regulation and citizen participation in decision-making over the penetration of market forces, community disintegration and the reduction of participation to the implementation of a few projects. By organizing communities and building small-scale enterprises more of the income generated stays in the community in the form of higher wages, social benefits and capitalization. Economic sustainability also depends on the formation of cooperatives to pool resources and know-how, and of linking them to local, regional, national and world markets. By the same token, environmental sustainability is better served by small-scale use, because, with appropriate technology, it offers a better opportunity to mimic natural processes (Hartshorn, 1989). Because of the interconnectedness of nature, excessive human intervention in any one area (as occurs with large-scale development) damages the whole web of life.

The *Plan Piloto Forestal* of Quintana Roo, Mexico, is a good example of the grass-roots development approach. An alliance of forest peasant communities (*Ejidos*) wrestled control of their forests from private interests and government corporations. The *Plan Piloto*, with help from the government and international aid agencies, began to market its own timber (mahogany). Member communities received better prices for the timber than before, employed more personnel at higher wages, trained personnel in management, began to add industrial value to the timber instead of just selling whole logs and redistributed portions of the profits to member communities. There are many such cooperatives in Mexico (for an introduction, see Paré et al., 1997).

A significant strand of the grass-roots development perspective takes a different stance from current neoliberal trends with respect to the role of the state and social participation. The state has an important role to play in the crafting of industrial and extension policies favoring grass-roots development (Lipschutz and Conca, 1993). Thus, the strengthening of state institutions is vital to carry out increased functions. Otherwise, community enterprises, networks of cooperatives and links to markets are not likely to flourish beyond a few individual instances. Nongovernmental organizations (NGOs) are considered central to this process, as well as the inclusion of social groups in the policy-making process. By contrast, a more civil society-centered strand of the grass-roots development approach argues against deeper involvement of state institutions. Instead, it emphasizes the nexus between NGOs and organized communities (Browder, 1989; Leonard, 1989; Ekins, 1992).

Regardless of the position versus the state, at its irreducible core, participation is about a focus on organized communities as a vehicle for the self-determination of subordinate class and ethnic-based social groups (Friedmann and Rangan, 1993: 1–10; Ghai, 1994: 1–12; Ghai and Vivian, 1992: 1–19). Moreover, participation is about more than just helping to implement policy. It extends to broad deliberation by organized civil society in defining policy agendas, prescribing solutions and formulating policy.

This was the experience of forestry NGOs in Costa Rica, such as Aguadefor and Codeforsa in the 1990s. They worked both independently and through political organizations such as Junta Nacional Forestal and the Cámara Forestal Costaricence to influence the formulation of Costa Rica's new forest law (Carriere, 1991; Silva, 1997b; Silva, forthcoming). These organizations tirelessly lobbied ministries, foreign agencies and legislators. They also took part in countless committee meetings, formal workshops and official commissions.

Both critics and supporters of the grass-roots development approach to sustainable development recognize several difficulties with its implementation. First, in relation to its emphasis on small-scale development it is capital intensive, especially in terms of human capital. Where will financing come from, given that the private sector will not invest in such efforts? Up-front costs are significant: from planning stages involving the leadership of local communities, to the assignation of experts in the field, to coordination with government offices and to the purchase of equipment. Second, the ideal timeline for such projects is a long one: 5–10 years. The learning curve to use technology and to learn organizational skills can be long as well. Third, such projects are plagued by potential collective action problems within the community or between the governmental, international, NGO and community components of a project. Fourth, the techniques themselves may be experimental, unproven.

Much can go wrong. The planning process may be too top-down, thus alienating the community that was supposed to benefit. Given financing constraints, the project timeline may be too short, thus dropping crucial extension support before the community has sufficiently matured to take over the project. Irreconcilable or debilitating conflicts may erupt between the different participants in the project. Experimental techniques may fail. All of these difficulties may be exacerbated by the institutional weakness of state agencies involved in such projects. They frequently lack the personnel, equipment, training and authority to fulfil their role adequately. Not surprisingly this contributes to friction between the partners in a project.

The politics of natural resource extraction in the context of sustainable development

Ideally, a well-rounded policy framework for sustainable development would integrate both the large-scale, market-orientated and the grass-roots development approaches (Silva, 1997b). Too much emphasis on the former does not solve the problem of rural poverty and associated environmental degradation. By the same token, an overemphasis on grass-roots development would probably deprive a nation of necessary resources for healthy economic growth.

But the world is not ideal. Moreover, selecting the trade-offs between the two models for sustainable development is especially difficult to achieve in one of the most popular approaches to decision-making: the rational actor model, where policy-makers are appraised of a problem and offered a list of options best suited to solve it. The principal obstacles for the effective use of this technique are the scarcity of financing and the logic of globalization. These problems ensure that politics – the authoritative allocation of value – plays a significant role in determining policy outcomes: whether the market and grass-roots approaches will be integrated or whether one will dominate the other (usually market over grass-roots). In other words, politics will influence the agenda from which the choices offered to policy-makers will be drawn. Whether effective grass-roots development solutions get on that agenda is not guaranteed.

A political economy approach to public policy offers a good starting point to understand the politics of reforming policies for the sustainable development of natural resources. It helps us to identify the main actors, interests and power as defined by their location in international and domestic economic and state structures. Such approaches also emphasize the role of knowledge and coalitional behavior among actors as key to understanding outcomes (Box 7.2).

> **BOX 7.2 Extractive reserves in Brazil**
>
> The creation of extractive reserves in Brazilian Amazonia constitutes the most famous example that runs the gamut of actors: international, state, private sector, peasant and indigenous peoples (Schwartzman, 1991). In the 1970s and early 1980s the establishment of large ranches ignited a struggle over land between large-scale ranchers, on the one hand, and small-holders, rubber tappers and indigenous peoples, on the other. Initially, large-scale landowners easily prevailed because traditional rivalries divided these subordinate social groups. Ideas, however, brought them together. A Brazilian NGO came up with the concept of the extractive reserve: areas of land that would be set aside for low-impact extraction of natural resources. With this idea the environment became a new issue capable of uniting social groups that had been in conflict with each other. All could identify with the need for land and the preservation of the resources necessary for their livelihood and/or cultural survival (Keck, 1995). However, even when local peoples formed alliances they could not prevail. Local and state governments (Brazil has a federal form of government) generally backed large-scale landowners. Moreover, alliances between organized local peoples and Brazilian national unions and political parties were also ineffective. The federal government was adamantly in favor of large-scale development. This was when international actors tipped the balance in favor of local peoples. First, the Brazilian NGOs were in contact with powerful US-based international NGOs. These took the fight to the Congress of the United States, which brought pressure to bear on the World Bank to postpone loans to Brazil. At that point, the Brazilian federal government took note and decreed the establishment of extractive reserves. Unfortunately, this heroic effort has not had the success it deserved, for the extractive reserves had run into many economic problems (Assies, 1997; Hall, 1997).

International structure has both a political and an economic dimension. International actors include governments, transnational corporations, multilateral development banks and international organizations (Haggard, 1990). Their interests regarding environment and development depend on the position of the political parties in control of the executive branch, the country's situation in the international division of labor, the logic of international business and the balance of power among member states. Depending on the circumstance, these actors – especially when they are from developed countries – possess significant political and economic power, which they can bring to bear in national policy debates.

Domestic political and economic structure define state actors and social groups (Gourevitch, 1986; Frieden, 1991). State actors include the presidency, relevant ministries and agencies and political parties represented in the legislature. Social actors run the gamut from large-scale economic interests to peasants, smallholders and native peoples. In this schema the structure of state institutions (their cohesion, the tightness of policy-making teams, the hierarchy of ministries and their porosity to social forces) is crucial for shaping the power resources of state actors and social groups (Skocpol, 1979). By the same token, the economic and organizational capabilities of social actors affect their strength or weakness in relation to state actors (Migdal *et al.*, 1994).

Some social actors, however, are not defined by domestic or international structure. These are the environmental nongovernmental organizations (NGOs) (Fisher, 1993a). They can be important advocates of market-friendly or grass-roots development approaches to forest policy in the policy formulation process. When they are professional organizations, their policy stances generally derive from the intellectual and scientific ideas of their middle-class staff. When they have a peasant base, their economic interest often leads them to advocate grass-roots development ideas about combining environment and development (Keck, 1995). In addition to these characteristics, some of the large NGOs of the developed world are important international actors in their own right. Domestic NGOs can also be significant actors in the policy process. Their power often depends on their financial and organizational capabilities and the quality of the expertise available to them.

Ideas and knowledge about the environment and their relationship to development also plays an important role in outcomes. As Peter Haas (1990) has argued, they inform the policy stances of the state agencies charged with formulating policies related to environment and development. Of course, this view assumes that the state is not a unitary actor (Migdal et al., 1994). Knowledge, as previously mentioned, also provides NGOs with their rationale for action.

Mapping configurations of actors, interests, ideas and power provides a good starting point for uncovering how state institutions, social groups, intellectuals and the country's relationship to the world political economy influence policies regarding environment and development (Hurrell and Kingsbury, 1992). But this static approach only offers a partial account. In the end, policy outcomes generally depend on the dynamics of coalition formation between social, state, international and nongovernmental actors on the one hand and the historically specific international and domestic structures they are enmeshed in on the other. These alliances and structural conditions define the sum of power that competing coalitions muster in support of alternative policy stances during the policy formulation stage of the policy process (Gourevitch, 1986; Frieden, 1991; Rueschemeyer et al., 1992).

Reforming natural resources policy

Reforming natural resources policy in the mold of sustainable development poses a significant challenge. The range of policy options is well known and broad. It includes infusing large-scale resource use with an environmental component to ensure either sustained-yield practices or a sharp reduction in pollution or both. But it is also recognized that these practices alone will not provide conditions sufficient to alleviate pressure from impoverished rural populations. Nor will they protect the welfare, cultural heritage or survival of native peoples. This is why reform of large-scale resource use must be accompanied by efforts to promote grass-roots development. Integrating both approaches to the sustainable development of natural resource use, however, is not an easy task (Silva, 1994). Strong political conflicts may stand in the way of including grass-roots development components in policy reform. Nevertheless, these conflicts can be overcome. What it takes to do so depends on the initial disposition of state and dominant class actors towards such programs, the degree of local social conflict, whether or not local groups are organized and the role of international actors.

Where government and dominant social classes are cohesive in their resistance to grass-roots development, it takes high levels of social conflict and very broad alliances of local, national and international actors to force their inclusion. As the case of extractive reserves in Brazil suggests (Box 7.2) high levels of social conflict fused a strong coalition between well-organized regional subordinate social groups with national political and institutional affiliations. This brought the issue to the national political arena. However, domestic groups by themselves are usually not strong enough to prevail. They may agitate but they cannot win. Greater pressure from international sources is required. Here, international NGOs, linked with those of the developing country in question, can begin an awareness campaign in developed countries and lobby their governments and international institutions to take action. Threatening to suspend loans and other sanctions usually gets the attention of policy-makers in developing countries.

Where some key sectors of government actors and dominant social classes support sustainable development at the grass-roots for one reason or another, local actors may succeed largely on their own, but international actors and local NGOs make important contributions to policy-making. Tension between dominant and subordinate social groups is often a catalyst for organization in rural communities, which provides the drive from within to demand policies favorable for grass-roots development. Knowing they cannot act alone, communities often actively seek allies. When they find them, and especially if they are government actors, the organized community will strive to infuse policy content with its interests. International actors, often the development agencies of more social democratic governments, provide critical support for local communities in their efforts to shape policies emanating from relatively sympathetic government offices. They legitimize the demands of radicalized communities and help with project design and management.

Where most relevant government actors are largely indifferent or weak, international actors may be the most important catalysts for the inclusion of grass-roots development dimensions to resource use reform. The efforts will be project-orientated and the specifics of the grass-roots development orientation will depend on the goals of the lead international agency involved. It cannot be sufficiently stressed that success demands involvement of local communities and a very long-term international presence. Moreover, one must always keep in mind that scaling up projects – expanding their application to other areas – requires active government assistance. International presence or pressure alone cannot do the job.

Inclusion of a grass-roots development component to natural resource use never has been an easy task, and globalization makes it even more difficult. The emphasis on free markets by international agencies and governments impedes consideration of the non-market special needs of rural populations in developing countries. Those needs focus on the redistribution of national wealth via subsidies (credit, inputs, price supports) and involve the promotion of an industrial policy to create markets for their products. For these same reasons, globalization favors large-scale industry and concentration of wealth rather than small-scale production and more equitable distribution of wealth. As a result, many international agencies now focus on strengthening institutions for environmental management of pollution and parks for preservation only. Neither threaten the large-scale market development perspective.

Nevertheless, social tension in the countryside and lack of employment in cities to absorb displaced rural populations still provide a fulcrum from which to apply pressure for the inclusion of the grass-roots development dimension of sustainable development. The rural poor continue to organize and increasingly frame their demands with reference to environmental problems and conservation. This provides social and political allies domestically and internationally. It also supplies a cadre of professionals capable of working together with organized communities to devise plans for attaching a grass-roots development component to carbon offset agreements, pollution tax credits, energy taxes and effective parks management. Equally important, although it is not clear that this is occurring, new international institutions must build in participation and support for grass-roots development (Young, 1994; Haas *et al.*, 1993).

THE ENVIRONMENTAL QUESTION IN THE CITY

The city provides a very localized scale in which to study the interactions between environment and people – in contrast to the study of natural resources. Furthermore, it is argued that the environmental question in the city should be framed within the context of social relations. To what extent should society (and governments) prioritize investment in unhealthy environments, where polluted water and air dramatically reduce the life chances of the local populations? In this context, environmental sustainability should concern not just the issue of resource sustainability but also whether the basic needs of local populations can be met.

The major problem in Latin American cities is that of trying to alleviate poverty and of ensuring that planning at both the national and the urban scale significantly reduces the number of people living with unmet basic needs. If the environmental question concerns providing people with access to basic environmental standards, it must be emphasized that urban dwellers now predominate in Latin America. In 1998 the population of the 17 mainland countries of Latin America totaled about 450 million; of these 340 million (just over 75 per cent) were living in urban areas. Perhaps as much as 100 million of these urban dwellers lived in environments which posed considerable threats to their lives. Threats include unsafe and insufficient water, poor quality and often overcrowded shelters, inadequate provision for waste and sanitation, unsafe housing sites and a lack of adequate health-care services.

Thus, government environmental policy should also focus on the actual environments experienced by the poorer sections of society, particularly evident in urban areas where differential access to clean water and air is marked. In these areas, governments (in cooperation with other private sector and NGOs) should be involved in setting (and meeting) targets to improve access to such basic environmental needs as clean air and water. To provide focus, the following examines the measures which could be adopted by political actors within the public and private sectors and within the sphere of NGOs. After that the nature of urban environmental problems will be examined at the scale of the household, neighborhood and city before an assessment of the constraints involved in finding solutions is made.

Political action and city environments

Public sector action

The public sector has traditionally had the primary role in improving city environments. This has taken place both at the national level in terms of relevant ministries (public works, housing, health, education) and at the metropolitan or city government level. However, with the shift to the neoliberal economic model, public resources to improve city environments have been under continual downward pressure as ministerial and city budgets have normally been reduced in real terms.

Nevertheless, efficient public sector action requires a competent city and municipal government, efficiently run public enterprises and support from the national government. One area in which public sector action has normally predominated is in the supply of water. The governments of the State of Mexico and the Federal District, for example, have implemented policies designed to increase the proportion of clean water imported from external sources, including three water-gathering systems: the Cutzumala, Tecolutla and the Amacuzal systems (Schteingart, 1989). Pumping an increasing proportion of fresh water from outside the Valley of Mexico is very costly in terms of both energy and finance. The cost of providing water has increased to above what is received in water charges (Ezcurra and Mazari-Hiriart, 1996) and the level of capital and energy required to maintain the water pumps is dangerously high.

In contrast, Hardoy and Satterthwaite (1989) discuss a range of options which require a comparatively low level of capital investment in order to expand the provision of water, and which place a far lower level of demand on local resources. For example, local water resources can be used for small independent networks in particular areas of the city. The installation of a piped water system can often replace water vendors (Fig. 7.2) and thereby provide poorer households with a more cost-efficient, and safer, water supply. Although the initial installation cost is high, once established the cost of extending the system to more households is comparatively low. In Lima, the cost of water per unit volume bought from private vendors was 17 times higher than the price of equivalent volumes consumed through the mains network (World Bank, 1988).

In addition, if services such as piped water, drains and efficient waste collection are installed simultaneously, large cost savings can be achieved. This is because the services are interdependent. If drains are provided, but there is no waste collection service, the drains can become blocked: the cost of unblocking them, or indeed the risk of disease if the drains are unattended, is extremely high. This demonstrates the potential value of public sector action in providing coordinated investment to provide environmental improvements, particularly in the poorer areas of the city.

FIGURE 7.2 The water supply system in San Juan de Lurigancho, a poor squatter settlement of 100 000 inhabitants outside Lima, early 1980s; water fuelling point and tanker supplying (costly) water to poor households.

Private sector action

As in the case of Brazil (Box 7.3), there is considerable scope for public–private partnerships in tackling environmental problems. Furthermore, where local government is weak, private companies can invariably offer a better and more effective level of service provision. In particular, services which are not a natural monopoly, such as waste collection, may be improved at an affordable cost, with appropriate encouragement and regulation by local government. As Hardoy and Satterthwaite (1989: 169) comment, 'the aim is to seek a compromise between guaranteeing a basic level of service to everyone and maximizing cost recovery. Privatization is only valid if considered within the more important long-term goal of strong, competent and representative local government'.

Pacheco (1992) illustrates this point by suggesting that the environmental problem of litter pollution in Bogota can be reduced by achieving greater commercialization of recycled products and increased status of privatized waste collectors in order to improve working conditions and increase job satisfaction. In addition, changes in industrial production processes and the promotion of 'environmentally friendly' technologies must be promoted within private and public sector companies to reduce the

---

BOX 7.3 Public–private action and environmental sanitation technologies in Brazil (source: ECLAC, 1991: 97)

The Environmental Sanitation Technologies Company (CETESB) is a mixed public–private enterprise carrying out functions delegated by the government of the state of Sao Paulo for pollution control and environmental conservation. The government of the state of Sao Paulo owns 99.8 per cent of the shares; the management and the members of the board of directors own the rest. The company operates independently in all legal and financial matters and, with regard to its substantive operations, is linked to the Department of the Environment of the state of Sao Paulo.

A noteworthy project which CETESB has been implementing for a number of years is aimed at restoring the environment of the industrial zone of the Municipality of Cubatao (with World Bank financing, alongside that of the industrial firms concerned). The Cubatao industrial park, which covers 100 km^2, is located in a region where unfavorable climatic conditions, from an environmental standpoint, prevail. In 1985 the area generated 3 per cent of Brazil's gross domestic product (GDP) while at the same time releasing nearly 1000 tonnes of pollutants daily, 25 per cent of them in their solid state. In 1984, a five-year program for controlling environmental pollution went into effect under CETESB management.

The success of the project is evident. Since 1986 no emergency situation due to air pollution has arisen in the Municipality, and states of alert declined steadily until 1989, when none were declared. In complete fulfillment of the targets set, air and water pollution have been reduced by almost 90 per cent and soil pollution has disappeared altogether. Sources of pollution are systematically monitored and a substantial program of tree planting is hoped to restore forest cover to the surrounding slopes of the basin. The United Nations Environmental Programme (UNEP) now believes that the Cubatao experience should be repeated in other regions of the world affected by similar problems.

volume of waste and facilitate further recycling. The Mexican government has turned to the private sector to spread the cost of supplying water to the growing metropolis. In 1989, a group of companies in the Vallejo area of the city formed a collaboration in an attempt to combat rising water prices and potential shortages. A new company was formed, Aguas Industriales de Vallejo, which rehabilitated a former municipal wastewater treatment plant. The project has contributed to a reduction in pollution from untreated sewage and has enabled shareholder companies to purchase water at a price which is 25 per cent below that previously charged by the government.

The role of nongovernmental organizations

Nongovernmental organizations (NGOs) and organizations established by the residents of particular city communities can often provide cheaper services than can private commercial enterprises, since no profit is made. They are also more accountable to their customers because control of some services rests largely with representatives of the consumers of those services (Peattie, 1990). Indeed, as Hardoy and Satterthwaite (1989: 169) argue, 'there is tremendous potential in new partnerships between local governments and local community organizations – which could be regarded as privatization in another form'. Joint initiatives can be established, for example, to drain stagnant pools, to install drains, pipes and access roads, to clear land for schools and health clinics and to locate and destroy disease vectors within individual households.

A case study of Lima illustrates the ways in which NGOs can work with local governments and local communities to tackle environmental problems. In the late 1980s, a project was undertaken in the La Rinconada region to the north of Lima's center by the Institute for Local Democracy in Peru (IPADEL). This involved working with Save the Children and a low-income community to design, implement and maintain systems to improve water supply and services for sanitation, washing, and the collection and disposal of household waste. As Dawson (1992: 96) explains, 'the idea of developing a prototype environmental health system is to provide local governments with a model to assist them in public health improvement in squatter settlements'.

The project has largely been successful, generating local employment at a time of a depressed economic climate and ensuring optimal use of local resources. The scheme has ensured that even the poorest families in La Rinconada have access to the newly installed system of clean water and units consisting of combined toilets and showers. The project has targeted local needs and encouraged a positive response to the city's worsening environmental problems, which are particularly intense within the squatter settlements.

Overview

Clearly, then, there are a number of policy options which Latin American cities must consider when aiming to alleviate environmental problems. Often, simple and cost-efficient methods can be adopted to ensure the provision of adequate services for the city inhabitants and, in particular, the poor. It is important to distinguish between environmental risk and environmental degradation amongst the poor. Illegal or informal settlements generally have high levels of environmental risk because of the inadequacies in provision for piped waters, sanitation, drainage, garbage collection and other services. But most of these settlements contribute little to environmental degradation in the sense of degrading natural resources or sinks.

Overall, sustainable development cannot even begin to be achieved without strong, competent local government, joint initiatives between municipal governments, NGOs and private enterprises and support from the national government. However, this is not always the case. In fact, many governments can still be intent on adopting a 'Western model' of slum clearance rather than on upgrading existing squatter settlements, where the scale of environmental problems is intense. Indeed, as urban expansion continues in Latin American cities, environmental problems proliferate and the ultimate goal of 'sustainable cities' becomes ever more distant. What, however, is the nature of these environmental problems?

Environmental problems and the city

Owing to the diversity and great number of environmental problems in existence, analysis will focus on the problems associated with water supply, sanitation and waste disposal, although reference will also be made to other prominent factors. The analysis will be framed in terms of three different spatial scales: namely the household, the neighbourhood and the city. Environmental problems are considerably interlinked between these scales but such a spatial framework aids in highlighting the priorities for action at different geographical scales.

Household level

In most Latin American cities, between 30 per cent and 60 per cent of their population is housed in poorly serviced and structurally unsound housing units. It is in these settlements that the majority of environmental problems are concentrated. The most significant of these can be categorized under four headings: poor water supply, poor sanitation, insufficient waste disposal and overcrowding.

Poor water supply
Problems with the water supply in many Latin American cities include (Hardoy et al., 1992: 38): a lack of available drinking water; sewage seepage into groundwater and rivers; pollution from industrial effluents. Millions of urban dwellers are exposed to life-threatening diseases as they have no option but to utilize contaminated water. There are no reliable regional statistics concerning those of the urban poor that are fortunate enough to be serviced with piped water to their homes. Despite the fact that in most nations there have been significant increases in the proportion of the urban population served with water piped to their homes, there are still many millions of households who only have access to standpipes to collect their daily requirements; this often involves lengthy treks, carrying bulky water-carriers.

However, the plight of some is even more wretched. For those who are not served by piped water or standpipes, water from streams or other surface sources represents one of the few alternatives. These sources are often little more than open sewers. As a result, diarrhoeal diseases, typhoid, polio and cholera cause high morbidity and mortality rates in the poorer sections of Latin American cities (Hardoy et al., 1992).

In addition to the quality of the water, quantity is of major importance. Access to water from sources such as standpipes is greatly influenced by the time and energy required to collect it. The distance one has to carry the water, and the weight of the load, in addition to time spent waiting in queues at the source all affect the consumption. This limited access to water inhibits its use for washing and personal hygiene, in addition to washing food, cooking utensils and clothes (Hardoy *et al.*, 1992: 43). Its scarcity therefore represents a further risk to health.

Poor sanitation
Sanitation systems for the urban poor are generally of extremely poor quality, if they exist at all. For the majority, channels such as rivers and streams exist as the only possible facility for the disposal of sewage, thereby polluting the same source used for water consumption. In addition to the contamination of water supplies, the lack of drains and sewers results in waterlogged soil and stagnant pools, breeding grounds for diseases such as hookworm. In most cities in Latin America, such inadequate drainage often results in damp walls and thus damp living environments, increasing the incidence of respiratory diseases.

For those cities with sewers, the services generally serve the richer residential areas and government and commercial districts. In larger, richer cities higher proportions of homes are connected. One example is Mexico City where it is estimated that 70 per cent of the metropolitan area has access to the sewage system. However, three million people are still not served (Hardoy *et al.*, 1992). The increased population densities in cities, and the resultant increase in the demand for water and waste disposal, has made it impossible to segregate the disposal of waste with the drawing of water from impure sources. The resultant contamination of water supplies represents a common cause for disease.

Insufficient waste disposal
In 1985, only 41 per cent of the urban population of Latin America had access to sewage systems, and over 90 per cent of all wastewater was discharged directly into other water without any kind of treatment (Bartone, 1990). Subsequently, a decade of crisis and recession has reduced the amount of resources which the region can allocate to sewage and water treatment systems. There has been some improvement in the 1990s (UNCHS, 1996), particularly in access to sewage systems. Furthermore, in countries where cholera broke out in the late 1980s (such as Peru and Chile), considerable investment has been made in treating wastewater.

The problem is worst in squatter settlements, where little or no room is left for roads and public spaces. This results in high densities, which have a negative impact on environmental and health conditions. In particular, high densities make nonsewer sanitation environmentally hazardous as the removal, treatment and final disposal of refuse is generally very limited within Latin American households. International experience has shown that nonsewer sanitation starts to have adverse effects on health when densities exceed 100–150 persons per hectare (Ferguson, 1996). Threats to health include the diseases contracted from pathogens of typhoid fever, paratyphoid, hepatitis and dysentery (Bakacs, 1970). These have been proven to remain active in poorly treated refuse for months.

Overcrowding
The high density of population in Latin American cities exacerbates health problems. The highest densities tend to be found in poor slum areas of the inner city, such as the Rimac area of Lima (Fig. 7.3). The high density in these poor urban areas, combined with the lack of ventilation, results in the increasing likelihood of contraction of airborne diseases; respiratory infections, tuberculosis (still the single largest cause of adult death worldwide), mumps and measles are spread rapidly in such environments. Overcrowding has also been linked with high rates of death, disablement and serious injury from household accidents, especially where children and toxic chemicals are involved.

FIGURE 7.3 Overcrowded slum housing by the edge of the heavily polluted Rimac River, Lima, early 1980s.

Neighborhood level

The rapid growth of the poorer sectors of Latin American cities has often been associated with their uncontrolled expansion into the worst locations in terms of physical stability. Squatter settlements are frequently constructed on hillsides prone to landslides or on land prone to flooding, as in Rio and Caracas (Diaz, 1992). However, landslides represent only one of the hazards associated with unstable housing sites. The lack of available land has resulted in settlement on solid-waste dumps, adjacent to open drains and sewers, or close to quarries and factories. Households in these poor and poorly-located neighborhoods have even higher incidences of disease than those recorded in the 'average' poor household.

Inadequate provision for sanitation disposal results in excreta and wastewater being dumped into open drains. This, despite originating as a household problem, represents health dangers for the whole of the neighborhood. In addition, the insufficient garbage removal system represents a further health hazard as it is generally left to rot in small tips, canals and sewers or on the streets. This, in turn, creates further problems by blocking drainage channels, which then overflow to create open sewers, prime locations for pathogens to breed. Finally, because of the settlement processes in poor urban areas, little or no consideration is given to public open space. Consequently, children commonly play on waste land used for rubbish disposal, and these areas represent breeding grounds for pests and diseases.

City level

The pollutants contaminating the water supplies of many Latin American cities include sewage, industrial effluents and contaminated run-off. Owing to the lack of any form of

consensus-building before regulations are issued, which increases the efficiency of policy implementation. Of course, all agree that institutional strengthening is required for effective formulation and implementation of environmental policy. That is one reason why both the Inter-American Development Bank and the World Bank have focused on the problem. They, along with the United Nations Environmental Programme, are also encouraging the creation of environmental bureaus within line ministries (mining, transportation, agriculture, industry and so on).

Additional questions suggesting trade-offs between economic growth, social justice and environmental quality abound. What relationship between public sector and private sector institutions serves policy formulation and implementation best? Are market incentives or is regulation the most effective way to ensure results? How can fragile lands, forests and remaining wilderness be protected from the onslaught of economic development and migrant poor? Do national systems of protected areas work best under a strict preservationist regime? Or, in the absence of concern for the livelihood needs of the rural poor, do these systems succumb to land invasions and environmental degradation?

The answer to these questions, as expressed in policy, must necessarily involve politics: the process of authoritatively allocating value. Exhortations of the need for political will do not suffice. Existing socio-economic systems, the manner and beneficiaries of natural resource extraction, who receives urban benefits and who does not – all of these are sustained by coalitions of private interests, government actors and international agencies. Change requires the construction of countervailing coalitions. How much change and in what direction – how the questions raised above are answered – depends on the exact nature of such coalitions and the compromises they entail.

ACKNOWLEDGEMENTS

The authors would like to thank David Satterthwaite and Tom Klak for comments on drafts.

FURTHER READING

Collinson, H. (ed.) 1996 *Green guerillas: conflicts and initiatives in Latin America and the Caribbean*. Latin America Bureau, London. From diverse viewpoints, this book shows how environmental issues are central to the human rights struggles of urban and rural communities.

Hardoy, J.E., Mitlin, D., Satterthwaite, D. 1992 *Environmental problems in Third World cities*. Earthscan, London. Reviews environmental problems in cities of all Third World regions but gives many useful references to problems of health and water provision in Latin American cities. Wide-ranging and comprehensive in coverage of issues.

Kolk, A. 1996 *Forests in international environmental politics: international organizations, NGOs, and the Brazilian Amazon*. International Books, Utrecht. This book offers an excellent introduction to the dynamics of international environmental politics, with a special focus on the Brazilian Amazon.

MacDonald, G.J., Nielson, D.L., Stern, M.A. (eds) 1997 *Latin American environmental policy in international perspective*. Westview Press, Boulder, CO. This edited volume assesses Latin America's progress toward sustainable development, analysing the role of democratization, NGOs and international pressures.

Porter, G., Welsh Brown, J. 1996 *Global environmental politics*, 2nd edn. Westview Press, Boulder, CO. A first-rate introduction to contemporary trends in the ever-changing world of environmental politics.

Satterthwaite, D., Hart, R., Levy, C. *et al.* 1996 *The environment for children*. Earthscan, London. Focuses on urban environments as they affect children, particularly in terms of health and education. Reviews environmental hazards that threaten the life and health of infants, children and their parents and provides some ideas as to how these problems might be addressed.

reliance on both the private sector and NGOs. However, the reductions in public spending consequent upon neoliberal reform have seriously impacted upon Latin American city administrations and national governments. They have limited financial resources with which to tackle environmental problems and provide the urban poor with basic services.

In a normative sense, there are a number of cost-efficient measures which could be adopted, provided that the political will to identify and challenge environmental problems exists. Leitmann *et al.* (1992) argue for a long-term urban environmental management strategy which comprises four stages (Fig. 7.4). They emphasize that such a model can only be achieved if Latin American cities target four distinctive action areas. First, there is the need to strengthen governance. This involves mobilizing public support and encouraging local-level participation and awareness in environmental issues. Clearly defined institutional initiatives should be presented to local communities to demonstrate state support.

Second, policies and their implementation must be improved, particularly in the area of environmental health. Third, the management and delivery of environmental services and infrastructure should be enhanced. Finally, there is the need for more information and improved communications. Analytical frameworks should be used to identify, understand and prioritize environmental problems and design appropriate policies to manage them. There is some evidence that improvements are gradually taking place in all these areas during the 1990s. Improvements are, however, slow and patchy. Ultimately, as Stren *et al.* (1992) argue, the problem of sustainable cities does not rest within the issue of the failure to preserve natural resources but in the inability to satisfy basic needs such as food, housing, health, education and sanitation.

Conclusions

Population growth in Latin America and the drive for renewed economic growth in the context of globalization and free-market economics has significantly increased pressure on natural resources. These problems have been augmented by rising concentration of wealth among the rich. This has swelled the ranks of the poor whose migration patterns overwhelm the carrying capacity of fragile marginal lands and shantytowns lacking basic services. Taken together, these events deplete vital natural resources, expand pollution and, in general, degrade the environment. In short, the quality of life for humans in rural and urban settings declines.

The concept of sustainable development – struggling to become a paradigm – was born to address these trends. It recognizes the connections between economic development, poverty and environmental degradation. As a result, it does not ask policy-makers to give up the goal of economic growth for the sake of the environment. From this perspective to do so would be to destroy it. But it does urge them not to ignore the livelihood needs of the poor and the integrity of the environment as they pursue their economic ends.

Accomplishing these aims is a difficult proposition. Scarce funding and the divergent interests of experts who advise policy-makers have generated many competing policy prescriptions to advance towards the goal of sustainable development. Virtually all of them call attention to trade-offs between the major components of the concept. A number of key questions policy-makers must ask frame the trade-offs they face. The emerging environmental agenda in Latin America and among multilateral lending banks reflects the choices made to date.

Where is the problem of poverty and environmental degradation more severe, in cities or the countryside? For the moment, town has won over country. Intensifying urbanization, air pollution and deficiencies in sanitation services have driven the decision. More people live in cities and the public health hazards of pollution are apparent. Also, urban populations have the most immediate impact on politicians. As a result, policy debates focus on emissions from industry and automobiles, waste treatment and water supplies. Rural areas receive comparatively little attention except in areas of severe soil erosion and desertification. This neglect poses immediate health hazards to farm workers (direct contact and run-off from pesticides and agrochemicals) and longer-term damage to city dwellers. Soil and water may become polluted, affecting peasant villages. Finally, failure to address rural incomes and land reform increases pressure on fragile lands and their resources, which policy-makers, for all intents and purposes, have abandoned to their fate (Chapter 11).

At what scale do projects deliver the most efficiency? Proponents of large-scale projects argue they reach the most people or largest area with the most efficiency. Attention to large-scale business, farms, water treatment, mining complexes and the like has immediate effects. Their implementation is easier to monitor (less firms to control). Since fewer organizations are involved, there is less opportunity for project failure due to miscommunication or conflicts of interest. Proponents of smaller-scale projects argue that large-scale ones frequently break down due to equipment failure, unforeseen side-effects (for example siltation of water plants) or fiscal problems in the agencies involved. Smaller-scale projects have less start up costs and use more inexpensive and environmentally friendly inputs (for example recyclable materials and compost). Moreover, simpler technology, such as more efficient ovens and modest plantation groves in rural areas, can relieve pressure on fuel wood collection, a major source of deforestation. Active participation by local peoples with clear benefits fosters a stakeholder outlook crucial for project success.

By the same token, the question of how to address the basic needs of the impoverished is also linked to the issue of scale. At a very basic level, one camp argues that relatively unfettered market forces spearheaded by large-scale domestic and foreign enterprises will drive economic growth. This will increase employment, which is what people need. Others maintain employment alone is not sufficient. Market forces in Latin America may generate employment, but at low wages which make it impossible to access necessary services – health, education, sanitation. In many cases, penetration of market forces increases poverty. As a result, additional policies are required that focus on small-scale economic projects with active participation of local peoples.

What public institutions must be created to handle environmental issues? One debate has focused on whether a ministry of environment (Venezuela) or a coordinating agency (Chile) are the best forms. Supporters of ministries argue that cabinet rank is essential to be taken seriously by administration officials and to participate in policy-making in the executive; not to mention to have the ability to enforce statutes. They also claim that coordinating agencies cannot set sufficiently broad policy agendas because they must generate consensus between agencies. Proponents of coordinating agencies maintain that ministries only foster interbureaucratic rivalries in which, because they are new and relatively weak, they will be marginalized. Coordinating agencies ensure

> **BOX 7.4 Local freshwater resources in Lima, 1990** (source: ECLAC, 1991: 36)
>
> The city of Lima provides an example of the problems of water supply and the availability of farmland which are created by uncontrolled urban growth. The city's explosive growth and the loss of prime farmland in the latter half of the twentieth century have been accompanied by alarming changes in the area's surface water and groundwater systems as a result of the heavy demands placed upon them. For example, between 1970 and 1979, Lima's consumption of surface water climbed from 6.8 m³/s to 11 m³/s. At this time, the valleys' agricultural zones required 6.5 m³/s; these zones had previously served as a recharge area for the groundwater aquifer, but had ceased to perform that function by the 1970s. Since the Rimac River provided scarcely more than 15m³/s, by 1979 there was already a deficit of 1.5 m³/s. This shortfall was covered by pumping as much as 9.5 m³/s of groundwater when the river was low. By 1990, the average level of the water table was 20 m lower than when this practice began. In consequence, Lima suffered from serious water shortages.

authorities are still constrained in their potential to combat environmental problems and reduce inequalities both by national governments and by development assistance agencies (Mitlin, 1996). Focusing on only the 'needs of the poor' diverts attention away from the comparatively more important issues of the structural causes of poverty and environmental degradation. It is therefore apparent that the majority of 'solutions' may simply provide 'quick-fix' remedies that fail to address the critical issues of poverty and inequality which are, essentially, the root cause of environmental problems.

Nevertheless, there is growing evidence of considerable innovation in many urban areas of Latin America in the 1990s, much of it associated with democratization at national and city levels. There seems to be a new generation of mayors who have come to power because local elections have been introduced. Examples include: the innovations in Cali under the mayorship of Rodrigo Guerrerro; participatory budgeting in many urban centres in Brazil, most of it associated with municipal governments led by the Workers' Party; the record of elected mayors from the Frente Amplio in Montevideo that have instigated policies which improve conditions for low-income groups; the innovations of Cardenas as mayor of Mexico City.

National level

Whilst most of the solutions to environmental problems involve improved local practice, they also necessitate competent national governments. In most of Latin America, the recent orientation towards neoliberalism has meant that the state has reduced its role in welfare provision at the same time as economic restructuring has caused significant increases in unemployment. In the short term at least, neoliberal reform has been associated with increasing inequalities in income distribution. Structural adjustment programs have normally been associated with large reductions in fiscal expenditure with investment in urban infrastructure (such as for increased water supply and sewage systems) especially hard hit. As a result, the poorest and most vulnerable sectors of the population only have access to inadequately funded government services, and they are becoming increasingly marginalized in society.

In order to combat environmental problems successfully, the issue of inequality must be addressed. However, in Latin America, governments do little to tackle such inequalities and, in many circumstances, actually reinforce it (Mitlin, 1996). Inadequacies in environmental legislation mean lower costs for most companies but increased environmental health risks for much of the urban population. Similarly, inadequacies in occupational health and safety legislation results in lowers costs for the employers but higher health burdens for employees. It is only under international pressure (as with NAFTA and US government pressure on the enforcement of the Mexican government's environmental legislation) that environmental concerns seem to become more important at the level of national governments in Latin America.

Basic needs and sustainable cities

Neoliberal reform has changed the institutional framework within which environmental problems of the city are being approached. Although the public sector maintains its dominance in strategic action to alleviate problems, there is an increasing

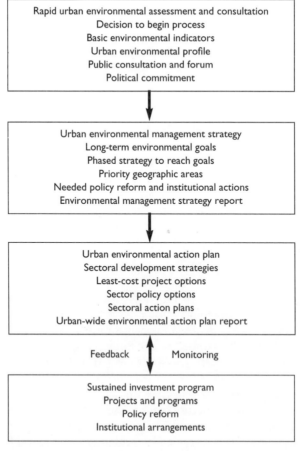

FIGURE 7.4 The urban environmental planning process. *Source:* Leitman *et al.* (1992: 139).

efficient policy controls, many hazardous wastes are dumped untreated into sewers and rivers. Alternatively, they are buried and can eventually contaminate the groundwater. The problem of sewage disposal has already been considered at the household and neighborhood levels. Its implications are considerable on a city-wide scale, as a large proportion of the population drinks water from sources that are contaminated with untreated sewage.

At the spatial scale of the city, air pollution, as in Mexico City and Santiago, becomes a major problem for people's health. Most of the toxins in the air over Latin American cities originate from the combustion of fossil fuels. The main source of such air pollution is sulphur dioxide, which is emitted from power stations, industrial plants and domestic stoves. Sulphur dioxide provides irritation for eyes and respiratory systems, particularly severe for those with genetic respiratory problems (such as asthma). Furthermore, the increasing ubiquity of the motor car has resulted in rising levels of airborne oxidizing and photochemical pollutants (Hardoy et al., 1992).

In a number of cases, the spatial location of Latin American cities in interior mountain basins has exacerbated this problem. Mexico City is located in a small mountain basin, 2200 m above sea level and surrounded by mountains. Such a location 'provides the conditions for thermal inversions which trap pollutants within the valley. In addition, high altitude also means a lower concentration of oxygen in the air which causes higher emissions of hydrocarbons and carbon monoxide from motor vehicle engines' (Hardoy et al., 1992: 64). Santiago is another city whose topography and climate favor the accumulation of polluting particles and gases over the city, particularly during the winter months.

Clearly, the cities of Latin America are afflicted by a significant number of environmental problems. These problems exist at all geographical scales within the urban environment, but their effects are most devastating in areas which are inhabited by the urban poor, because of lack of resources. What are the constraints involved in finding solutions to these problems?

The constraints to sustainability

There exist a number of constraining factors that inhibit the effective implementation of proposed solutions to environmental problems. These constraints can be witnessed on a number of geographical levels and this section examines these not only at the individual, community and city scales but also at the national level, where neoliberal reform is most dominant.

Individual and community level

Few improvement schemes will be successful unless they are backed by those for whom they are intended to serve. In the case of slum clearance, perhaps the most important consequence is the damaging effect on the community network. The lack of water, electricity and waste disposal ensures that residents establish support networks and mutual aid links. Such links are not only of great significance for day-to-day survival, but they also provide the basis for the community to mobilize and request public investments in the area from local and national authorities (Gilbert and Ward, 1984). If these networks are broken down, so too is the community attempt to diminish local environmental problems.

It is for this reason that research interest on community organizations has intensified in recent years (Akin Aina, 1990; Cuenya *et al.*, 1990). This reflects the important influence of collective solutions and responses to the problems of urbanization and the environment. These are problems which municipal and national governments have failed to solve, thereby forcing ordinary people and their communities to confront the challenge. Some have argued that, on their own, community organizations are incapable of instigating change. They are often seen as 'too weak and their actions too sectoral to have much political influence' (Akin Aina, 1990: 6). However, in some countries, such as Mexico and Brazil, urban social movements have proved strong and influential, whilst there has been a long history of community organization in Lima (Arevalo Torres, 1997). Nevertheless, whilst the potential importance of community actions should not be disregarded, their influence is often determined by the level of cooperation with municipal and city governments, and, in this sense, their effectiveness can be considerably undermined.

City level

Budgetary constraints have resulted in most municipal governments lacking the trained personnel, financial base and autonomy to provide the needed investments. This weakness of city government also makes other changes difficult to implement – from the enforcement of environmental legislation to the efficient collection of waste and the management of solid-waste sites (Hardoy and Satterthwaite, 1989).

The insufficient financial base represents the most significant constraining factor for alleviating environmental problems. This is invariably a result of the dramatic physical expansion of the city, demographic growth and the inability of municipal governments to expand their revenue base concurrently. The result of this has been insufficient extension of sewer networks combined with insufficient sewage treatment, and inadequate provision of city-wide piped water, waste disposal and health-care services. However, the implementation of such services is essential for achieving sustainable development (Hardoy *et al.*, 1992).

The severe shortage of water supply for much of Latin America's urban population is a serious environmental problem, but rarely is its cause environmental. In most cities, it is not a shortage of freshwater resources that is the problem but the government's refusal to give a higher priority to water supply. In some cities, there are serious constraints on expanding freshwater supplies, since the size of the city and its production base has grown to exceed the capacity of local freshwater resources to supply it on a sustainable basis (Box 7.4). It is common for richer residential areas and the main commercial and industrial concerns to receive good quality water supplies, sewers and drains, while 30 per cent or more of the population of poorer residential areas receive limited service provision or none at all. National and city governments frequently claim that extending piped water, sewers and drains to poorer areas is too expensive (Hardoy and Satterthwaite, 1989).

The idea that privatization can improve housing and service provision in areas of poor housing is greatly overstated (Hardoy and Satterthwaite, 1989). The lack of infrastructure and services supplied in most illegal settlements, the level of tenant exploitation and the very poor conditions in most of the rental housing stock do not support the idea that privatization will necessarily improve standards for the urban poor. However, it is important not to cast too much blame on municipal governments or their privatization policies. The privatization of national industries does generate much-needed revenue, and this can then be reinvested in low-income settlements. But many local

BOX 8.2 The double aspect of nineteenth-century democracy in Chile

J.S. Valenzuela has thrown light onto the duality of progress and restriction in the democratic process by showing how nineteenth-century Chilean elections excluded the great majority of the population because they were women or illiterate, but at the same time the majority of the voters came from middle and low strata and the electoral campaigns deeply involved many more people, including women (1997).

terms, because of its restricted character. Two features of this stage must be underlined. First, during this time liberal ideas were adopted, lay education expanded, a free press was established, a republican state was built up and democratic forms of government were introduced, but all this with *de facto* extraordinary restrictions to the wide participation of the people (Box 8.2). Second, contrary to the European trajectory, industrialization was postponed and replaced by a raw material exporting system which kept the backwardness of the productive sectors.

Thus Latin American modernity during the nineteenth century was more political and cultural than economic and, generally speaking, very restricted. In spite of these limitations, the modernizations introduced went hand in hand with a reconstitution of cultural identity in which the values of freedom, democracy, racial equality, science and lay education experienced a considerable advance *vis-à-vis* the prevalent values of colonial times. The eighteenth century French Enlightenment, British liberalism and, especially, the positivism derived from Auguste Comte played a very important ideological role in this process. The old values had been heavily influenced by Catholic religion, closely related to political authoritarianism and not very open to scientific reason, thus justifying slavery, racism, the Inquisition and religious monopoly. Just as much as the Creoles wanted freedom to trade with Britain and the rest of Europe, they also wanted cultural freedom from the tutelage of the church. Not that the new enlightened values and practices of the new republics totally displaced the Indo-Iberian cultural pole, but at least they modified and transformed it in important ways. The scientific rationality of nineteenth century Europe became quite influential among the Latin American ruling classes and university academics. They thought it was the only hope to bring about 'order and progress' to the newly emerging republics.

Pro-modern thinkers of the time felt that modernity could only be achieved as long as the Indo-Iberian cultural pattern was totally replaced by a new one, but they were not able to recognize how deeply influenced by the old racist prejudices they still were. At the same time, their vision of modernity was shaped by the naive wish to become a true image of the United States or Europe. Latin America was still to be civilized and its barbaric cultural features eradicated. Sarmiento, for instance, explicitly argued that the real struggle in Latin America was one between civilization and barbarism. The former was represented by Europe and the United States, the latter was the result of racial inferiority. According to Sarmiento Latin Americans were born of a mixture of three races – Spanish, Indian and black – which by their very nature were opposed to the spirit which has made civilization possible (quoted in Zea, 1990: 60–1). He pleaded 'let us be the United States of South America' in the same way as Alberdi proposed to form practical men, 'Yankees of the South'.

This vision was more or less shared everywhere in Latin America by other positivists such as Gil Fortoul, Prado, Bunge and Ingenieros. Prado maintained that 'the principal obstacle has come, necessarily, from what is the first social factor: *race*. . . . I cannot but recognise the pernicious influence which inferior races have exercised on Peru' (1983: 126–7, emphasis in original). Gil Fortoul, in his turn, argued in similar vein that some races, such as the European, have better aptitudes than others for civilization (1983: 104). Carlos Octavio Bunge (1926) sought to study the Latin American political organization through a psychological analysis of its three basic races. He defends the white race of Spanish origin, individualistic and superior. The Indian oscillates between passive fatalism and vengeance (Bunge, 1926: 133) and, at any rate, tends to disappear as a result of the degeneration of its race. The black race, in its turn, is characterized by its servility and infatuation (1926: 145). Hence racial mixes cannot be much better: mestizos lack moral sense and psychic stability; mulattos are false, impulsive and petulant (1926: 157 and 166).

It is not therefore surprising that one of the policies they proposed to modernize Latin America consisted in improving its race by means of European white immigration. Thus Prado, evoking Alberdi, argued that

> it is *necessary to increase* the number of our population, and what is more, to *change its condition* in a sense advantageous to the cause of progress. *In America to govern is to populate*; and the population must be sought in the spontaneous immigration, attracted by the action of laws, governments and individuals, of superior and vigorous races which, by mixing with ours, bring practical ideas of freedom, work and industry. Let us not promote, let us oppose the immigration of inferior races (1983: 133, emphasis in original).

Another way of compensating for Latin America's racial inferiority was education. Sarmiento rhetorically asked: what can South America do in order to achieve the prosperous destiny of North America? And he answered in the typical enlightened manner: 'Instruction, education diffused in the mass of the inhabitants', 'to level itself up; and it is already doing it with the other European races, by making up for the Indian blood with modern ideas, by finishing with the Middle Ages' (quoted in Martínez Estrada, 1968: 134 and 137). In the same vein Gil Fortoul argued that 'the intellectual and moral influence of the more civilised peoples has begun to neutralise or modify the primitive influences of the race' (1983: 115). Modernization, therefore, depended upon Latin Americans being able to replace their colonial and racial heritage by means of immigration and/or scientific education. For nineteenth-century Latin American authors there was clearly a need to achieve modernity by destroying the colonial cultural identity. But obviously, it was not easy to dismantle such an identity and they themselves unwittingly shared its racism and elitism.

Still, their projects of modernization are also projects of a new cultural identity with opposite characteristics to the Indo-Iberian cultural pattern they detested. They wanted to construct it upon the basis of the values of the Enlightenment: political and religious freedom, science and reason. We have seen how they adopted many of these values in theory only, and how the democratic progress they brought about was restricted to the ruling classes, but in spite of that there is no doubt that a major cultural change has been produced and that a renovated identity begins to emerge in which, nevertheless, many of the old values still remain.

> **BOX 8.1 Other classifications of trajectories to modernity**
>
> Therborn propounds four routes: European, New Worlds (including North America and South America), colonial zone (Africa and South Pacific) and externally induced modernization (Japan) (Therborn, 1995: 5–6).
>
> Marín distinguishes at least seven trajectories: Western Europe; North America and Australia; Eastern Europe and the Soviet Union; Latin America; Japan and South East Asia; India and finally Africa (1997; p. 3).
>
> The problem with Therborn's classification is that North and South America can hardly be located within the same trajectory. With regard to Marín's, one has to be aware that Eastern Europe is only a subgroup, started in 1945 of a common European trajectory spanning more than four centuries.

The North American trajectory to modernity is historically the closest to the European one and is the result of a veritable cultural transplant to another land. The idea of a cultural transplant or of 'transplanted peoples' has been developed by Darcy Ribeiro in order to account for the settlement of European migrants who wanted to resume their style of life and culture in a different continent with more freedom and better prospects (1992: 377). But it differs from the European route in that its initial progress was hindered by the English colonial power until independence. Once independence was achieved, the process of construction of modernity continued to be different from the European one because the United States began without the burden of the 'old regime' and, therefore, it almost does not know restrictions to political participation, and the 'social question' only arises in a very attenuated way (Wagner, 1994: 53).

The African trajectory to modernity is very different because it begins with a colonial imposition of capitalism by the end of the nineteenth century. The expansion of the British Empire destroyed by means of force the traditional modes of production and tribal modes of life. Whilst the Latin American modernity commenced with independence at the beginning of the nineteenth century, African modernity started with its colonization and developed under the colonial powers until the second half of the twentieth century. It suffers therefore from all the traumas and instabilities which stem from a very close colonial situation. An important problem of African modernity is that many African countries are artificial creations which arose by adding territories to the convenience of the colonial powers and without taking into account important tribal and cultural divisions which still subsist.

Japan has also a special trajectory to modernity which was pushed forward by its own traditional dominant class as a way of avoiding the colonizing designs of the West. The process started well into the nineteenth century with the Meiji restoration of 1868. This new elite wanted to keep a traditional mode of life but to construct a modern economy and state. The transition from a semi-feudal system to a modern economy was considered by the Meiji elite a necessity for national survival. Without modernization the Europeans would end up conquering the country and creating a new colony, as was happening elsewhere in Asia. The former isolationist policy of the Tokugawa regime had been successful for a while, but, by the middle of the nineteenth century, European countries were already aggressively 'opening up' the whole of Asia to international trade and had forced Japan to sign some treaties which conceded commercial privileges to foreigners. The Meiji reaction was to try to oppose foreign penetration by adopting the same foreign methods and instruments.

European modernity began incipiently out of endogenous processes around the sixteenth century and was consolidated with the Enlightenment in the eighteenth century. It could be said that the European trajectory to modernity has evolved in five stages. From the beginning of the sixteenth century to the end of the eighteenth century one finds a founding stage where modernity existed more as the ideas of some philosophers and where material and political progress as much as levels of general consciousness were low. The second phase, starting from the end of the eighteenth century revolutionary wave, covers most of the nineteenth century. The economic front was characterized by the industrial revolution and it was this process and the organized struggles of the working class that led to the political opening up of the system. In this period the Enlightenment ideas configure more precisely the meaning of modernity. Political life began to democratize and a wider public shared the experience of living in a new and revolutionary epoch. However, the distance between the project of modernity as an organized discourse which establishes a true imaginary of modernity, and the modern social practices and institutions which each society has actually developed and implemented, is still huge (Wagner, 1994: 4).

It is partly because of this that the third stage, from 1900 to 1945, was a phase of crisis and transition. The ambiguities of the modernizing process, with its rhetorical promises and practical exclusions, and the very criticisms which these inconsistencies attracted, led to a process of refurbishment of modernity in which the 'social question' assumed a fundamental importance (Wagner, 1994: 58). Liberal principles were subjected to critique and the creation of a welfare state for all citizens was on the agenda. These ideas were practically consolidated in a fourth stage which encompassed 1945 to 1973. This is what Wagner has called 'organized modernity', the golden epoch of capitalism (1994: 73). Nevertheless, this phase of stability and economic growth finished towards the end of the 1960s, and modernity entered once more a period of crisis. At the root of this second crisis of modernity there is an economic problem of slack accumulation and too heavy a state expenditure. In the 1990s the contours of a neoliberal new stage have begun to appear in which a post-Fordist regime of accumulation is entrenched. Universal state-supported welfare and full employment are no longer considered to be fully achievable goals and are partly abandoned.

THE LATIN AMERICAN TRAJECTORY TO MODERNITY

From independence to 1900: oligarchic modernity

Latin American processes of modernization started later than those of Europe and North America, commencing at the beginning of the nineteenth century with the process of independence. Prior to that, Spain and Portugal, the colonial powers, prevented their expansion for at least three centuries. The ideas of the Enlightenment exercised an important influence in these modernization processes but they had to confront the resistance of the Indo-Iberian cultural pole which had been consolidated during the three centuries of colonial life. This is why the first stage of modernization during the nineteenth century might be called oligarchic, in spite of the obvious contradiction in

of modernization in the 1950s believed in an ineluctable transition to modernity through a series of stages which would eventually overcome the traditional cultural pattern. In many contemporary neoliberal positions in Latin America is implicit the idea that the application of appropriate economic policies is a sufficient condition for an accelerated development which will lead Latin America to a modernity similar to that of Europe or North America.

But also those who oppose enlightened modernity do it because of Latin America's supposed true religious, Hispanic or Indian identity. For Indianists, modernity has destroyed Latin America's true identity which remains in the Indian traditions forgotten and oppressed since the conquest. For hispanists, Latin America's true identity can be found in the Spanish medieval cultural values which have been forgotten by the modernization processes since independence. For religious currents which emphasize the Christian or even Catholic nature of the Latin American ethos, the true identity has not been recognized by the enlightened Latin American elites, but can still be found in popular religiosity. All of them believe that Latin American identity was formed in the past once and for all, and that it was subsequently lost in the alienated pursuit of modernity. All of them believe that as long as modernity takes courses of action which are against the true Latin American identity, it cannot succeed and will lead to failure. Hence they propose that the only way out of this dilemma is to recover the lost essence of Latin America by going back to the Indian cultural matrix or to the values of medieval Hispanic culture or to Christian religion.

Between these two extremes are those like Octavio Paz, Carlos Fuentes and Claudio Véliz, who, despite adhering to modernity, try to show how difficult the process of Latin American modernization has been, as a result of the Spanish baroque legacy. For Fuentes 'we are a continent in desperate search of its modernity' (1990: 12–13) and for Paz, since the beginnings of the twentieth century we are 'totally installed in pseudo-modernity' (1979: 64), that is to say, for Paz Latin American modernity has never become really genuine. More recently, Claudio Véliz (1994) has argued in a similar vein that Latin America's stubborn baroque identity has been a major obstacle for its modernization and that only in the 1990s, bombarded by all sorts of consumer artefacts, has it begun to crumble to give way to an Anglo-Saxon kind of modernity. Somehow Latin American identity would have delayed the search for modernity or would have allowed reaching only a semblance of modernity.

It is interesting to verify that, in spite of the many differences amongst these authors and currents of thought and their favorable or unfavorable positions with respect to modernity, in all of them modernity is conceived as an eminently European phenomenon which can only be understood from the European experience and self-consciousness; which means that it is supposed to be totally alien to Latin America and can only exist in the region in conflict with its true identity. Some oppose it for this reason and others want to impose it in spite of this reason. The former believe that modernity cannot succeed in Latin America; the latter believe that Latin America's identity has to be dismantled. Both recognize the existence of a conflict which has to be resolved in favor of one or the other. Modernity and identity are conceived as absolute phenomena with opposite roots.

Contrary to these absolutist theories which present modernity and identity in Latin America as mutually exclusive phenomena I would like to show their continuity and interconnection. The same historical process of identity construction is, from independence onwards, a process of construction of modernity. It is true that modernity was born in Europe, but Europe does not monopolize all its trajectory. Precisely because it is a globalizing phenomenon, modernity is actively and not passively incorporated, adapted and put in context in Latin America in most institutional and value dimensions. That in these institutional and value processes there are important differences from Europe there is no doubt. Latin America has a specific way of being in modernity. Latin American modernity is not exactly the same as European modernity; it is a mixture, a hybrid, a product of a process of mediation which has its own trajectory; it is neither purely endogenous nor entirely imposed from without; some call it subordinate or peripheral (Brunner, 1994: 144; Parker, 1993).

The objective of this chapter is to show historically how within five distinct stages of modernization Latin America has been simultaneously constructing its cultural identity and to show the way in which these two phenomena, in spite of being intimately interconnected, are frequently perceived as opposite alternatives. I would also like to explore the reasons why if modernity and identity are not mutually excluding phenomena there has been such a marked tendency in Latin American history to consider modernity as something external and opposed to its identity. At the root of this mistaken perception one can find, on the one hand, oversimplified conceptions of modernity which totally conflate its different trajectories in a single European or North American model which has to be reached and, on the other, essentialist conceptions of cultural identity which freeze its contents and do not consider real cultural change. The former are more prevalent in times of accelerated development and economic expansion. The latter have emerged with greater force in periods of crisis in which economic growth and general welfare stall or go down.

In Latin America's independent history there have been roughly five alternating stages: from independence to 1900, expansion; from 1900 to 1945, crisis; from 1945 to 1970, expansion; from 1970 to 1990, crisis; and from 1990 onwards, expansion. In describing these stages of modernization I hope to be able to show what are the central elements of the Latin American trajectory to modernity in contrast to the European one, and what is the specific Latin American way of being in modernity. But this requires a small detour about the various trajectories to modernity.

HISTORICAL TRAJECTORIES TO MODERNITY

From the point of view of its historical evolution, modernity is a complex process which follows various routes (Therborn, 1995; Wagner, 1994). It is frequently believed that modernity is an essentially Western European phenomenon and it is forgotten that its very globalizing tendency makes it expand all over the world thus being forced to connect with different realities and to acquire different configurations and trajectories. No doubt, modernity was born in Europe and became a necessary point of reference for the processes of modernization in the rest of the world, but follows different routes in Japan and South East Asia, in North America and Australia, in Africa and in Latin America. Thus at least five routes (see also Box 8.1) to modernity can be distinguished which diverge, especially at the beginning, but which, as globalization expands, commence to converge. To make a full analysis of these five trajectories is beyond the scope of this chapter. That is why, after mentioning in a very brief and general manner some characteristics which distinguish the North American, Japanese, African and European trajectories, I shall concentrate on the Latin American route.

Part 5

Cultural change and modernity

8

MODERNITY AND IDENTITY: CULTURAL CHANGE IN LATIN AMERICA

JORGE LARRAIN

INTRODUCTION

The theme of modernity and culture in Latin America is full of historical paradoxes. Latin America was 'discovered' and colonized at the beginning of European modernity and thus became the 'other' of European modern identity. But Latin America was deliberately kept apart from the main processes of modernity by the colonial power. With the process of independence from Spain, Latin America enthusiastically embraced the Enlightenment's ideas, but more in their formal, cultural and discursive horizon than in their political and economic institutional practice, where for a long time traditional and excluding structures were kept in place. When finally political and economic modernity began to be implemented in practice during the twentieth century, cultural doubts began to emerge as to whether Latin America could adequately modernize or whether it was good that it modernized by following European and North American patterns. While in practice modernizing processes were widened, disquieting questions arose as to whether they could be carried out in an authentic manner. Hence, it could be said that Latin America was born in modern times without being allowed to become modern; when it could become modern, it became so only partially, in the realm of programmatic discourse, and when it began to be modern in practice then doubts emerged as to whether this conspired against its identity.

This chapter will try to show how, from the beginning of the nineteenth century, modernity has been presented in Latin America as an alternative to identity, as much by those who are suspicious of Enlightened modernity as by those who badly wanted it at all cost. Examples of the latter are plentiful. Nineteenth century Latin American positivism, for instance, believed that 'order and progress' could be provided by Enlightenment's ideas, and precisely because of this it strongly opposed the prevalent Indo-Iberian cultural identity. In the same way, the optimistic North American theories

stratum of this identity, secularization is not just a threat to the church but, more fundamentally, a threat to the very Latin American culture. This irreligious threat succeeded in converting the Latin American elites, the creoles, to instrumental reason, but it did not succeed against the popular religiosity of the mestizos, which has resisted all attacks to remain until today the most spontaneous and genuine expression of the cultural ethos.

Yet by the end of the 1980s and in spite of the strength of these attacks against modernity, the project of rapidly advancing to modernity even at the cost of identity was becoming dominant in Latin America supported by the increasingly overwhelming success of neoliberalism. At this time the antagonism between supporters and detractors of modernity reached its peak, but both sides seemed to share the idea that modernity was something external, which either had to be prevented from expanding in order to preserve identity or had to be brought about at all cost in order to change the old identity.

From 1990 onwards: the neoliberal stage

The stage which opens up after the end of dictatorships continues with and accelerates economic and political modernization under the influence of an already consolidated neoliberal ideology. Once more the concerns about identity recede as the neoliberal optimism gets the upper hand everywhere. But this time the modernizing wave has the opposite sign to the post-war effort. Instead of protection to industry, economies open to the world market are advocated. Instead of an interventionist state, the tendency is to reaffirm reduced-size states with lower expenditure and greater control of macroeconomic variables to fight inflation; instead of welfare state, the privatization of health education and social security. Dependency is no longer blamed for lack of development but rather excessive state intervention in the economy is the new culprit which has smothered growth. Capitalism is still dynamic if it is allowed to work according to the market forces. What has failed in the past is not capitalism but socialism.

The free-market and open-economy policies produced in the first instance a significant diminution of industrial production and industrial employment. Some countries such as Mexico and Brazil managed, after a while, to expand their industrial exports thus compensating for the flood of imports from foreign manufacturers. The rest, on the contrary, follow a more radical *laissez faire* model which, although diversifying the exports of primary products, make more permanent the low level of industrial production and employment. In this the Latin American trajectory to modernity (with the exception of Mexico and Brazil) is very different from the Asiatic one where the state assumes a very important role in the acquisition and adaptation of first-class technologies and in the promotion of industrial exports.

These economic processes occur now in a new political context which values democracy, participation and respect for human rights. This new stage continues with open-economy policies, but in a manner different from Europe it has to begin by modernizing and democratizing the state. This process has progressed but many problems still remain. The rebuilding and modernization of the political structures which had collapsed is one of the characteristics of the Latin American trajectory to modernity in the 1990s. This process still goes on.

Some authors openly propound the idea that Latin America must abandon its old identity in order to be able to enter modernity fully. Claudio Véliz (1994), for instance, maintains that the main problem that modernization has in Latin America is the cultural resistance which the very Latin American identity has opposed to it. Véliz agrees with Morandé that the Latin American identity is baroque, but he sees it as an obstacle to development. The failure of Latin America to modernize until the 1990s is a result of its own baroque identity, of its aversion to risk and change, of its distrust of the new, of its preference for stability and central control, of its respect for status and old loyalties. Nevertheless, after many centuries of resistance to change, the magnificent baroque dome has begun finally to deteriorate and crumble under the impact of thousands of banal cultural artefacts coming from the Anglo-Saxon world, from tennis shoes to toasters and jeans. Hence Claudio Véliz advocates in the 1990s an Anglo-Saxon kind of modernity. This goes to show that in his view the process of modernization in Latin America is externally-led and antagonistic to its cultural identity (Véliz, 1994: chapter IX).

Still, as in other times of accelerating development and economic expansion, a new kind of identity seems to be implicitly advocated and discursively constructed by the neoliberal project, only this time it is very different from the developmentalist identity which responded to the populist matrix in the 1950s and 1960s. Now a new kind of identity is being constructed which is embodied in the figure of the successful and innovating entrepreneur and which offers widespread (credit card) consumption as the linchpin which could deliver the masses. The former values of equality, state-sponsored welfare, fairness and general austerity which most ideologies propounded are now replaced by individual success, conspicuous consumption and privatized welfare. The point now is no longer justice, full employment or industrial development but rather to become winner nations comparable to the Asian tigers.

Hence, after this brief historical exploration about the ways in which modernity and identity have been constructed in practice and theoretically presented by Latin American intellectuals, one can clearly see the oscillation between the prominence given to one or the other in alternate periods: modernity in periods of expansion; identity in times of crisis. Yet this does not mean that at times of expansion identity ceases to have a presence. It is still being constructed even if it is not mentioned. True, people and authors do not seem to refer to it very much, but in the major modernizing projects of the day there is implicit the project of a new identity which will replace the old. In the very critique of the old identity there is the seed of the new identity.

SOME SPECIFIC ELEMENTS OF LATIN AMERICAN MODERNITY AND CULTURE

The combined construction of modernity and identity in Latin America which I have explored historically in five stages has produced certain specific features and characteristics which can be presented more systematically. These cultural features should not be essentialized; they are the result of history and they can change, be modified or even disappear altogether. But they still have an important presence today and are the results of a specific historical evolution. I have selected those characteristics which seem to me most relevant and which mark a contrast to other trajectories to modernity. I make no claim to the approach being exhaustive.

The first feature I would like to refer to is clientelism or cultural and political personalism. As I mentioned in the above section, this feature arose from very precise historical circumstances but its effects have remained until today. I differ in this from Manuel

new identity, a kind of developmentalist identity whose goal was economic development of an industrial kind, in which the state played a central role and in which the value of equality was still very important. Political struggle in this period was about how to achieve development and welfare for all. The economic system continued to be capitalist but the modernizers wanted to humanize it, and following populist policies they wanted to protect the workers and redistribute national income in their favor. The chosen path of protected industrialization determined that the ruling industrializing classes had some common interests with their workers. The new identity had therefore a populist matrix which combined industrial development with state support and workers' rights. Paternalism, personalism and clientelism are its hallmarks.

From 1970 to 1990: dictatorships and the lost decade

By the end of the 1960s a new crisis started which coincided with the second crisis of European modernity. The processes of industrialization and development lost their dynamism, and social and labor agitation became widespread. While in Europe right-wing governments were elected which sought to limit trade union power and state expenditure, in Latin America the challenge of the Chilean socialist experiment and the exhaustion of other populist experiences of the left precipitated a wave of military dictatorships which did the same, only in a more drastic and authoritarian way. This shows the precariousness of Latin American modern political institutions compared with those of Europe. They are incapable of channelling and absorbing the political turmoil within a framework of stability.

Some argue that at least dictatorships opened the way to a new globalized stage of development and economic modernization in Latin America. It is true that in many cases, and most definitely in Chile, they changed the direction of the economic policies by opening up their countries to foreign investment and foreign goods. They implemented in practice the first neoliberal ideas according to which it was necessary to export, to abolish tariffs and subsidies, to abandon an inefficient industry and to let the market forces produce development by means of a more rational allocation of resources. But it took many years before a new stage of expansion began to yield some economic fruits. In the case of the Chilean dictatorship it took the first four years (1973–77) for the harsh economic policies to begin to have some acceptable results, only to plunge again into a deep financial crisis in 1982 (Moulian, 1997: 201–12).

It has to be remembered that during the 1980s, the so-called 'lost decade', Latin America suffered negative growth. Besides, from the point of view of political and social modernity, dictatorships meant an important regression insofar as they abolished democratic institutions, systematically violated human rights, dismantled forms of social participation and consistently sought to destroy social organizations representing the poorest sectors of society. The exclusion of wide social sectors was increased as unemployment levels soared and salaries plummeted.

This second crisis of modernity explains in part and goes hand in hand with a profound identity crisis which is marked by pessimism and renewed doubts as to whether the road to modernity which had been followed could be wrong. Thus in the 1980s forms of neo-Indianism, religious fundamentalism and even of postmodernism emerged which were very critical of modernity. Many critical social scientists, some disillusioned Marxists, some influenced by Catholicism or postmodernism, initiated a process of radical deconstruction which entailed a forceful and bitter critique of

Western instrumental rationality and a reappraisal of a different, supposedly original, kind of cultural identity which could provide new and fresher arguments to oppose the increasingly successful wave of neoliberal ideas which was beginning to sweep Latin American countries. Some, the neo-Indianists, resorted once more to exploring the Latin American origins and the forgotten cultural patterns present in the Indian communities in the hope of finding there the elements of a new alternative way ahead which included community-orientated and ecological dimensions.

Luis Guillermo Lumbreras, for instance, argued that Latin America's 'tropical and mountainous lands were not necessarily suited for the procedures of the prairies and cold forests'. So the whole process of development in Latin America had been misdirected from the beginning and the only solution is to recover 'the knowledge of our ancestors' and 'to make use of that knowledge' (Lumbreras, 1991: 22). Anibal Quijano in his turn makes a critique of instrumental reason and dreams of a utopia constructed upon the basis of an alternative reason (1988: 62; 1991: 34–8). This reason comes from the past and was cultivated by the Indian communities, but also has its reality in the present. This different rationality, based on solidarity, collective effort and reciprocity, remains alive in the mass of the urban poor, in their popular kitchens, in their cooperatives and in their forms of organization to survive. Similar ideas can be found in Galeano (1991), Rojas Mix (1991) and Stolcke (1991). In all these contemporary Latin American writers one still finds the idea that Latin America's future depends on its being true to some age-old Indian traditions or principles which were forgotten by instrumental reason, alienated enlightened elites and neoliberal modernizing attempts.

Morandé (1984), following a religious point of view, criticizes the modernizing efforts in Latin America for they would negate its true identity. Modernization, as it has occurred in Latin America, would be antithetical to its most profound being insofar as it has sought its ultimate foundation in the European enlightened rational model. The Latin American cultural ethos, on the contrary, was formed before the Enlightenment, it has a necessary Catholic underlying structure, it prefers sapiential to scientific knowledge and is best expressed in popular religiosity (Morandé, 1984: 144–5). According to Morandé, Latin America's leading intellectual elite was unable to recognise its deepest cultural roots and, because of that, led their countries into modernizing experiments which, by ignoring Latin America's true identity, could only fail.

The attempts to repeat Weber's rationalization process were bound to fail, whether they took the form of an evolutionary transition to modernity, the form of a democratic or revolutionary road to socialism or the form of a 'shock treatment' which suddenly introduces the self-regulating free market. Contrary to the Protestant ethic and the need to save and invest as a proof of salvation, Latin American culture puts an emphasis on work as sacrifice and on religious festivities as ritual squandering. Basically, Latin Americans are not supposed to be motivated by technical progress, and the subordination of their ethos to instrumental rationality was a form of alienation, a mistake punished by chronic failure.

According to Morandé this does not necessarily mean that the Latin American cultural identity is fundamentally anti-modern. What he argues is that it was constituted before modernity arrived, or rather that the Latin American cultural identity was constituted within a different kind of modernity, the Catholic, Counter-Reformationist, Spanish, baroque modernity. So what threatens Latin American identity is not just any kind of modernity but the modernity which entails a process of secularization, the modernity stemming from the Enlightenment. It follows that, given the Catholic sub-

the twentieth century wanted to recover an old identity and could not recognize the important new changes and modernizations which were occurring in practice with the end of the oligarchic period.

From post-war to the 1970s: industrial expansion

The third stage, from the end of World War Two, consolidated democracies with a wider participation and important processes of modernization of the Latin American socio-economic base. Amongst them a growing industrialization and expanded patterns of consumption, education and urbanization should be mentioned. The expansion of the mass media (including television) and of radical political movements which seek profound structural reforms are also noteworthy (García Canclini, 1989: 81–2). Most states developed interventionist and protectionist policies which controlled most of the economic life but also introduced some aspects of a welfare state in health, social security and housing. In spite of all this, the benefits of modernity continued to be highly concentrated and the mass of the people continued to be excluded.

Even though this stage coincides with the phase of organized capitalism in Europe and one could find some common features, there are also important differences. In the first place, the role of the state in the promotion of industrialization in Latin America was much more accentuated than that of private initiative. Second, the participation of foreign capital in the Latin American process of development was becoming increasingly more important than that of national capital. Protectionism benefits more international than national corporations. This leads many authors to put forward theories of dependency. Third, the elements of welfare state which populist governments introduced and the progress of industrialization did not cover or reach most of the population as in Europe, and a sizeable number of excluded and marginalized poor people accumulated around big cities.

The comparison with the Asiatic trajectory to modernity is interesting in this respect. While in Asia, Japan and others built sophisticated and highly automated and flexible technologies, strongly supported by the state, with a view to exporting industrial products in the international market, in Latin America the process of industrialization was happy with second-rate technologies, partly because its horizon was only the protected national market, and partly because the state failed to assume the role of promoting a national technological capacity. This is why the success of industrialization has depended, to a great extent, upon the size of the internal market. Thus in the cases of Brazil and Mexico, which are the countries with the largest internal markets in Latin America, competition and economies of scale allowed international levels of competitiveness (Gwynne, 1996: 228–9 and 220). In the rest of Latin America industrial production had high costs and little demand.

Be that as it may, the process of modernization and change occurring in Latin America was accompanied and promoted by ideas and theories which propounded the dismantling of the traditional agrarian cultural identities and their replacement by modern values and institutions. First, it was the ideas of the North American sociology of development, usually called 'modernization theories'. They put forward the idea that Latin America was in transition from traditional society to modern society and that the very advanced (North American or European) industrial societies were the ideal model which backward countries would inevitably reach. The modernization process was conceived as a historical necessity which, following a transitional route, repeated the same stages previously gone over by advanced societies. The obstacles presented to this transition by a traditional culture underpinned by an oligarchic system controlled by the old landowning aristocracies would inevitably be overcome. In Germani's analysis, what remained to be dismantled in order to reach true democracy were the traditional cultural values both of the elites and of the masses in Latin America, which presented stiff resistance to modernity especially through what he called traditional and ideological authoritarianism (1965: chapter 4).

Then, in the early 1950s, the pioneering economic work of the Economic Commission for Latin America (ECLA) focused on the existence of a center–periphery world system which favored the central industrial countries. According to ECLA's analysis those countries specialized in the production of industrial goods grew faster than those specialized in the production of raw materials and therefore the gap between central and peripheral economies was becoming increasingly wider. This is why it propounded the idea that the countries of Latin America had to modernize their societies by switching from a raw material export-orientated economy to an industrial-led economy in order to lessen their dependency on the external demand for raw materials and to substitute for it the expansion of the internal demand. This meant for ECLA a change from a model of development 'towards the outside' to a model of development 'towards the inside'. The political and economic initiative to bring about modernization and industrialization had to be mainly in the hands of the state. ECLA contended that given the many difficulties which a process of industrialization had to face it was crucial that the state took the initiative of organizing, promoting and supervising all the industrializing efforts in order to guarantee the continuity of the process.

Then, the ideas of imperialism and dependency, the resurgence of Marxism and the hope in socialism arose in the 1960s and early 1970s. The disillusion with the results of import substituting industrialization (ISI) processes, the lack of dynamic economic growth and the increasing number of contradictions surfacing mainly because of widespread poverty and the destitution of growing sectors of the population gave rise to a powerful critique of the capitalist system as being unable to deliver economic development in the conditions of the periphery. The failure of the capitalist modernization process was blamed mainly on imperialism. Capitalism did not work in Latin America because it was dependent upon the main industrial centers. This was the time of the resurgence of Marxism and socialist projects whose main objective was to struggle against dependency and bring about national development. But the critique of capitalism and imperialism did not prevent the adoption of new foreign models prevalent in the socialist countries.

Even with its deficiencies and problems, the post-war advance of modernity is notable and showed the continued cultural importance in Latin America of European and North American rationalistic and developmentalist ideas. In spite of the differences between these modernizing theories, their basic premise continued to be development and modernization as the only means to overcome poverty and/or dependency. Nevertheless, in all these positions the idea remained that modernity was something essentially European or North American which Latin America ought to acquire by repeating the historical experiences of Europe or North America. The idea also remained that the traditional agrarian cultural pattern in Latin America would have to be dismantled in order for modernization processes to succeed. This is why agrarian reforms were considered so crucial to the modernization process.

Implicit in the various modernizing approaches there was nevertheless a project for a

From 1900 to 1945: the crisis of oligarchic modernity

The second stage during the first half of the twentieth century historically coincided with the first crisis of European modernity and, in a way, that crisis was reflected in Latin America. But the consequences of the crisis were specific to Latin America: the oligarchic power began to crumble, the so-called 'social question' came to the fore, new populist regimes emerged which widened the franchise and incorporated middle classes to government, and processes of import substituting industrialization were initiated. Thus, while in Europe a crisis of liberal industrialism was experienced, in Latin America it was the prevalent oligarchic and aristocratic export-orientated system which entered into its terminal phase and incipient industrialization processes started with some success. As Mouzelis has argued, this means that the end of the oligarchic regime occurred in a pre-industrial context and that therefore the new openness of the political system did not include the active participation of organized working classes, as in Europe, and only tended to incorporate the middle classes to the power structures (1986: xvi). This peculiarity helps explain the emergence of populist regimes and the survival of paternalist and clientelistic political forms.

This stage of crisis and change in Latin America was accompanied in its beginnings by the emergence of an anti-imperialist consciousness, by a new revaluation of *mestizaje*, by a new Indianist kind of consciousness which criticized the discrimination against the Indian communities and by a growing social consciousness about the problems of the working class. In general most of these trends showed a renewed interest in Latin America's specific cultural identity and opposed the kind of modernity offered by the North American or European models. At the beginning of the century, in the context of North American expansionism, a series of intellectuals raised their voices against the United States and its hegemonic aspirations in relation to Latin America. José Martí of Cuba, Rubén Darío of Nicaragua, José Vasconcelos of Mexico, Rufino Blanco Fombona of Venezuela and Manuel Ugarte of Argentina joined their critical voices to that of the Uruguayan José Enrique Rodó. The latter achieved an enormous influence with his book entitled *Ariel*, which was published in 1900. Rodó (1976) started a critique of what he called *nordomanía*, the Latin American inclination to copy foreign models, especially North American, and advocated a return to its own reality. Rodó celebrated the feelings and virtues of the Latin race and maintained that Latin America possessed a greater cultural sensitivity and a more idealist sense of life than did the United States, which was materialistic and utilitarian to an excess.

Against the positivist idea that *mestizaje* degenerated the race, Vasconcelos (1927) celebrated the values of *mestizaje* and of the Latin race and opposed them to the characteristics of the Saxon race. The attitude towards the Indians is crucial. Whereas Saxon colonizers 'committed the sin of destroying those races ... we assimilated them, and this gives us new rights and hope of a mission without precedent in history' (1927: 14). This mission is the formation of a new fifth 'integral race', 'cosmic race' or 'synthesis race', made of the fusion of whites, blacks, Indians and Mongols. In this fifth race all the peoples of the world will eventually fuse, and the honor to accomplish such an integrating mission belongs to the Latinos (1927: 15). In a similar vein, the works of Indianist authors such as Valcárcel (1925 & 1972), Mariátegui (1976), Aguirre Beltrán (1976) and others, advocated a return to Indian values and customs in opposition to the European cultural heritage. Indianism oscillated between the absolute and essentialist affirmation of the Indian race (Valcárcel) and its assimilation to the national culture (Aguirre

Beltrán), but on the whole, and with the exception of Mariátegui, it wanted to affirm identity against modernity.

Later on and in the context of the big depression this difficult period seems to have promoted very pessimistic discourses which underlined the negative features of the Latin American identity or tried to rescue the Hispanic features of the Latin American character (Eyzaguirre, 1947; Lira, 1985). Thus, for instance, Martínez Estrada (1946) focused on the idea of resentment as best expressing the Latin American ethos; Alcides Arguedas (1975) described the duplicity of the Bolivians; and Octavio Paz (1959) analysed the double personality of the Mexicans who show a face which conceals a deep emptiness and resentment. Although these harsh self-criticisms did not seem to leave room for any pride in the Latin American identity, they still wanted to emphasize the peculiarities of the Latin American cultural identity as against the European pattern. The point was to explain why Latin America was different and why modernity could not succeed.

Later still, in the mid-1940s, hispanist currents of thought emerged which tried to respond to the questions about identity raised by Indianists and essayists of the 1930s and regretted the conscious neglect of the Hispanic culture. Eyzaguirre (1947), for instance, argued that Iberoamérica (notice the intentional use of the word against the prevalent 'Latin America') would not have existed without the presence of Spain, and the first sense of the Iberoamerican identity must be found in the Hispanic cultural roots. In spite of this, Latin America would have turned its back to its true 'collective soul' and would have run after other cultures, sometimes antagonistic to its own, to imitate alien political and social models which could not be successful in such a different context (Eyzaguirre, 1947: 39), hence the failed attempts at copying North American federalism, French Jacobinism and British parliamentarianism, instead of using as an example the old Castilian *fueros*.

Osvaldo Lira, in his turn, tried to prove that 'Hispanic American nations constituted from the beginning and continue to constitute today a perfectly homogenous whole of culture among themselves and with Spain', and that 'all and every one of the perfections which the Hispanic American culture contains derive exclusively, as from its intrinsic essential first principle, from Spanish culture' (1985: 13 and 60). Lira considered the Indo-American cultures existent at the time of conquest as pseudo-cultures, incapable of contributing any central value or orientating principle. Following in the steps of Eyzaguirre, and indeed of all essentialists, Lira also lamented the treason committed by the majority of the leading sectors in Hispanic America, who 'far from keeping themselves irreducible ... to foreign influences, have allowed themselves to be seduced by them, forgetting and in many cases reneging on the very values which engender them to historical existence' (1985: 55).

In this way, a period of crisis and of important economic and political changes is accompanied by new forms of social consciousness and by a search for identity which tries a variety of avenues but which in any case has abandoned the nineteenth century certainties and in some significant instances attempts to affirm a Latin American identity against modernity. Nevertheless, the main thrust of modern industrialization, widened political participation and social rights which populism brought about continued to be the center around which the great national debates revolved thus influencing in practice the construction of cultural identity. Just as nineteenth-century positivists wanted modernity at any cost but remained trapped in some forms of the old identity they wanted to abandon, thus the essayists and critics of modernity in the first half of

9

CIVIL SOCIETY, SOCIAL DIFFERENCE AND POLITICS: ISSUES OF IDENTITY AND REPRESENTATION

SARAH A. RADCLIFFE

The 'high modernization' period in Latin America (1970–90) was broadly characterized by a weak civil society and the depoliticization of society (Chapter 8). Yet the period of the 1970s and 1980s also oversaw the mobilization of diverse groups of Latin American society into social movements and other forms of collective political action. How can we explain the paradox of apparent depoliticization together with mass mobilization around an innovatory set of political repertoires and new agendas? From environmentalist groups in Venezuela strategically using mass media to convey their message, to the soup kitchens in Lima's shantytowns, to black peasant land-titling movements in Colombia, and the indigenous march on La Paz in Bolivia in the run-up to the Columbus quincentenary, Latin America appears at first glance to be politicized and to have active citizens. Groups of women, gays, Indians, shantytown dwellers, colonists and peasant producers appear to capture the political spaces of their societies. A period of modernity with limited civilian mobilization transforms, through the actions of diverse subjects, into a modernity in which questions of development, identity and modernity are being addressed by a great variety of voices.

This chapter examines the ways in which the latest phases of Latin American modernity have engaged with political and cultural struggles. In the 1990s, the complex and evolving sense of Latin American identity and modernity focused on the reconstruction of political and civilian life, especially in the region's engagement with transition from military rule and democratization (Chapters 2 and 3) and demands of new social subjects. To answer the apparent contradiction, we need to examine the ways in which social differences have been produced and remade in Latin America, and the differential political insertion of social groups into the state and political agency.

Latin American social relations are profoundly influenced by hierarchies in which social difference – along lines of race, class, gender, sexuality and other differences, separately or in combination – affect the formation of subjectivity and citizens. Larraín

(Chapter 8) links this to the existence of a long tradition of authoritarianism in the region. Although shaped by the baroque, Enlightenment ideals, Larraín argues, were more completely embraced with regard to cultural and discursive formations than in political structures. Yet as many scholars have now pointed out, the cultural and discursive formations of the Enlightenment were not in themselves unauthoritarian or free of implicit hierarchies. In Enlightenment ideas of the subject, personhood was declared to be Universal, that is, 'unmarked' by gender, class, race/ethnicity or other social difference. However, alongside such ideals existed an implicit normative comparative structure in which the model citizen and subject was in practice implicitly masculine, white/European, urban and bourgeois. In Latin America too, the ideal citizen-subject was expected to be male, white (later mestizo) and urban bourgeois. Such subjectivity was built into Latin American conceptions of citizenship, development and state–society relations, with authoritarianism displayed towards those whose characteristics denied them full citizenship under this schema. It was as if, drawing on Laclau (1985: 39), there was an internal frontier between the 'European' Latin American political world, into which the subalterns could not be assimilated, and the fringe of subaltern groups who were treated to domination. Latin America's problematic relationship with modernity suggests that the Enlightenment goals of equality and inclusion are still to be struggled for by a wide range of institutions and actors (Schuurman 1993: 187). The social movements and the flourishing forms of civil organization represent in this context the latest forms of resistance and accommodation from those designated as Others, what Elizabeth Jelin (1990: 5) characterizes as 'struggles from below'.

DEFINING SOCIAL MOVEMENTS

From rural and urban areas, from diverse (and often mixed) class groups, and involving a range of ethnic and racial groups, social movements at first sight seem to have little in common. How do we then map and understand the diversity of social and political movements in the region? The sheer heterogeneity of civil, political and social forms defies simple categorization, together with a diverse social and political repertoire and range of discourses and practices. In comparison with other socio-political groups, however, they are characterized by their self-reflectivity or self-knowledge, which is socially confirmed (Melucci, 1995): it can be seen as the 'work that society performs on itself' (quoted in Escobar, 1992: 92). It is this dimension of reflexive modernity that provides a common theme, and links with Scott's definition of the social movement as a 'collective actor constituted by individuals who consider themselves to have common interests and, for at least some significant part of their social existence, a common identity' (Scott, 1990: 6).

Social movements are distinguished from other political groups such as unions or political parties on a number of criteria, although the boundaries between styles of politics can be blurred (as the examples below illustrate). First, social movements are often cyclical, moving through a pattern of greater and lesser visibility, and even disappearance once aims have been achieved. Nevertheless, the formation and transferral of collective memory of practices, discourses and meanings may serve as a resource for later social actors and their movements (Foweraker, 1995: 100; Alvarez et al., 1998). Second, through their use of social mobilization as power, social movements differ from the formalized proposals and institutionalized channels utilized by political parties and

and renewed interest in, political democracy and human rights by intellectuals and popular majorities. Paradoxically, this has contributed to the relative depoliticization of society because it has led to wide agreements between formerly antagonistic political forces and has refocused the interest of most social scientists. In spite of the above mentioned weaknesses of democratic institutions in Latin America, in spite of corruption, terrorism and human rights violations, the democratic system has recently emerged as the only legitimate framework for political action. Even in Central America, where authoritarianism, open or otherwise, has been endemic for many years, there has been a strong movement toward democracy in the 1990s.

Conclusions

Modernity is not absent from Latin America, nor is it inauthentic or the same as European modernity. It has its own historical trajectory and its specific characteristics although it shares many general features. The Latin American trajectory to modernity is simultaneously an important part of the process of identity construction: it does not oppose an already-made-in-the-past, essential and immovable cultural identity, nor does it entail the acquisition of an alien identity (Anglo-Saxon, for instance) either. Modernity and cultural identity are processes which are being historically constructed and which do not necessarily entail a radical disjunction even if there may be tensions between them. The features of Latin American modernity which I have explored, as much the general ones as the specific ones, constitute, for better or worse, important elements of the Latin American cultural identity today. But, of course, nothing prevents their critical appraisal or their change in the future.

The question arises as to why, if modernization processes have been closely intertwined with processes of identity construction in Latin America, there has been such a manifest tendency to consider modernity as something external and in opposition to identity. No doubt, as I advanced in the introduction, part of the problem is a result of the failure to distinguish different trajectories to modernity and to consider cultural identity as historically constructed and changing. But these failures have been prompted by historical facts. The first fact which may be of importance is the postponement of the beginning of modernity for three centuries because of the Spanish and Portuguese colonial blockades which established cultural barriers which surrounded their domains. This means that when the forerunners of independence began to get soaked in modern ideas by means of trips to Europe and the contraband of books, modernity could not but present itself as something external which others had developed outside Latin America. This left a lasting mark in the social imaginary which tends to associate modernity with Europe or the United States.

The persistence of this idea was reinforced during the nineteenth century by an extraverted economy and a cultural orientation which continued to look to Europe as the source of all culture and civilization. When in the first half of the twentieth century the oligarchic regime was in its terminal crisis, and currents of thought arose which questioned Latin America's economic and cultural extraversion, modernity appeared once more as something external, but this time as a negative imposition bound to destroy Latin America's supposedly original and unchanging Indian or Hispanic identity. The attempts to recover and reaffirm Latin America's own identity in times of crisis led to a critique of foreign cultures, and modernity had been considered until then as belonging to foreign cultures. Hence up to World War Two, and for opposite reasons, modernity was considered something external.

In the past 50 years the situation has changed, but not entirely. Soon after the war modernization theories confirmed once more the external character of modernity by arguing that Latin America had to repeat the same historical process that Europe and the United States had gone through. In the 1970s and 1980s, in times of crisis, once more, modernity was put in question because it came from without and was not supposed to be compatible with Latin America's identity. In the 1990s neoliberalism has succeeded in imposing an extraverted strategy which again looks up to the developed world as the model and source of all progress. Thus the polarity between modernity and identity has continued to be present in the social imaginary while in practice Latin American modernity and cultural identity continue to be constructed in close connection with one another.

FURTHER READING

Larraín, J. 1996 *Modernidad, razón e identidad en América Latina*. Editorial Andrés Bello, Santiago. Develops a general review of identity issues in Latin America.

Canclini, N. G. 1990 *Culturas híbridas*. Grijalbo, Mexico. Explores differences in Latin American culture.

Brunner, J.J. 1994 *Cartografías de la modernidad*. Dolmen, Santiago. Provides useful debates about Latin America's road to modernity.

Véliz, C. 1994 *The new world of the gothic fox: culture and economy in English and Spanish America*. University of California Press, Berkeley, CA. Proposes the thesis that the arrival of modernity in Latin America means the necessary dismantling of its old baroque cultural identity.

> **BOX 8.4 Racism denied**
>
> Flores Galindo has observed that
>
> > In Peru nobody would define himself or herself as a racist. Nevertheless, racial categories not only tinge but sometimes condition our social perception. They are present in the configuration of professional groups, in the messages transmitted by the media or in the call for beauty contests . . . racism exists notwithstanding racial terms, suppressed in the procedures of public identification. Yet a masked and even denied phenomenon, does not cease to be real (1994: 215).
>
> Raúl Béjar has argued that in the case of Mexico 'it is a commonplace to say that in this country there is no racial discrimination', yet it is possible to affirm that 'prejudice has grown in the cultural history of Mexico' and that this affects 'especially the Indian or quasi Indian . . . blacks . . . and Chinese' (1988: 213–14).

gated, oppressed by mestizos and subject to special laws and forms of administration. Nevertheless, the very fact of *mestizaje* and the fact that in many cases social class overlaps with gradations in skin color (the darker the skin the lower the class) leads frequently to a denial of racism.

This has even a base in the social sciences, which have often underlined the differences between the Spanish treatment of Indians and blacks and the British treatment of them. Gilberto Freyre, in his classic book *Casa Grande e Senzala* ('The masters and the slaves', 1946) argued that the treatment of slaves in Brazil was softer than in North America, especially because of closer, even sexual, relationships between masters and slaves in the hacienda. Many historians and social analysts have subsequently noted that whereas in North America the white settlers imposed their separation from Indians and blacks, in Latin America a wide process of *mestizaje* took place, thus causing to emerge a continuum of racial gradations. From this the myth arose that in Latin America a 'racial democracy' exists and that racism is a problem of foreign countries (Cubitt, 1995: 122–6). This idea continues to be believed today and shows its prevalence in that, with the exception of some degrees in anthropology, there is a significant absence of courses and studies on Latin American race problems in social science degrees.

A significant phenomenon which differentiates the Latin America modernity from others is the lack of autonomy and development of civil society. In Latin America civil society (the private sphere of individuals, classes and organizations regulated by civil law) is weak, insufficiently developed and very dependent upon the dictates of the state and politics. This is one of the consequences of the absence of strong and autonomous bourgeois classes to develop the economy and culture of society independently from politics and any state support. Brunner adequately argues that in contrast to the modernity of central countries, Latin American modernity suffers from a 'voracity of politics which swallows everything and behind which everyone seeks protection or justification: equally entrepreneurs, intellectuals, universities, trade unions, social organisations, clerics, the armed forces' (1988: 33).

For instance, it is significant that some universities, institutes and even the media could lose a good number, or the best, of their members each time there is a change of government, and recruitment of civil servants takes place in order to replace those who are leaving. At the same time, it is not rare to see the functionaries of the outgoing government using their power to prepare in advance places of work in some universities or institutes which in this way are further 'colonized' by certain political tendencies or power groups which recruit only members or supporters of their own sector. Neither is it rare to find that a good number of research and consultancy institutions depend almost exclusively upon services rendered under contract to various state organizations. Many cultural centers are directly created by local governments and managed by the political majority which controls them. Hence politics exercises a disproportionate influence upon civil society and cultural institutions.

It is also necessary to refer to the fragility of the political institutions in Latin American countries. Since independence Latin America has appeared before the eyes of the world as a continent of revolutions and *caudillos*, military coups and conspiracies, where the institutional order is permanently under threat of being surpassed, so much so that important scholarship has been devoted to explaining the Latin American systematic political instability (Kling, 1970; Huntington, 1968). The wave of military dictatorships which began in the 1960s and covered the 1970s and part of the 1980s did not even respect countries such as Chile which had a reputation for institutional stability. True, Latin America is now going through a period of return to democracy, but the symptoms of institutional weakness remain quite evident throughout Latin America, especially in Argentina, Venezuela, Colombia, Peru and almost all Central America.

A relatively recent feature of Latin American modernity, especially the Chilean one, is the relative depoliticization of society. Military dictatorships sought to depoliticize society by eliminating elections, abolishing political parties and closing down parliaments. Their policies of exclusion and violations of human rights, however, got in the long term the opposite result: society got more intensely politicized against military governments. This led to a search for crucial agreements and coalitions which would allow the return to democracy. One of the conditions for this search for democratic consensus was to make autonomous the economic sphere in order to protect it from the ups and downs of everyday political discussion. From now on the economic system was to be self-regulating according to the laws of the market, and a consensus economic policy on macro-economic variables was to be introduced.

As Cousiño and Valenzuela have put it, 'once the economic system is made autonomous, politics loses the capacity to observe and intervene upon the economy and, therefore, abandons its pretension to place itself in the perspective of totality' (1994: 17). The consequence of this is that politics itself is converted into another functional self-centered system which refuses to intervene in the basic course of the economy. Hence, what had been an immense area of disagreement and political disputation is left out of discussion. It can be concluded from this that Chile's democratization, mediated by the autonomy of the economy, has resulted in a considerable and significant depoliticization of society. The military dictatorship in Chile started the process of reorganizing the economy, but this could be consolidated only with the redemocratization of the country by the end of the 1980s: the price of the new stability was the increasing autonomy of, and the loss of political control over, the economy.

Finally, a very recent feature of Latin American culture has been the revaluation of,

Barrera who has argued that with the kind of state that emerged out of years of authoritarianism and neoliberalism clientelism has disappeared (Barrera, 1996). I think his arguments show only a probable diminution of clientelism but by no means its disappearance. Recruitment of civil servants, university lecturers and mass media journalists continues through clientelistic or personalist networks of friends and supporters. The processes of public contest for a job are absent, scarcely developed or work in a purely nominal fashion when procedures are 'fixed' to favor a pre-selected individual. Clientelistic recruitment flourishes in Latin America and shows as much the absence of normal channels of social mobility as the narrowness and high competitiveness of political and cultural environments.

Education, acquired skills and personal achievements are not enough to secure access to certain political or cultural jobs. Well-placed 'contacts', 'godfathers' or 'friends' are required to facilitate entry. Because the system depends on the patronage of certain individuals who exercise institutional power, it secures the personal loyalty of the recruited and favors institutional immobility. Thus veritable institutional fiefdoms are created which because of their discriminatory character are almost impenetrable to those who do not belong to the group that controls them. By paraphrasing Habermas (1989: 164) in a slightly different sense, one could speak of a true refeudalization of cultural and state institutions (Box 8.3).

A second feature could be called 'ideological traditionalism'. In putting forward his theory of transition to modernity Gino Germani spoke in the 1960s of the 'fusion effect' by means of which modern values could be reinterpreted in contexts different from developed societies, with the result that traditional structures were reinforced (1965: 104). A particular form of this fusion effect was ideological traditionalism in which leading groups accepted and promoted changes necessary for development in the economic sphere but rejected changes required for such a process in other spheres (1965: 112).

In late modernity a similar phenomenon takes place in which certain leading groups advocate total freedom in the economic sphere but appeal to traditional moral values in other respects. Thus they emphasize almost a religious respect for authority and order, the traditional family and the national heritage, or they cast doubts about democracy and oppose, for instance, divorce laws or the decriminalization of adultery for women. A good example of this is Chile where adultery was a crime for which only women could be punished until 1995 and where until this very day a divorce law has not been passed because of Catholic and conservative opposition. Renato Cristi has argued convincingly that conservative thought in Chile never opposed liberalism as such but rather 'the democratic element which takes over its reservoir of ideas from the 19th century onwards' (1992: 157). These fusions are not exclusive to the developing world. The emergence of a New Right in the United States and the United Kingdom has also been

---

BOX 8.3 Habermas's 'refeudalization of the public sphere'

By means of this concept Habermas refers to the loss of public space for rational critique and discussion of state affairs which had emerged at the beginning of modernity. This public space, as a result of state interventions and the commercialization of the media, was subsequently replaced by the manipulation of the masses as a new 'feudal' means to avoid genuine discussion, thus legitimating public authority (1989: 164).

characterized by the way in which it has combined traditional conservative attitudes about authority, Victorian values, internal order and external security with a new emphasis on free markets (Levitas, 1986; Hall and Jacques, 1983).

However, traditionalism in Latin America has stronger institutional bases than it has in Europe or the United States. One of them is the extraordinary power and influence of the more traditional Catholic church over political and legislative matters. This can be explained by the privileged role which the Catholic church has played since the colonial times in the maintenance of social and political order. As I will show next, the church and religious mechanisms played a central role in the exercise of authority and the political control of people.

A cultural aspect which has survived from colonial days, at times in a moderate form, at other times in an exacerbated form, is authoritarianism. This is a trend which persists in the political field, in the administration of public and private organizations, in family life and, in general, in Latin American culture which concedes an extraordinary importance to the role of and respect for authority. Its origin is clearly related to three centuries of colonial life in which a strong Indo-Iberian cultural pole was constituted which accentuated religious monopoly and political authoritarianism. As Imaz has put it, 'for three centuries there existed a clear relationship between political authoritarianism and the legitimating role of Inquisition' (1984: 121).

Flores Galindo has documented how the seventeenth century religious congregations' persistent struggles against idolatry in the central sierra of Peru had a connotation of political control: 'the relative precariousness of the military system forced an apparent hypertrophy of religious mechanisms, so that, in that way, through fervor or more frequently fear, control over men could be secured' (1994: 66). In spite of the democratizing influences of Enlightenment thought, which certainly achieved some partial moderation of the authoritarianism of the Indo-Iberian cultural pole from independence onwards, its cultural force has not easily been extinguished in Latin American socio-political life.

In the particular case of Chile, various authors have highlighted the historical crucial role of Portales's strong and authoritarian government in the formation of the Chilean state (Edwards, 1987; Góngora, 1981). Portales's conception consisted in that, because of a lack of republican virtues, democracy had to be postponed and unconditional obedience to a strong authority had to be established. The action in favor of the public good of such authority could not be hindered by laws and constitutions. It divided the country between the good (men of order) and the bad (conspirators to whom the rigor of the law had to be applied) (Góngora, 1981: 12–16). It is not surprising that general Pinochet's regime should have frequently invoked such a conception.

Another important feature is masked racism. The existence of racism in Latin America is well documented even though it is a relatively neglected area of social sciences and generally it is not perceived as an important social problem. It is clear though that from very early days there has been in Latin America an exaggerated valuation of 'whiteness' and a negative vision of Indians and blacks (Box 8.4).

It is well known that various governments attempted to 'improve the race' by means of 'whitening' policies which favored European immigration. There also exists spatial segregation whereby Indian areas are the poorest and most abandoned and whereby the shantytowns in cities contain a bigger proportion of people with darker skins, be they Indians, mestizos, mulattos or blacks. There is no equality of opportunity for them. Some surviving Indian groups constitute true internal colonies, geographically segre-

unions. However, with the recent transformations of the state, the boundary between groups over their use of institutional or mobilizing actions has become increasingly blurred. Third, in whatever context, social movements need to define their terms and agendas, and provide some closure around their projects (Scott, 1990: 9). Such closure is provided by the movements' cultural resources, as well as by the political, economic and social context in which they operate. The type of state under which they operate, the broad political economic context of poverty and marginalization, as well as locally specific social relations and cultural capital shape – and individuate – social movements.

The characteristics of social movements influence in turn the ways in which they are studied and analysed. Social movements research is often defined by its broadly actor-orientated approaches (Schuurman, 1993) and the ethnographic documentation of the meanings, lifeworlds and contexts of participants. The focus on how and why different social (and often everyday) processes hold together lies at the heart of this project (Melucci, 1995). Why do actors initiate and participate in a movement, and how do the groups arise and perpetuate themselves?

Social movements are often focused around 'constellations' of issues, that is a specific topic or grievance, compared with political parties' elaboration of manifesto-style programs addressing a wide range of issues (Scott, 1990: 16–26). Nevertheless, one demand can in itself have multiple facets and be subject to change over time. The dynamic and fluid nature of the 'agendas' of social movements thus focuses attention on the ways in which social actors continuously create and recreate a 'community' around a certain topic in an ever-changing context. For example, the women's human rights movements in Central America and the Southern Cone that organized for the return of 'disappeared' relatives from military regimes found that their concerns grew to encompass issues of male violence against women, their roles as mothers in a Catholic society and the juridical–political context for human rights (Fisher, 1993b; Schirmer, 1993). In other words, the self-reflectivity of social movements can prompt consideration of a wider 'astronomy' of issues.

In the political sense, social movements often arise in the context of the 'failure and inadequacy of institutions of interest intermediation' (Scott, 1990: 9). In this light, social movements have specific goals and achievements, namely the incorporation and then satisfaction of their interests and needs through institutional channels. In Latin America, development interests are at the crux of movements' agendas, as the promises of modern development have been delayed or highly uneven (Escobar and Alvarez, 1992). Paradoxically, this may be combined with a sense of disillusionment with the prevailing political system, either in its practice or in its ideals (Box 9.1). In summary, Latin American social movements have generally both social–civil *and* political dimensions, without being reducible to either identity politics or political economy. In this context, we find the overlapping and mutually influential presence of issues around social and citizenship rights, social difference, identity and use of politico-legal channels variously combined in social movements.

Social movements have an extraordinary range of practices, strategies and tactics through which to further their goals. Any one social movement may develop a particular 'repertoire' of actions and methods (Eckstein, 1989), and often finds a 'multiplicity of practices' (Calderón et al., 1992: 27) suits its mode of action and situation. How groups organize and operate and who they 'recruit' as participants are defined and remade in light of the types of community through which networks are initially created. The

> **BOX 9.1 The Mothers of the Plaza de Mayo, Argentina**
>
> Between 15 000 and 25 000 people died in Argentina in its Dirty War (1976–83) against 'subversives' and in defence of the 'last bastion of Western civilization'. In the closure of political activity, political activists and members of diverse organisations were perceived as a threat to stability and were 'disappeared' after imprisonment, torture and killing by the ruling military. The Mothers of the Plaza de Mayo were established in 1977 to trace their 'disappeared' sons and daughters, who had been abducted by the security forces or by unnamed groups. They processed weekly around the main square in downtown Buenos Aires, and, despite intimidation, they in effect initiated a public debate about the legitimacy of the authoritarian regime. After the return to civilian rule, the *Madres* were at the forefront of demands for the persecution of military crimes and the bringing to justice of those responsible for disappearance, continuing too in their campaigns to account for all of the disappeared.

innovative and exploratory use of tactics by social movements has been dramatic in Latin America, with social groups formulating new repertoires through which to change themselves and their society. For example, under economic austerity, poor women in Brazil organized carefully planned and nonviolent systematic raids on supermarkets to gain foodbasket items for their families. Fluidity between different goals are also common, with movements overseeing 'shifts between resistance, protest and proposals' (Escobar and Alvarez, 1992: 4). While telling us about the *how* of civil organization, organizational and managerial aspects miss the other side of the equation, namely, people's motivations, hopes, desires and meanings in their organization (Escobar and Alvarez, 1992; Scott, 1990).

The power of social movements often rests on their ability to mobilize in opposition to, and as external pressure on, local governments, states or powerful groups, suggesting that the maintenance of autonomy or independence from formal political institutions is a *sine qua non*. In theoretical terms, autonomy of social movements has been of central importance, defining the essential nature of social movements, *vis-à-vis* other forms of political and social action (Scott, 1990). Autonomy seen as a value and core strategy (Escobar and Alvarez, 1992) is not, however, so simple. Autonomy is not merely the condition for the social movements *per se* but depends upon circumstances. Two examples will suffice. Under the 1970s–80s military regimes, states were violently opposed to any mobilization of civil society, generating an insecure position of autonomy from which social movements could speak. Indeed, social movements for human rights and for the return of 'disappeared' during this period often functioned by default as *the* political. The Mothers of the Plaza de Mayo in Argentina paraded weekly around the central square of Buenos Aires' political district demanding the return of disappeared children [Box 9.1]. Moreover, under civilian regimes, autonomy may undermine movements' attempts to access resources while granting greater ideological and organizational freedom. Mexico's strong corporatist state has the resources and political will to co-opt social movements into its regime of power. Yet women's organizations respond differently to the resources on offer, generating distinct communities, depending on their pro-state or independent position (Craske, 1993). Autonomy too operates at

> **BOX 9.2 The peasant patrols of northern Peru: *rondas campesinas* and the state** (source: Starn, 1992)
>
> The peasant patrols, or *rondas campesinas*, of Peru, operating under conditions of state repression against the Maoist army of Sendero Luminoso (Shining Path), trod a fine path between the militarized elements, while forging alternative modernities for their rural locations (Starn, 1992). Started in the 1970s and widespread by the 1980s throughout northern Peru, *rondas* arose out of a concern over security, namely the theft of farm animals and a disillusion with formal judicial procedures. Political parties of the left as well as church catechists contributed to the nonclass, open organization. As *rondas* become embedded in the daily routines of villages, conflict resolution and women's *rondas* emerged to deal with a widening circle of social issues. In these actions, villagers were drawing on a wide range of institutional machineries, cultural capital and social relations, making a hybrid practice out of the state conflict resolution measures, local knowledges and dress. Identity in this case came *after* the organizational dimensions of the *rondas*, as participants started to identify as *ronderos* as well as Peruvians and *campesinos* or peasants.

a number of levels, from the individual participant to the group, and the group's autonomy *vis-à-vis* other movements (Scott, 1990: 19–21).

Nevertheless, as the contexts for social movements' operation change into the next century, so too the question of autonomy is being reexamined. Autonomy has often been bound up in spurious questions about the authenticity of social movements and their ability to fit into pre-given ideological frameworks. Rather the question is about what the community forming the movement wants to achieve, given its decisions about nonnegotiable matters. Arenas in which social movements might *not* negotiate may include cultural aspects [for example, the defence of tradition, such as the pan-Mayanists (Warren, 1998)], or independence from political parties (such as Guadalajara's women's groups) or their own process of politicization. Regarding the latter, Peruvian peasant patrols dealt with the state over local conflict-resolution procedures and stemmed the northward spread of the Shining Path, while retaining their own priorities and cultural frames of reference (Starn, 1992; Box 9.2). While European and North American social movements have arguably turned their back on the state's ever-extending reach and the technologies of state power (Touraine, 1981), Latin American social movements fall along a continuum between coexistence and cooperation with the state and other political institutions. As Starn makes clear in his analysis of the Peruvian *rondas*, participants did not want the state's overthrow, rather they wished the dismissal of corrupt officials and for the better functioning of judicial systems.

THE EMERGENCE OF SOCIAL MOVEMENTS: STRUCTURAL CONTEXT

Latin American understandings of social movements have been driven by the need to analyse the regional context within which social movements emerge. Latin America's insertion into modernity thereby lies at the heart of questions about the rise of new forms of political organization, just as it introduces new factors into groups' agendas and identities. The origin of social movements from the 1960s and peaking in the 1980s is attributed to both political economic and socio-cultural factors. The political economic context includes the failure of inclusive development, high rates of poverty and social marginalization and the structure of the state, with socio-cultural dimensions including struggles for identity and representation, and the definition of a new politics of needs (Slater, 1989; Escobar and Alvarez, 1992; Eckstein, 1989). The factors marking out the region's experiences as unique (in their combination if not individually) range from the political, economic and social dimensions of contemporary Latin American life.

In the political economic field, the context set by the state is central. The crisis in the developmentalist state, that is the states which premised their legitimacy on rapid industrialization and modernization in the 1950s–70s (Chapters 2 and 8) initiated a questioning of economic and political exclusion by diverse groups. When the state was not reliable in fulfilling its promises, self-help was often the next step (Lehmann, 1990: 150; Scott, 1990). However, self-help was not a romantic position of autonomy, not least because the rise of bureaucracies and technocrats under industrial modernization meant increasing public engagements over resource distribution. The crisis of distribution and growth faced by the clientalist and populist states of the 1960s and 1970s created new horizons for subjects, as well as generating the conditions for military juntas to step in to economic and political management during the 1970–80s (Chapter 2). Even before authoritarian states, however, popular organizations, trade unions and organized political parties had failed to consolidate democratization, leaving many groups outside the bounds of a centralized state system premised on social exclusion. In this sense, civil and political rights were *new* to Latin America (Foweraker, 1995: 28), representing a new dimension to the formulation of modernity by protest groups. Under conditions of excessive centralization and corrupt and inefficient management, the notion of a public sphere had to be constructed often in an anti-centralist impulse. The 'terrible tension' (Calderón *et al.*, 1992: 25) between state and civil society was thus premised on the fact that while the state was the main interlocutor or referent for social movements, everyday issues were being brought into an unmapped and unconstructed public sphere. Precisely at the moment when authoritarian states were invading the space of domesticity and family life, so the public realm was invaded by 'private' concerns, emerging from the submerged experiences of disempowered citizens (cf. Sheffner, 1995; Foweraker, 1995). Ambivalence and tension around public–private issues was found also in European social movements, yet the 'decisively political character' (Touraine, 1981) of the gulf between state and citizen is particular to Latin America.

At another level, the state's promise of development created an imagined horizon for all subjects, although the processes of exclusion exacerbated the mismatch between private experience and public provision of services. Urban social movements arose in light of the failure of municipal and national governments to provide poor populations with the means of collective consumption (Castells, 1983). With marches on government offices or the creation of self-help construction groups, urban poor organize to leverage more resources out of class-based states, as well as mobilize their own small level of construction materials and labor. In Brazil by 1982 there were an estimated 8000 neighborhood associations. In the *favelas* of São Paulo neighborhood associations organized marches under the military dictatorship to demand assistance, forming strategic links with popular leaders as municipalities were weakened.

Whatever the regime in Latin America, economic inequalities lie at the heart of the region's situation, informing in turn the debates in civil society and the urgency of political mobilization. In the immediate term, harsh economic policies of adjustment quickly followed on the heels of 1980s debt crises, the failure of agrarian reforms and associated colonization programs in the 1960s and 1970s. Yet, over the longer term, broader changes associated with the geographical mobility of firms and the feminization of worldwide economic restructuring further repositioned the state *vis-à-vis* its promises of modernity and development. In Mexico, austerity in the 1980s undermined the state's ability to buy out opposition, hence boosting the social movements. The new international division of labor and the rapid increase in numbers relying on the informal sector (Chapter 10) made class a more problematic category for political organization while further reducing the state's liabilities towards its citizens. Failure of development to give what it promised generated interest in 'better' types of development and more equitable distribution of its benefits, combined in some cases with a questioning of the peripheral Fordist, urban and technologized path to development.

However, it is not just political-economic dimensions of modernity that inform the contestation and remapping of civil society and the state. Globally, social movements have been linked to the encroachment of the state or technology on a core set of social values (Touraine, 1981; Scott, 1990) against which defensive mobilizations occur. In this perspective, social movements are primarily socially defined civil groupings – their political engagements are secondary to the emergence of identities and culturally specific repertoires of action (Lehmann, 1990; Escobar and Alvarez, 1992; Alvarez et al., 1998). In Latin America, late modernity has taken a different path, while large segments of society remain outside the reach (benign or controlling) of state-led projects of modernization. Nevertheless, ethnic difference (primarily of indigenous and black Latin Americans), social hierarchy and deeply embedded ideologies of gender and sexuality continue to act as significant blocks to full participation and citizenship for many people. In the populations composing these subalterns, face-to-face relations work to affirm a sense of identity and community, although these groups are also often riven by divisions around class and so on.

Affected by the broad structural modernist changes of Latin America's mid-century – migration, education, nationalism – subjects were politicized. Social experiences of migration into cities, insertion into a growing circle of meaning resulting from education and military service, as well as state appeals to nationalism have all contributed to the political-*cultural* (rather than political-economic) dimensions informing social mobilization and conflict. Migration and migrant populations generated new political dynamics incorporating former subalterns as the new interlocutors to interact with nation-states and local governments, demanding resources as well as incorporation into the imagined community of nationhood. Rapid urbanization in Latin America during the mid-twentieth-century transformed civil society, political consciousness and practices of social interaction. Urban growth together with attempts at integration of 'citizens' through nationalist school curricula and military conscription paradoxically gave a language of citizenship rights just as the reality failed to match. With increasing levels of (especially urban) education, a new political culture arising with wider franchise (advertising, demonstrations, campaigning) drew people into a new relationship with the political sphere (Calderón et al., 1992: 25). In this context, social movements began to question the nature of the states and political systems within which they were embedded.

The recovery of nonviolent and meaningful quotidian existence also lies broadly behind the impetus to civilian organization. Particularly for those groups whose labor has been at the forefront of economic restructuring, and whose rights have been the first to be dropped under neoliberalism, alienation together with concerns linked to their violent social incorporation were central. While development as a modernist machinery can arguably produce fragmented individuals (Escobar, 1992), it also creates new spaces for social actors and identities.

The effectiveness of social movements can thus be gauged in relation to a number of variables, both measurable and more nebulous. Institutional changes and the emergence of new elites or social mobility can be identified as material and significant outcomes of social organization. Yet the symbolic challenges to existing discourses and practices of (mis)representation may be equally significant to participants, although the gains are more tenuous. Depending on the criteria used to define social movements, success can comprise a wide range of outcomes, from subtle shifts in 'cultural codes' (Melucci, 1995) or political culture (Alvarez *et al.*, 1998) to specific resources (land-title, legislative change, recognition of a social group, etc.) (Sheffner, 1995). Existing within the ambivalent spaces between civil and political society, social movements' gains are inevitably going to be highly varied and multifaceted. While being more than the 'weapons of the weak' (Scott, 1986) in a material sense, social movements are not merely struggles over production and consumption, but are also about meaning, communication and representation (Melucci 1995).

Post-Marxism's discussion of social movements suggested that social movements were a sign of an 'excess of meaning', generating a proliferation of new social actors and a rainbow of political-social relations through which to rethink social cohesion (Laclau and Mouffe, 1985; cf. Geras, 1987). The context of Latin American late modernities suggest that institutional histories and prevailing political cultures have a more central role in the emergence and direction of social movements, that is Touraine's 'decisively political'. Whether in terms of the constraints placed on movements by disappearance of leaders under military regimes, the bias of judiciaries or the profound hierarchies of race, class and gender, the material impact of the machineries of dominance are direct acts impacting on the bodies, subjectivities and strategies of citizens.

In concluding this section, we have seen how the broad structural features of Latin America which lie behind the emergence of social movements can be summarized as the political economy of exclusionary social relations and states, together with the particularly acute processes of marginalization (indeed invisibility) of subaltern groups during political and economic crisis in the 1980s. Moreover, transformations in the lifeworlds of poor and marginal populations generated new communities and agendas, under broad processes of differential incorporation into modernizing political economies. While identity and community are by no means new (neither are authoritarian states), from the 1960s and particularly from the 1980s the context was set for 'new forms of organizing against changing political or economic oppressions' (Sheffner, 1995: 604). The process of social movement formation and their active dynamic can be simplified. Political economy sets the following structural conditions:

- grievances,
- social problems for particular social groups,
- interpretation by a self-defining community,
- self-conscious community (new social movement),

- strategic actions,
- success or failure.

The stage between the interpretation of the grievances and the formation of a self-defining, self-reflective community is thus a question of individuals' choice (Hobsbawm, 1996) although not under conditions of their own choosing. Community as well as specific (often material or more broadly discursive) outcomes can thus be the goals of social movements.

BETWEEN POLITICS AND THE EVERYDAY

After outlining the main structural factors behind the emergence of civil and political mobilization in the previous section, we turn to examine the *specific* contexts and social relations within which these movements emerged in Latin America. Although the macrolevel factors lay behind the experiences of most subaltern and marginal groups, they do not explain the particular networks, social relations and cultural resources through which organizations came into being.

Just as the national urban hierarchies were being reconfigured, so too were networks within and between rural and urban areas expanding. Neighborhoods became the location for personal and group networks (such as migrant associations in Lima and other cities) linking populations with churches, rural villages with outlying migrant *paisanos* (countrymen) and various state institutions. The quotidian networks provided for a microstructure of organization (Evers, 1985: 44), a local social pattern through which civil society could be imagined (cf. Melucci, 1989; Scott, 1990). With localized networks, daily life becomes the means of reproducing and extending grass-roots meanings into a wider group (Melucci, 1989). In this sense, social movements are intermediary between political power and the networks of everyday life, placing them at a level which is 'structurally ambivalent' (Melucci, 1995: 115). This is particularly true in Latin America where the public space for engagement and negotiation is either limited or lacking altogether (Fowerraker, 1995: 30). Thus, while public space is determined largely by the state, alternative publics may arise in the 'counterpublics' of nondominant groups (Alvarez et al., 1998). While supportive of local meanings, ambivalent and weak networks risk keeping movements parochial and fragmented (Alvarez et al., 1998). However, the degree to which social movements remained at the local level with a cyclical pattern of activity and quiescence varied greatly, as did the scope of action. For example, the indigenous agrarian regional movement of Unión de Comuneros Emiliano Zapata (UCEZ) in Michoacan, Mexico, transcended locality by coordinating with other social movements such as the miners' union (to minimize state repression) as well as using the judicial apparatus to engage with the nation-state (Gledhill, 1988). As recent trends have shown, however, new contexts can lead to expanded networks (socially and geographically) permitting 'activism at a distance' and a greatly expanded scope of operation for some movements (see below).

With the reformulation of Catholic theology and the new 'option for the poor', founded on Liberation Theology, spaces for the organization of social movements opened up in the 1970s, especially where the state had been discredited (Lehmann, 1990). Bible-reading and adult literacy groups founded under this theology provided a focus for meetings, as well as cultural meanings (of oppression, exploitation and redemption) through which to reformulate world-views among subaltern and marginal groups (Lehmann, 1996). Examples abound in Latin America of the role of CEBs (*comunidades eclesiásticas de base*, or Christian base communities) in fomenting and supporting social movements, particularly in their early days, whether in agrarian or women's organizations. In the mobilization of agrarian groups in Chiapas during the 1970s, catechists formulated petitions for land and basic services for rural communities. Similarly, peasant women in southern Peru were influenced by catechists and bible-study groups which sprang up in rural areas throughout the Andes in the 1970s. Drawing on strong female Biblical role models, catechists encouraged *campesina* women to learn how to run meetings and organize petitions to local municipalities, giving them confidence, skills and a new language of protest (Radcliffe, 1993a).

Tensions between church officials and civil groups can arise once mobilization takes off or, in other cases, the process of critique put in train by consciousness-raising turns participants against the precepts of the Catholic church. *Campesina* women in Peru for instance found themselves in disagreement with the church's representatives over core ideas of femininity and women's rights. By contrast, Archbishop Romero in El Salvador gave strong support to human rights movements during the military regime, despite female members' increasingly sophisticated analyses of the country's gender regime (Schirmer, 1993).

As the nature of the public sphere, broadly defined after military-era containment of social movements, switched to a civilian public sphere of the late twentieth century, so too the types of networks operating within society and influencing social mobilization have changed. Rather than being bounded by the microstructures of everyday life and the heavily policed terrains of authoritarian regimes, civil society operates in a wider and multilayered set of spaces. Mobility of capital and labor, together with the transnational connections offered by media and new communication technologies, provide new horizons, leading to what Calderón *et al.* characterize as a 'new diaspora of collective action' (1992: 23). At the end of the twentieth century, networks – encompassing but not coterminous with social movements – have rapidly emerged, between and across diverse institutions, movements and social actors, creating 'social movement webs' (Alvarez *et al.*, 1998). The notion of webs fruitfully suggests a method for tracing the personnel, tactics, ideas and information across space and between social movements. At the end of the twentieth century, increasing evidence of these spaces of flows are emerging, whether in the Central American and Andean indigenous movements or among feminists and nongovernmental organizations (NGOs), either separately or in combination (for example, Saporta Sternbach *et al.*, 1992; Warren, 1998).

Whether in the urban social movements or the most recent cyber-culture movements, social movements have been embedded within broader horizontal linkages, which define their constituency (at least initially) and shape their political culture. Between organizations these horizontal linkages may lead to greater extension of networks, or alternatively may show up the limits to social movement activism. Cooperation and conflict are both possible outcomes, illustrated by the ways in which the streams of women's movements in Peru at times successfully combined middle-class feminist concerns with popular women's organizations, while in other cases relations were fraught by accusations of hierarchical behavior and authoritarianism (Vargas, 1991).

Claims to specific rights often lie at the centre of mobilization, especially after failures of modernization (Calderón *et al.*, 1992). While this may include material and economic rights, such as access to secure title to land or housing, or assistance in agricultural extension and bilingual education, there is also a broader set of 'rights'-based issues to do with the construction of the subject under Enlightenment notions of difference. As discussed above, the premise of Latin American modernity was on the constitution of hierarchies of value around 'full' subjectivity, defined in relation to those 'lacking' certain attributes. In the impulse to organization, the attainment of recognition of subalterns as subjects in their own rights constitutes a crucial element of Latin American politics. The feminist movements of the region, as well as the organization of black Latin Americans (Minority Rights Group, 1995) and indigenous populations all speak to the broader terrain of rights, that of the 'reconstitution of the regime of rights' (Calderón *et al.*, 1992: 29; also Foweraker, 1995). Again, globally available resources may be inserted into social movements' agendas: the United Nations Convention on the Elimination of All Forms of Discrimination against Women was utilized widely by popular and peasant women's organizations in Peru during the 1980s as it provided them with a language in which to negotiate with governments, international agencies as well as feminists. While rights provide a mobilizing and 'open' political arena around which groups can coordinate, the networks between individuals and groups can vary greatly in their strengths and weaknesses.

Given that mobilization occurs simultaneously with a definition of interests that are multifaceted and unfixed, so too rights and interests are defined and acted upon in light of the gender, ethnicity and class of movement participants. Patriotism as the state's attempt to mobilize citizens in the interests of their country, *patria*, has failed to reduce interests to this single dimension (Radcliffe and Westwood, 1996). Rather a complex socially embedded politics of interests exists in which interests become incorporated into policy and which attempt to make citizenship in a more active and participatory way.

During the transition to civilian regimes and the process of redemocratization, social movements as individual actors and as institutions have necessarily seen a change in their engagement with other machineries of development and power. In Chile under democracy, social movements were *not* strengthened by the re-emergence of diverse political parties, which drew movement leaders into their remit while marginalizing the movements themselves. After 1985, leaders steadily moved into political parties, with negative consequences for popular groups' access to decision-makers (Schuurman, 1993). Simultaneously, middle-class professional women became 'femocrats' in the restructured state agencies (such as the National Women's Bureau SERNAM), whereas popular women found themselves reinscribed into a passive recipient role (Schild, 1998).

UNDERSTANDING SOCIAL MOVEMENTS

In the discussion so far we have suggested that social movements can be seen usefully as organizations and networks which are attempting (successfully or otherwise) to reconfigure their access to resources and the representation of their identity. In other words, social movements require both strategies and creativity, an agenda for seeing where they are going and a sense of how they are going to present themselves to get there. These ideas are drawn from two fields of theoretical writings and case studies, known as resource mobilization theory (RMT) and identity theory (sometimes known as new social movements theory).

RMT focuses on the actions of movements in the political sphere, analysing how they mobilize resources for their goals within the context of kinds of participation and the rationales for action. To RMT, structural change and how groups mobilize in response are the key issues to be addressed, reflecting as it does a valid recognition that nonformal politics can be both rational and important. Social movements are apprehended as a constant process of organization requiring resources and inputs from its members, therefore raising questions about how leadership operates under specific circumstances (Foweraker, 1995). Nevertheless, beyond its important focus on the structural dimensions and the mobilizing role of leaders, RMT has been grounded in an overly economistic interpretation of participants' actions, viewing them merely as individual *'homo economicus'* (Sheffner, 1995). With the cultural turn associated with post-structural social theory, increasing emphasis is now placed on the intrinsic meanings and values of movements and on the ways in which cultural, ideological and noneconomic interest-related dimensions feed into mobilization.

New social movement (NSM) theory, by contrast, comes more directly from the post-structural social theory, with a focus on culture and identity lying at the heart of its concerns. In his discussion of the importance of identity, Evers (1985: 48) asks whether power is the only or most important potential social dimension for understanding social movements. Rather, he suggests, social movements are driven by the need to express an identity, a way of 'doing society' in which realization of subjecthood can take place. In this theory, NSMs are not new organizational structures (as they are in RMT) but rather the outcome of submerged social networks. From the emergence of identities in these networks, social movements are identified as a means to democratize authoritarian political cultures through the slow, persistent transformation of meaning. Formal (often exclusionary) politics, relying on patron–clientage and the silencing of subaltern groups, gives way to newly empowered social actors, unconstrained by the given social categorizations of their national society (for example, Escobar and Alvarez, 1992: 326). Mediating between communities and the party system, NSMs provide a forum for the creation of identities, around a system of meaning and cultural-economic equivalences known as moral economies (Foweraker, 1995; Sheffner, 1995). Social movements act in a broad terrain encompassing the negotiation over meanings – meanings of the state, citizenship and so on – with the (un)intended effect of transforming wider social discourses and national narratives (Alvarez et al., 1998). However, it is clear that the transformatory potential of social movements was often celebrated too soon, as the material basis for their emergence and the persistence of social difference complicated a simple cultural project (Roberts, 1997). Another weakness of NSM theory was its focus on identity at the cost of understanding the material bases of protest, together with a noted failure to define (and predict) the success of movements (Sheffner, 1995).

In the 1990s, social movements have increasingly been viewed synthetically in ways that draw on the strengths of RMT and NSM theory, while attempting to avoid their pitfalls. Rather than judging civil society and its conflicts with the state and social actors in make-or-break terms of resource mobilization or path-breaking social relations, it is possible to make contextualized, multifaceted analyses of factors as diverse as the state, political cultures, questions of representation and networks as well as flows of resources and cultural capital, and how these engage (or do not) with specific social movements and developments in civil society. In other words, it is a question of looking at 'community, ideology, structural factors and strategy' (Sheffner, 1995: 604). Identities are not given automatically but must be constructed, in light of the types of communities in which participants are embedded (Melucci, 1995: 342). Moreover, the identity or community through which participants engage in action must be realigned constantly (Foweraker, 1995: 12), responding as much to external as internal factors. Social movements, then, are both expressive and instrumental, and, to quote Jean Cohen, civil society is the 'target and terrain of collective action' (quoted in Foweraker, 1995: 19–20). In Latin America in the late twentieth century, when civil society is being widely debated and discussed, such material and identity-related issues are of vital importance. Moreover, three particularly urgent questions require such an approach: the role of social movements in the transition to democracy, the significance of social movements in transforming political cultures in Latin America and, last, the ways in which social movements may shape current economic restructuring.

Despite arising under conditions of political authoritarianism, social movements are the basis for more democratic and participatory politics, through their attempts to reopen a debate in political culture and their efforts to make new organizational forms. In this sense, social movements have acted at crucial moments, such as in the movement 'schools for democracy', focusing on political rights, an expanded notion of citizenship and on making political demands on the state. The transition to democracy from military rule was influenced by social movements' insistence on human rights, holding rulers to account and their rejection of the closure of public space (Foweraker, 1995). The Mothers of the Plaza de Mayo, for instance (Box 9.1), through their constant demands on the Argentine military juntas, initiated a process of reoccupation of public space that eventually contributed to the loss of the military's legitimacy and the return to civilian rule (Radcliffe, 1993b). In Chile too, submerged associations and social movements pushed the country towards a civilian return. In their basis in community and their focused agenda, social movements impact on transitions in the broad sense of political culture (a point returned to below) as much as in the process of negotiation that led to the end of authoritarian regimes. While they can contribute to transition, their impact depends upon the form and timing of their entry into a multifactored process that also includes elites, business representatives, political parties and others. In a context where the transition is driven behind closed doors with negotiations between political elites, social movements do not often have direct access to the negotiating table (Foweraker, 1995).

The significance of social movements for Latin America's political culture is undeniable, although their role as innovative breeding grounds for egalitarian and inclusive social organization has been tempered by a realization of the persistence of strong legacies of authoritarianism and exclusion. From an identity perspective, social movements appear to offer the ideal location from which to speak of different political and social relations. From this perspective, the 'view from the margins' is necessarily critical of dominant patterns, fostering 'alternative modernities' as models (Alvarez et al., 1998: 9). Yet many groups are ambivalently and deliberately placed on the delicate frontier between rejection of given modernities and the formulation of others. Ecuador's indigenous movement engages with the state over land-title and political presence (Chapter 2) while using the machineries of state modernization (land-reform agencies, legislation on Indian communities, etc.) for its own ends. In this sense, it has played a double game in which hybrid practices are created, combining (state) modernity's promise of resources and a distancing from modernity's racisms and authoritarianism (Bebbington et al., 1992; Radcliffe and Westwood, 1996). With the strategic choice of the label 'indigenous nationalities' as both a resource and as an identity, Ecuadorian indigenous groups have mobilized in what Santana terms 'a sustained – if difficult and sinuous – march towards ... citizenship' (Santana, 1995: 6). The goal of citizenship then necessarily engages with the modernist institution of the nation-state, while the meaning of modernity that entails is struggled over. Basing their movement on ethnic identity offers both cultural capital (Alvarez et al., 1998) and a means to get political leverage.

The clear separation of society and the state, common for much of Latin America's experience of modernity, is thus becoming more blurred as it moves into the twenty-first century. Similarly, at a general level, what were once a social movement's demands are permeating the political field in a 'web-like, capillary fashion' (Alvarez et al., 1998: 16). In this context, the long-term impact of mobilization is still to be mapped out, in the 'legal-institutional terrain linking civil society and the state' (Foweraker, 1995: 103). While citizenship is not a particular interest promoted by the state in the post-transition era, there is also a shift to state recognition of difference within society. For example, Bolivia under a neoliberal president created a subsecretariat of ethnic, generational and gender affairs with transversal powers at a ministerial level, while Ecuador's ministry of indigenous and black nationalities reflects the recent ethnic mobilization in that country. Perhaps what is being seen in the late 1990s is a set of 'informal, discontinuous and plural public spaces' (Alvarez et al., 1998: 18) rather than the early modernist public–private split.

Another question that arises in the late twentieth century is the significance of social movements in challenging the economic and social restructuring associated with neoliberalism. While much attention was focused on social movement's role in regime change, less has been found out about resistances to economic neoliberalism (Roberts, 1997). Manuel Castells (1997), among others, argues forcefully that the network society is disenfranchising large populations around the world, who organize social movements in opposition to the new world order. With neoliberal rules of redistribution and access, social movements and civil groups have entered energetically into the political arena, bringing certain repertoires and languages while learning the new ones. Groups such as the Zapatistas in Chiapas, Mexico, offer in this view potential alternatives to new forces of marginalization through the creation of a 'project identity' (Castells, 1997: 65), especially in the Zapatista case – the first informational guerrilla movement – formed around the uncharted possibilities of telecommunications. Other writers are less optimistic about the room for maneuver under neoliberalism, arguing that in the post-transition period social movements have not been shown to be successful in affecting resource distribution (Foweraker, 1995: 104). Given the partial and decentralized picture of neoliberalism (Chapter 1) it could be argued that restructuring opens out new avenues just as it closes down others.

Having examined some of the theoretical background to social movements, we now turn to see how they operate in Latin American society. How extensive are they? How do they organize between different language or ethnic groups? What makes a women's movement different from a feminist movement?

UNDERSTANDING THE (POLITICAL) SPACES OF SOCIAL MOVEMENTS

This section addresses the ways in which social mobilization uses and attributes meanings to space and place. A recurrent theme in the literature on social movements is that they are characteristically geographically limited or localized, tending to function at the level of rural communities or urban neighborhoods, running alongside a theme of fragmentation and lack of widespread coordination, implicitly bringing the social movements into contrast to nationwide politics that are the 'norm' of democratic political life. Within this characterization of social movements, especially in the early literature, the underlying measure of comparison was with national-level unions, themselves often the means by which national politicians co-opted political support and minimalised opposition from working populations. Particularly in the Southern Cone under mid-century industrializing corporatist–populist regimes, civilian political institutions mirrored the national territory, extending their patronage, influence and structures across the country.

The widespread use of spatial metaphors in the field of social movements additionally contributes to confusion over their geographies and use of places. To give examples of these spatial metaphors: Escobar and Alvarez (1992) speak of the 'fragmented social and political space' of social movements, while Foweraker (1995) talks of the terrain of politics. While metaphors assist our apprehension of the issues at stake, they can function as substitutes for analysis of the geographies involved.

Geographical processes constitute social movements and reinforce the politics of social difference. The uneven spread of development in Latin America, with its extreme patterns of urban and regional bias, generates differences that are expressed as much in social terms (for example placing blame on certain populations) as in the distribution of resources. Grievances are then expressed over the security or precariousness of development gains. Mobilization for gaining title to housing in urban shantytowns and title to agricultural land in rural areas illustrate these demands. Expressions of demands are often intimately bound up with the meanings attributed to the land, or place, within which people find themselves struggling. They might identify with land through cosmologies, or by means of narratives of invasion and squatting. A self-reflective community may thus be grounded in a particular place, while its agendas will be shaped by the geographies within which it is embedded. Success in achieving demands can often relate to the ability of negotiators to deal with the spatial concerns. Moreover, in the late twentieth century it is becoming increasingly clear that the 'state-centered' view of politics is losing its significance and explanatory power. States have spatial technologies of power to control territory, population and borders, yet these are becoming unbundled in the context of neoliberalism, expansion of globalizing economic structures and with the rise of international institutions. The late twentieth century in Latin America, then, just as in other world regions, is undergoing a reorganization of its spatial politics (or geopolitics), a fact that is recognized and utilized both by social movements and by states.

Rather than seeing a simple spatial division between, on the one hand, local social movement, and, on the other, national political institutions, analysis of the diverse geographies of social movements has become a focus. The once taken-for-granted geographies of social movements and states are understood in terms of multifaceted spatial imaginations and spatial practices. Social movements can be examined in terms of the diverse spaces in which and through which politics now operates, and in terms of the various connections between places which support and extend social movements' influence. Extending social movement theory with regard to the importance of resource mobilization and of identity issues it can be seen that spaces and places represent *resources* to be mobilized by social movements, while attachment to place provides a 'grounding' for *identity* and the imaginative geographies with which understandings of the world are made. Just as (formal) politics has its own spatiality, so too does resistance (Slater, 1998; Pile and Keith, 1997), with the practices through which resources and cultural capital can be mobilized. Three dimensions of the spatial aspects of social movements can be drawn out here, to highlight the multidimensional geographies into which social movement actors are actively embedded.

Place providing closure for a project

In the context of uneven development led by a centralized state, social movements have often based their mobilization around a sense of decentering, away from the failure of development in its centralist version. In their anti-hegemonic guise, social movements have long been engaged in struggles for the decentralization of power, and power's beneficial impacts at the local level. Regional grievances may therefore arise in which a sense of belonging to a regional culture provides strength to the movement. In Peru between 1968 and the late 1970s several regional movements arose in the context of economic crisis and the 'regional problem' that created internal differentiation of the national territory (Slater, 1989). Secondarily concerned with democratization, the regional defence fronts contained strong social linkages across rural–urban lines and class difference. New regional governments were eventually established in the 1980s, and in the case of Ucayali a new department was created. However, decentralization was achieved briefly, before disappearing under Fujimori's centralist tendencies.

Another aspect of the strategic use of place by social movements is closure provided by location as the basis of community identity. The strategic essentialism of the pan-Mayanists (Warren, 1998) lays claim to a spatial concentration of a community as the basis for its actions. That a place and a people is Mayan provides the basis for what is in effect a transnational network of activists. Local Maya identity serves as a mobilizing strategy, calling up coalitions between language, class and political groups and the generation of cultural capital.

Social movements reconfiguring space

Social movements highlight the spatial nature of (state) power through their transgression of the spatial rules organizing Latin American geographies. If ethnic groups have been seen as geographically bounded 'internal colonies' within a mestizo state (as in much of Central America and the Andes), then indigenous social movements have reconfigured that economic, political and social isolation through the extension of networks across rural–urban divides and also by carrying out demonstrations in the heart of mestizo nationhood. For example, black movements in Brazil have held visible demonstrations in downtown São Paulo, while the 1992 'March for Life, Dignity and Territory' to Quito from the Ecuadorian Amazon brought ethnic diversity to the center of the capital.

In this light, transnational flows of ideas and strategies are illustrations of how subaltern groups attempt to reconfigure the geographies of power, through accessing resources (material or cultural capital) from groups outside the nation-state. The Peruvian peasant patrols are embedded within transnational networks of ideas and support, located in 'larger networks of ideas, exchange and authority' (Starn, 1992: 94; Box 9.1). Within a globalizing world, spaces of flows (Castells, 1997) extend across Latin

America, giving wider reception to social demands from diverse audiences (think, for example, of the media coverage and websites dedicated to the Zapatista uprising in Chiapas, Mexico, since their emergence in January 1994[1]) and to the nature of political and social strategies adopted by the social movements themselves. Technological advances have expanded the possibility of 'activism at a distance' (Ribeiro, 1998: 325). Owing to the liberalization of regional markets, as well as NGOs' provision of computers to organizations and late modernity's emphasis on the image, more movements are diversifying their range of communication tools. Expanded networks of personnel, support and strategies transform the movements into transnational actors, working at a regional or increasingly global stage. New networks and patterns of political insertion in turn inform new cultures of organization as well as new 'complexes of brokering' (Mato, quoted in Yudice, 1998: 370) through which negotiations take place between NGOs, states, associations and movements and local populations.

Contesting boundaries

In their anti-hegemonic impetus, social movements challenge and negotiate the meanings attributed to subaltern groups and at times attempt to generate alternative languages through which to speak about social difference, development and modernity. In this process, terminology used by dominant groups to designate, and in some cases separate, populations has been taken, questioned and reconstituted, thereby reworking the boundaries between social groups. The exclusionary meanings attached by dominant society to ethnic labels, such as Indian or black, have been challenged by ethnic groups throughout the region. The different Indian populations of Ecuador have mobilized around the term 'indigenous nationalities' which to their mind entitles them to mark their distinctive heritage, while claiming the citizenship rights associated with nationhood (Box 9.3), while black Latin Americans around the region have contested the racial hierarchies that long made them invisible (Minority Rights Group, 1995).

Other movements around issues of sexuality and gender provide further illustration of this remaking of social boundaries through civil mobilization. One boundary that has come under pressure and negotiation with the social movements is the public–

BOX 9.3 Ecuador – land and indigenous identity

Demanding full citizenship rather than ambivalent inclusion in the imagined community, indigenous groups have contested the territorial boundaries of their land. Ecuador's large indigenous population, comprising 30–40 per cent of the Andean republic, has long dealt with the nation-state over issues of land and identity. The 1930s law legislating for indigenous communities provided, however precariously, a territorial basis and a grounding for indigenous identity. While class-based union and party politics prevailed in the 1960s and 1970s, the 1980s saw a radical shift in focus from agrarian issues to territorial questions articulated through the new cultural language of 'indigenous nationalities' (Santana, 1995). Representing both a necessary resource for survival as well as a support for a territorially-defined identity, land was the focus of struggles for land-titles and state recognition of indigenous territories (Radcliffe and Westword, 1996).

private divide, between what is historically seen as the masculine space of the street and public arena (including politics) and the domestic sphere, historically feminine and defining the heterosexual and patriarchal family relations that underpin the wider political structure (Dore and Molyneux, 1999). While these gendered divisions have long been breached by subaltern, peasant and indigenous groups (whose dependence on female wages and work breaks down the domesticity of women, or whose distinct gender relations offer other models of gender relations) (Melhuus and Stolen, 1996), the political sphere has long been defined by the masculinism of its culture and the exclusion of women. Female citizenship arrived late in Latin America, with voting rights conceded slowly over several decades in the mid-twentieth century,[2] while entry into formal representative positions has proceeded at a snail's pace. Women make up between 4 and 15 per cent of parliamentary representatives in the post-transition governments of the early 1990s.

Yet despite this highly gendered pattern of formal politics, women made up the great majority of participants in many social movements through the 1970s and 1980s, precisely when their formal avenues were being closed down. For instance, women made up 99 per cent of members of Chilean neighborhood associations, most of them involved in self-help groups purchasing and cooking cheap food and calling for housing tenure (Lehmann, 1990). In Lima's shantytowns, women organized communal kitchens in which they took turns to buy and cook food, thereby freeing their neighbors to work for wages (Vargas, 1991; Barrig, 1994). In these movements, the gendered impacts of structural limits on participation are clear – female reproductive and domestic work was increasingly difficult under conditions of economic austerity, prompting women to organize themselves. Whether in the shanties of Buenos Aires, Lima or Mexico City, new forms of organization breaching the public–private divide in divisions of labor and expectations of behavior emerged.

The authoritarian regimes raised different gendered issues. In addition to work pressures, the male public sphere of politics and the institutions (unions, political parties, municipalities) were repressed or severely restricted. As in relation to domestic work, women stepped into the breach in an unprecedented way, to carry out forms of resistance and organization under military governments. Although women were often reluctant to define what they did as political (Jelin, 1990), these activities contributed to a reconfiguration around the political sphere not least in its classic separation of the public and private. In the Brazilian women's movement (comprising neighborhood, consciousness-raising and feminist groups), women were engaged in 'deliberate attempts to push, redefine or reconstitute the boundary between the public and the private, the political and the personal' (Alvarez, 1990: 23).

The insertion of women into social movements in the 1970s through to the 1990s raises questions in turn about the long-term influence of participation on their social position and autonomy. Motherist groups, organizing around the identity and networks of a gendered maternal role, lie on one side of a complex picture of women's groups. The extent to which motherhood provides the grounds for an original and emancipatory politics has been much debated (Fisher, 1993b), not least in the case of the Mothers of the Plaza de Mayo, Argentina. The Mothers founded their women-only organization on the basis of a well-embedded gender regime with a long history of church and state endorsement of a sacrificial maternal role. In contrast to other mothers' groups, the *Madres* in Argentina failed to initiate a wider self-reflexivity regarding their role, thereby reproducing rather than challenging their political position. As Jane

Jaquette (1994: 225) asks, 'if interests define citizens, can women be citizens if they always act in the interests of *others*?' (emphasis in original).

In the differentiation of gendered interests, the distinction between practical and strategic interests has long been useful. Formulated by Maxine Molyneux in her analysis of the Sandinista gender politics in Nicaragua, the givens of female work and identities were seen to be expressed in terms of *practical* interests (for cheaper food, better health care for families), whereas critical engagements with the gender regime were associated with women's *strategic* interests (Molyneux, 1985). Feminist movements in Latin America long demanded strategic changes in the legislative, familial and marital relationships structuring women's lives. Their activism and ability to lobby governments for change, as well as forging strong international linkages, put Latin American feminist movements at the heart of civil and political life (Saporta Sternbach et al., 1992). Initially comprising urban and middle-class women, feminists increasingly intersected and interwove with the popular women's movements that arose initially around practical issues in the shantytowns (Vargas, 1991; Alvarez, 1990). In the Centro Femenino 8 de Marzo of Quito, Ecuador, popular women outlined an agenda of interests and needs in which strategic and practical interests were intertwined, informed by women's recognition of their disempowerment. Through consciousness-raising exercises, marches, meetings with feminists, lobbying of urban government and the collective provision of family needs, it is clear that poor women are not solely concerned with practical issues, as they 'negotiate power, construct collective identities and develop critical perspectives on the world in which they live – all factors that challenge dominant gender representations' (Conger Lind, 1992: 137). Quiteña women's politics of needs was transformed through organization – out of the house and in interaction with a wide range of social actors – and as the cultural repertoires available to them engaged with an increasingly questioning agenda. Rather than question whether strategic or practical needs are being met, social movements offer the possibility that gender regimes can be critiqued, alongside a critique of the other dominant institutions in participants' lives (Burgwal, 1995).

In summary, social movements are not restricted in their geographies nor are they passive inhabitants of place (neighborhood, lowland environments and so on). Space is not a mere background to their activities but is integral to their operation. Social movements' spaces are not passive containers or limits on activities and identities but are integral to their operation. Contesting boundaries between social categories or between spaces designated by dominant groups is – as shown by reference to the women's movements, indigenous movements and mobilization around issues of sexuality – an integral part of the civil histories of Latin American late modernity. Boundaries may, as in the case of Ecuadorian and other indigenous groups, be primarily material and legal, yet their cultural weight attributes them a wider significance within the cultural politics of mobilization. We can say that social movements can be viewed as existing in a more complex and dynamic relationship with place, and *using* space (as well as being fragmented by geography). We have seen that social movements mobilize space and the meanings around place in order to further their actions and their agendas. The 'archipelago of resistance' (Slater, 1998: 395) represented by specific social movements suggests that a mapping of civil and political mobilization has begun. Yet, as the discussion of reconfigured and increasingly globalized networks show, that archipelago effect may be rapidly superseded by an electronic circuit with key nodes linked through e-mail, visits, telephone and conferences.

Conclusions

After briefly sketching the conceptual and theoretical ideas behind subjectivity and society, this chapter analysed contemporary issues around the state, mobilization and social struggles. Present-day social movements can be understood in relation to participants' struggles for representation and their contests for material and cultural resources with which to construct their own paths through modernity. The range and diversity existing within the umbrella term of social movements and identity politics is astounding, yet the political economy of the region underlies many concerns, particularly the political influences of a centralist state on the public sphere. The chapter has also outlined briefly the trajectories of grass-roots politics and civil society at the end of the twentieth century and into the new millennium. With economic restructuring transforming identities, political economy and political parameters, the late twentieth century in Latin America is a time of rapid and exciting change. The picture of state–society relations traced here is a snapshot of what the future holds, in terms of increasing scales of operation of social movements combined with a radical shift in the nature of the state, both of which together can have unprecedented effects on the nature of citizenship, public cultures and political institutions. The paradox is, however, that just as the forms of agency are opening up, the state is being unbundled, weakened by its engagements with neoliberalism, a rapidly expanding global political economy and new networks of international NGOs and agencies across its borders. It remains to be seen how the continuing reworking of Latin American modernity engages with the civil and political needs of its citizens into the twenty-first century.

Endnotes

1 For more information on the Zapatistas, see Castells (1997) and the websites at www.stewards.net/chiapas/10.htm and www.physics.mcgill.ca/~oscarh/RSM/acuerdos.html.
2 Ecuador granted the vote to women in 1929, while Paraguay granted this right in 1961; most other countries extended female franchise in the 1940s and 1950s (Valdes and Gomariz, 1995).

Further reading

Alvarez, S., Dagnino, E., Escobar, A. (eds) 1998 *Cultures of politics, politics of cultures: re-visioning Latin American social movements*. Westview Press, Boulder, CO. A collection of 15 papers on the latest developments within Latin American social movements, plus an introductory chapter by the editors, and three short commentaries at the end.

Eckstein, S. (ed.) 1989 *Power and popular protest: Latin American social movements*. University of California Press, Berkeley, CA. A classic discussion of the initial force of social movements in the region.

Escobar, A., Alvarez, S. (eds) 1992 *The making of social movements in Latin America*. Westview Press, Boulder, CO. A strong collection of case studies drawing on the theoretical debates of earlier writers in a critical and informative way.

Foweraker, J. 1995 *Theorising social movements.* Pluto Press, London. Pluto Critical Studies on Latin America. The best review of the theoretical literature, and a clear analysis of the role of the state in social movements.

Radcliffe, S.A., Westwood, S. (eds) 1993 *Viva: women and popular protest in Latin America.* Routledge, London. A collection of nine chapters on women's participation in diverse social movements, including environmental, human rights, peasant and health groups, with an introductory chapter.

Part 6

The context of social change

10

POPULATION, MIGRATION, EMPLOYMENT AND GENDER

SYLVIA CHANT

This chapter attempts to provide an overview of recent trends in population, migration and employment in Latin America, with particular reference to the period of economic crisis and neoliberal restructuring in the 1980s and 1990s. Such a task is not easy, given diversity within as well as among countries, the varied timing and nature of liberalization programs in different parts of the continent and the fact that each of these topics is vast in its own right. Indeed, acknowledging that population, migration and employment in Latin America are influenced by an increasingly complex set of global and local forces, compressing analysis of these topics into a single chapter inevitably eclipses several perspectives. At the same time, it arguably allows a more ready appreciation of the ways in which demographic growth, the movement of people and the changing nature of labor markets are intimately interconnected. Given the importance of gender in all these issues, perspectives on gender are incorporated within the chapter's three major sections on population, migration and employment, as well as being the specific focus of a fourth section concentrating on gender and urban labor markets. In each respective discussion, an attempt is made to situate patterns and processes occurring in Latin America within the context of global influences and in relation to conceptual work pertaining to developing regions more generally. A major concern of the chapter is also to identify the implications of demographic and economic corollaries of restructuring in Latin American countries for people's lives and livelihoods at the grass roots. Thus, in addition to using large-scale secondary sources to piece together an idea of regional and national trends, the analysis also draws on microlevel case-study material. Some of the latter comes from my own field research on gender, households and urban poverty in Mexico and Costa Rica over the past 15 years.[1]

POPULATION

One of the most notable demographic changes in Latin America since 1980 has been its declining rates of population growth. This situation is in marked contrast to the period 1920–70 when the population in the continent grew faster than in any other world region, and even during the 1970s, when the growth rate was the second highest in the world (after Africa) (UNCHS, 1996: 42). With annual rates of population growth in most Latin American countries hovering between 2 and 3.5 per cent per annum in the postwar period, the regional population more than doubled between 1950 and 1980 alone (Safa, 1995b: 32). Yet, although by 1991 the population had increased to a total of 422 million (Cubitt, 1995: 110), Latin America contains less than one-tenth of the world's inhabitants and is among the regions with the lowest population densities.

Changing birth and death rates

The decline in annual population growth rates has been fairly steady in most Latin American countries since 1980 (Table 10.1), with upswings only in Nicaragua, and to a lesser extent Guatemala, in the first half of the 1990s owing to a combination of lower

TABLE 10.1 Population growth rates in Latin America, 1970–2000

Country	Population, 1994 (millions)	Average annual population growth rate (%)			
		1970–80	1980–90	1990–94	1994–2000
Haiti	7	–	1.3	1.9	1.9
Nicaragua	4.2	3.1	2.7	3.1	2.7
Honduras	5.8	3.2	3.3	3.0	2.8
Cuba	10.9	–	–	–	0.4
Bolivia	7.2	2.4	2.0	2.4	2.4
Guatemala	10.3	2.8	2.8	2.9	2.9
Ecuador	11.2	2.9	2.5	2.2	2.0
Dominican Republic	7.6	2.5	2.2	1.7	1.7
El Salvador	5.6	2.3	1.3	2.1	2.2
Paraguay	4.8	2.9	3.1	2.8	2.6
Colombia	36.3	2.2	1.9	1.9	1.7
Peru	23.2	2.7	2.2	1.9	1.8
Costa Rica	3.3	2.8	2.8	2.1	2.1
Panama	2.6	2.6	2.1	1.9	1.7
Venezuela	21.0	3.4	2.6	2.3	2.1
Brazil	159.1	2.4	2.0	1.7	1.3
Chile	14.0	1.6	1.7	1.5	1.4
Mexico	88.5	2.8	2.0	2.0	1.7
Uruguay	3.2	0.4	0.6	0.6	0.6
Argentina	34.2	1.6	1.5	1.2	1.3

Sources: UNDP (1997, Table 22) and World Bank (1995a, Table 25; 1996, Table 4).
Note: countries are listed in ascending order of gross national product (GNP) per capita in 1994.
– = no data.

mortality levels and repatriation of refugees in the aftermath of civil war (see Dunkerley, 1994: 48–9). Also worth noting is that the relatively wealthy countries of the Southern Cone (Argentina, Chile and Uruguay) were already experiencing low levels of population growth in the 1970s.

The major reason for declining population growth has been a massive reduction in fertility since the 1970s. In the region as a whole, the total fertility rate dropped from 5 in 1970–75 to 3.1 in 1990–95 (Roberts, 1995: 91–2), with most countries experiencing a 30–50 per cent reduction in their national rates between 1970 and 1994 (Table 10.2). Exceptions include Argentina and Uruguay, where total fertility rates had fallen before 1970.

Although figures for contraceptive use are not available for all countries, besides which data are rarely comprehensive or reliable (tending to refer only to married couples, for example), Table 10.2 indicates the existence of a broadly inverted relationship between contraceptive prevalence and total fertility rates in the 1990s. Thus in countries

TABLE 10.2 Fertility and contraception in Latin America

Country	Total fertility rate[a]				Contraceptive prevalence[c] rate, 1987–94 (%)
	1970	1980	1994	2000[b]	
Haiti	–	5.2	4.8	–	18
Nicaragua	6.9	6.2	4.9	4.2	49
Honduras	7.2	6.5	4.7	4.0	47
Cuba	–	–	1.5	–	70
Bolivia	6.5	5.5	4.7	4.1	45
Guatemala	6.5	6.5	5.2	4.6	32
Ecuador	6.2	5.0	3.3	2.9	57
Dominican Republic	6.1	4.2	2.9	2.7	56
El Salvador	6.3	5.3	3.8	3.4	53
Paraguay	5.9	4.8	4.5	3.7	56
Colombia	5.3	3.8	2.6	–	72
Peru	6.2	4.5	3.1	3.0	59
Costa Rica	4.9	3.7	2.9	–	75
Panama	5.2	3.7	2.7	2.5	64
Venezuela	5.3	4.1	3.2	2.8	–
Brazil	4.9	3.9	2.8	2.5	66
Chile	4.0	2.8	2.5	2.4	–
Mexico	6.5	4.5	3.2	2.6	53
Uruguay	2.9	2.7	2.2	2.2	–
Argentina	3.1	3.3	2.6	2.5	–

Sources: UNDP (1997, Table 22) and World Bank (1995a, Table 26; 1996, Table 6).
Note: countries are listed in ascending order of gross national product (GNP) per capita in 1994.
– = no data.
[a] The total fertility rate refers to the average number of children likely to be born to a woman if she survives until the end of her childbearing years and gives birth at each age in accordance with prevailing age-specific fertility rates.
[b] Estimate.
[c] Contraceptive prevalence refers to the proportion of women who are practising, or whose spouses are practising, any form of contraception. This is generally measured for married women aged between 15 and 49 years.

such as Colombia, Cuba and Costa Rica, where 70 per cent or more of the 'married' adult population use contraception, the average number of children born to women at a national level is less than 3. Alternatively, where less than half the population use contraception, as in Nicaragua, Honduras and Bolivia, the mean total fertility rate is nearer 5. By the same token, it is important to bear in mind that national averages often mask substantial variations between different classes and localities. For example, in the rural Department of El Paraíso in southeastern Honduras, where the primary economic activity is subsistence farming (corn and beans) and the bulk of the population is poor, the total fertility rate in 1988 was 6.6 (Paolisso and Gammage, 1996: 25). Although slightly lower levels were found in my own surveys of low-income urban settlements in Mexico and Costa Rica during the 1980s, they were still above the national averages (Box 10.1; Fig. 10.1). While limited knowledge of, or access to, family planning seems to play a part in influencing birth rates among the poor, another important reason for high birth rates is early motherhood. In Latin America as a whole, one third of women become mothers in their teenage years. Even in Cuba, where contraception and abortion are widely available, a survey of female workers showed that nearly half had their first child before the age of 20 years (Safa, 1995b: 43). It is important, therefore, to acknowledge the point that

> despite their zeal to reduce birth rates . . . population controllers leave many of the determinants of high fertility in place: the need for children as a source of labour or security, high infant mortality and limited economic opportunity for the poor (Hartmann, 1997: 83).

In demographic terms, fertility decline in the region has contributed to counteracting an upward pressure on population growth from two main sources: first, a rise in

Figure 10.1 Young mother, Puerto Vallarta, Mexico. Photograph by Sylvia Chant.

BOX 10.1 Fertility and family planning among the urban poor in Mexico and Costa Rica

The majority of young women I have interviewed in low-income urban communities in Mexico and Costa Rica in the 1990s express desires to have small families and are using a variety of family planning methods. However, this was not the case with many of their counterparts whom I interviewed in the 1980s.

From surveys conducted in low-income settlements in Mexico in 1982–83 and 1986, and in Costa Rica in 1989, most women over 30 years had given birth to at least five children, and their own mothers six or more. Many had had their first child by the age of 18 years when they got married or set up home with their *compañeros* (partners), with a substantial number having done so at the age of 15 or 16 years. Although many women of reproductive age in the 1980s were aware of artificial methods of family planning, the choice of contraception available through government health centres was extremely narrow, usually limited to the pill, the intra-uterine device (IUD) and sterilization. Although some women (particularly in Costa Rica) had opted for sterilization, since a husband's consent was necessary for the operation in both countries, this had lowered the take-up rate. Beyond the fact that most men were reluctant that they or their wives use contraception, hearsay about the side-effects of the pill and risks attached to IUDs was often enough to dissuade women from trying it themselves. Other barriers to family planning were presented by religious beliefs, particularly the idea that children were 'God-given' and that couples should not interfere with the course of destiny. In response to questions about how many children young women were planning to have in Mexico and Costa Rica in the 1980s, many responded '*Quién sabe? Sólo él sabe*' ('Who knows? Only He [God] knows'), or '*Los que el Señor me diera*' ('Those that the Lord gives me'). Women's refusal to use artificial birth control was backed (not to mention promoted) by the local priesthood.

Beyond the barriers attached to using family planning, many women perceived high-order births as positive or beneficial. One reason given was that they would have someone to look after them in old age. Another factor was that children were regarded as an economic resource rather than a burden. In Mexico, this sentiment was often expressed through the popular saying '*Cada niño trae su torta bajo su brazo*' (literally translated as 'Every child brings their own sandwich under their arm'). Women also talked about the importance of high fertility in a context where they were likely to lose at least one of their offspring through illness. Indeed, my survey of 350 low-income households in northwest Costa Rica in 1989, showed that at least one in every five to six births was likely to be a stillborn baby or a child who would die before they reached five years of age. Another reason why women in Costa Rica, in particular, did not restrict fertility was the prevalence of serial monogamy in Guanacaste. Here women would commonly have a child with each new partner in the hope of securing a long-term commitment and financial support from the man concerned.

The case of Melia, pictured in Fig. 10.1 in her home in a low-income settlement in the town of Puerto Vallarta, is typical of many young women I interviewed in Mexico during the 1980s and of the multiplicity of reasons embedded in fertility decisions among the urban poor. The photograph was taken in 1986 when Melia, a migrant from a rural area of Jalisco state, was 27 years old and had a total of six children ranging between three and nine years of age. The entire family lived in a one-roomed dwelling and their only source of income was Melia's husband's earnings from construction work. Although Melia was slightly built and had suffered a range of problems with her pregnancies, she had never used contraception and was not planning to. This was partly because she was a staunch Catholic who felt that her children were a 'gift from God', and partly because both she and her husband had come from large families. Both of them had happy childhoods surrounded by their brothers and sisters, and were still in close contact, even if some had stayed behind in the countryside and others had migrated to towns elsewhere in Mexico. At crucial moments of financial and family crisis, these grown-up siblings rallied round to help each other out. Aside from the practical advantages, Melia said that one could never feel lonely with a large family around, and that the beauty of having lots of children was that there was always noise and laughter in the house. Melia felt sorry for those women who could not have their own families and imagined that their homes must be unbearably quiet.

The main reasons for young women's desires to restrict their births in the 1990s seems to be based on two main factors. First, more mothers nowadays have jobs than their older counterparts and this presents problems of child care, even if there are some public and community facilities in both countries. The second, and related, factor is that with persistent pressure on household incomes, rising costs of living and the increased importance of education in Mexico and Costa Rica, it has simply become too expensive for poor families to have more than two or three children.

TABLE 10.3 Life expectancy and mortality in Latin America

Country	Life expectancy at birth (years)		Crude death rate (per 1000 population)		Infant mortality (per 1000 live births)	
	1970	1994	1970	1993	1970	1993
Haiti	–	57	–	–	–	–
Nicaragua	54	67	14	7	106	51
Honduras	53	66	14	6	110	41
Cuba	–	76	–	–	–	–
Bolivia	46	60	20	10	153	73
Guatemala	53	65	14	8	100	46
Ecuador	58	69	12	6	100	49
Dominican Republic	59	70	11	6	98	40
El Salvador	58	67	12	7	103	45
Paraguay	65	68	7	5	57	37
Colombia	59	70	9	6	77	36
Peru	54	65	14	7	116	63
Costa Rica	67	77	6	4	59	14
Panama	66	73	8	5	47	24
Venezuela	65	71	7	5	53	23
Brazil	59	67	10	7	95	57
Chile	62	72	10	6	77	16
Mexico	62	71	10	5	72	35
Uruguay	69	73	10	10	46	19
Argentina	67	72	9	8	52	24

Sources: Thomas (1996, Table 4.2) and World Bank (1995a, Tables 26 and 27; 1996, Table 1).
Notes: countries are listed in ascending order of gross national product (GNP) per capita in 1994.
– = No data.
[a] Figures rounded up to nearest whole number.

xpectancy from 61 to 68 years between the 1970s and the 1990s (Roberts, , second, a halving of mortality rates in many countries during the same cularly those in which crude death rates exceeded 10 per 1000 in 1970 The combination of falling birth rates and rising life expectancy in Latin begun to change the composition of dependency ratios in the continent. Whereas in 1975, persons aged 14 years or under were more than 90 per cent of the dependent population in Latin America, by the year 2000, persons over 65 years will represent at least one-quarter of this group (Sen, 1994: 8). Although some elderly people have no choice but to work (or beg) for a living, especially where relatives cannot support them and they do not receive pensions, in Mexico, almost 75 per cent of the population aged 65 or more are economically inactive (López and Izazola, 1995: 53). Although the continued growth of the working age population means that it is likely to be some time before the economic burden of this growing elderly cohort is felt, deteriorating conditions in the labor market may accelerate this process. Moreover, for women, who are generally responsible for caring for elderly people, the rise in 'Third Age' dependents may offset the fall in fertility and result in little appreciable change in their reproductive burdens (see Sánchez-Ayéndez, 1993).

Population, health and economic crisis

While infant mortality rates in all Latin American countries in 1993 were still way above the average for industrial market economies (7 per 1000), the decline in both general and infant mortality and rising life expectancy are generally attributed to the expansion of health care, better standards of living and diet and the control of major killer diseases such as poliomyelitis, smallpox and malaria (Gilbert, 1995a: 323; Portes and Schauffler, 1993: 33). The extent to which fertility and mortality will continue falling into the twenty-first century is open to debate, however, given the tendencies set in motion by economic crisis and neoliberal restructuring.

Medical services, for instance, have been weakened by cutbacks in health expenditure forced upon the majority of Latin American governments by structural adjustment programs (SAPs). Countries which have been particularly hard hit in this respect include Bolivia, Mexico, Ecuador, Argentina and Venezuela where declining allocations to health (and education) during adjustment were compounded by sharp falls in gross national product (GNP) (Stewart, 1995: 182–4). In Bolivia, for instance, central government health spending per capita plummeted by two-thirds between 1980 and 1984 (Asthana, 1994: 59). In Mexico, the real value of health expenditure per capita in 1987 was only 41 per cent of its 1981 level (Stewart, 1995: 203). Alongside general cuts in health expenditure, it is also significant that figures for the 1980s show that many adjusting countries (including Argentina, Bolivia, Brazil and Venezuela) attached lower priority to environmental health, primary health care, preventative medicine and programs for the poor (as opposed to hospitals and hospital equipment for example) (Stewart, 1995: 186). This also occurred in Nicaragua where, on top of the destruction of primary medical infrastructure wrought by the Contra War, economic adjustment in the late 1980s and early 1990s resulted in the erosion of several basic health programs. For example, immunization against major preventable diseases such as diphtheria dropped by 23 per cent between 1988 and 1994, tetanus immunizations by 44 per cent and measles by 61 per cent (CISAS, 1997: 88).

This constellation of events in the field of health care does not bode well for the urban poor whose risk of illness is frequently exacerbated by insanitary and overcrowded living conditions. Indeed, the incidence of 'diseases of poverty' commonly associated with rudely serviced residential environments (dengue, cholera, hepatitis, typhoid and tuberculosis, for example) has risen in a number of countries, including Peru, Chile and Costa Rica, during the 1980s and 1990s (Asthana, 1994: 59). In Nicaragua, too, alongside rising rates of infant, child and maternal morbidity and mortality, numbers of deaths due to respiratory and gastrointestinal illnesses have increased in the aftermath of adjustment (CISAS, 1997: 87). Besides shrinking government funds for urban environmental improvements, the combination of declining household incomes and rising food prices (often triggered by the lifting of government subsidies on basic staples) also presents a major threat to health and well-being.

Although many 'social costs' of adjustment are purportedly transitory in nature, several analyses point to deteriorating trends in health and nutrition in the past two decades. For example, between 1979 and 1986 numbers of low birth weight babies increased in parts of Brazil, Colombia, the Dominican Republic, El Salvador and Mexico (Stewart, 1995: 189). In Mexico, too, three diseases that became more important causes of death among children during the 1980s – dysentery, malnutrition and anaemia – are strongly linked with socio-economic deterioration, in the first case as a result of poor hygiene, and in the latter two to dietary insufficiency (Langer et al., 1991: 211). Indeed, annual per capita consumption of staples such as milk dropped from 112 to 101 litres per person in Mexico between 1980 and 1987, beans from around 18 to 14 kilograms and corn from more than 200 kilograms to only 142 during the same period (Cordera Campos and González Tiburcio, 1991: 32–3). In Nicaragua, the per capita consumption of food staples in general also fell by 25 per cent between 1989 and 1992 (CISAS, 1997: 87). Elsewhere in Latin America, such as in Peru and Uruguay, increased poverty has reduced the availability of calories per capita to levels below those of 1965 (Curto de Casas, 1994: 236). The extent to which such tendencies will impact on mortality is difficult to assess in the short run, although in seven countries of the region there were slowdowns in infant mortality reductions during the 1980s (Stewart, 1995: 187). This trend is especially worrying for low-income groups. In Nicaragua, although the national infant mortality rate is currently 51 per 1000 (Table 10.3), the level among illiterate (that is, poor) mothers is 138 per 1000, nearly six times higher than among mothers with primary, secondary or further education (CISAS, 1997: 88). Although fears of losing children may encourage people to revert to higher order births there is evidence from Mexico to suggest that couples have continued to reduce their fertility during the crisis years. As Selby et al. (1990: 169) argue (see also Box 10.1):

> Children cost more and they are not able to earn as much as they did in the past. The economic basis for high fertility rates has been undermined, at least in the cities of Mexico, and ... fertility rates are declining rapidly.

Family planning programs and fertility

Notwithstanding that reproductive behavior is affected by a multiplicity of factors such as culture, gender, education, incomes, migration and so on, we noted earlier that higher levels of contraceptive prevalence in Latin American countries are associated with lower fertility rates (Table 10.2). Given that family planning has been actively promoted by the international population establishment, and that, along with Asia, Latin

...a was the first developing region targeted for lobbying and donor pressure by ...ational agencies (Corrêa and Reichmann, 1994: 15), it is important to consider ...y planning in Latin America within the context of global strategies for fertility ...iction. Within this discussion, particular attention is paid to women, given their consistent position as frontline targets of family planning initiatives.

Global perspectives on population and population policy: a summary

Although there are numerous views on population, and it is possible to discern a divergence between population and development narratives over time, for much of the twentieth century, population thinking in the North has been dominated by the Malthusian notion that unchecked fertility contributes to poverty, underdevelopment and environmental degradation. Bearing in mind the view that this interpretation 'obscures the real causes of the current global crisis: the concentration of resources – economic, political, environmental – in the hands of an ever more tightly-linked international elite' (Hartmann, 1997: 84), the neo-Malthusian remedy for 'overpopulation' has been fertility control programs. Hardly surprisingly, such programs have been much more concerned with fertility reduction targets than with expanding people's own decisions over their reproductive behavior. As Furedy (1997: 65) summarizes: 'Since population control programmes are designed to reduce fertility, their aim is not to provide free choice but to influence people towards a particular outcome'. Moreover, given that the international population establishment has tried to impose Malthusian logic selectively upon the poor majority in the Third World, Hartmann (1987: 14) claims: 'It is not a giant step from this partitioning of Malthusian law to a similar partitioning in the realm of human rights'. In industrialized countries, the decision as to whether and when to bear children is a matter of choice; in developing countries, by contrast, individual rights are subordinated to the overriding imperative of population control (Hartman, 1987).[2]

With the third United Nations International Conference on Population and Development in Cairo in 1994, however, individual (and particularly women's) rights began to occupy a more prominent place on the global agenda. This accompanied a putative shift in the underlying rationale for population regulation during the 1990s. Accompanying the move away from a 'traditional economic development argument to a case for environmental balance' (Corrêa and Reichmann, 1994: 13), 'people' rather than 'numbers' has become the watchword for strategies of sustainable development (UNIRDAP, 1995: 8; Smyth, 1994: 3). Although some commentators see the shift from population control to 'reproductive rights' simply as a new means of promoting fertility reduction given the exhaustion of the 'family planning equals development' approach (Furedy, 1997), the Programme of Action 1994–2015 adopted at Cairo contains the letter, if not the bona fide spirit, of commitment to enhancing personal choice. More specifically, the 1994 Programme of Action made recommendations on six key issues:

- women's empowerment (including access to education to promote gender equality);
- reproductive and sexual health rights of men, women and adolescents;
- unsustainable patterns of production and consumption;
- population as an integral part of sustainable development and environment;
- provision of universal access to reproductive health services;
- family as the basic unit of society (UNIRDAP, 1995: 8).

Departing from the situation in which family planning was either coercive or overly persuasive, and building on the 'quality of care' framework developed by Judith Bruce and others at the Population Council, the Cairo Programme of Action also recommended that family planning programs be guided by imperatives of appropriate patient–provider interactions and informed choice

Feminist critiques of population policy

The Cairo conference marked something of a victory for the women's movement, and Corrêa and Reichmann (1994: 4) go as far to suggest that 'Official recognition of the reproductive rights and health approach to population may be one of the most significant achievements of contemporary feminism'. However, feminists have long criticized family planning programs because women's reproductive rights and general health have tended to be sacrificed to societal responsibility and the demographic rationale of reducing population growth (Smyth, 1994: 11). As Corrêa and Reichmann (1994: 4) have argued, the feminist framework critiques the attempts of states and markets to use science and technology to 'control women's bodies', and challenges the economic and demographic theories used to 'justify population policies that are harmful to women'. Even though the population establishment has increasingly adopted the language of 'rights' and 'choice' for women within their advocacy of family planning programs, they have not effectively acknowledged the fact that women's reproductive decision-making is mediated by regional and historical differences in social structures (Smyth, 1994: 12–13). Beyond this, it is also clear that appropriation of the language of reproductive rights can be a tactically expedient way of justifying anti-natalist agendas. As Furedy (1997: 126) observes: 'Since the seventies, the issue of women's position in society has provided the population lobby with a considerable degree of legitimacy to pursue its objective'.

Population policies, family planning and women's reproductive rights in Latin America

In Latin America, where birth control has been described as a 'very controversial issue' (Cubitt, 1995: 110), examples abound of cases in which consideration, let alone support, of women's rights has been absent from family planning initiatives. Some of the most striking examples include the interventions of North American and international development agencies, which, in the interests of alleviating population pressure on food and reducing poverty, have been 'aimed at direct population control rather than giving women the right and freedom to control their own fertility' (Cubitt, 1995: 110). In some cases, external agencies have tested unsafe contraceptives on poor Latin American women, or have enforced sterilization, the latter constituting a primary factor in the expulsion of the US Peace Corps from Bolivia and Peru. Although extreme instances of this nature are increasingly rare, it is important to take into account that resistance to family planning in the region from other interest groups has also played a major part in limiting women's reproductive choice and self-determination.

The most widespread, and powerful, source of formal resistance to family planning in Latin America has been the Roman Catholic Church. Resistance has taken a variety of forms, including outright blocking of state family planning programs and fomenting grass-roots opposition to artificial birth control. In a few cases, such as Chile, some of the lower-order clergy are now choosing to ignore the Vatican's teachings on contraception and supporting women's decisions to plan their families (Shallat, 1994: 149–50). Nonetheless, the strength of repeated papal pronouncements on the sanctity of human life is evidenced in the fact that, excepting Cuba, abortion is illegal throughout the region (Cubitt, 1995: 110). In Mexico and Bolivia, for instance, abortion is only permissible in cases where the mother's or infant's health is at serious risk, or if pregnancy results from rape or incest (Hardon, 1997a: 18). Moreover, in Mexico, people found practising abortions for other reasons can be imprisoned for up to eight years, and women who induce their own miscarriages may be detained for between six months and five years. Lack of abortion on demand has given rise to a situation where an estimated 4 per cent of Latin American women of reproductive age (15–49 years) have an 'unauthorized' ('unsafe') abortion each year, which is higher than any other region in the world. Even acknowledging the immense unreliablity of statistics, only one in 800 unsafe abortions in Latin America is actually estimated to result in maternal death (Hardon, 1997b: 24). However, this does not preclude the occurrence of numerous medical complications, with around 43 per cent of women inducing abortions in Brazil reported as suffering problems during or after terminations (Corrêa and Reichmann, 1994: 40).

Concern about rising numbers of illegal abortions and maternal deaths has prompted various Latin American governments to take a stronger stand against the Church over time, and to play a more active role in contraceptive provision. For example, 20 years ago the influence of the Church was so strong in Bolivia that the state referred to the country as 'dangerously uninhabited' and issued a Presidential Decree in 1977 forbidding public institutions to provide family planning services (Parras and Morales, 1997: 77). Although fertility continues to be high in Bolivia (Table 10.2), and its National Population Council still argues in favor of a larger population, in 1986 the government introduced a child-spacing policy on the grounds that it was the duty of the state to provide family planning information and services to high-risk groups. In 1993, the government took this further by introducing *Plan Vida*, which aims to widen access to family planning and reproductive health care and thereby reduce maternal, perinatal and infant mortality (Parras and Morales, 1997: 78). While birth control is not a goal in itself, the National Population Council is concerned to improve health standards, especially among women. Strong emphasis is placed on individual rights in using contraception, and concern to increase basic education for women is a national priority (Hardon, 1997a: 16). The absence of demographic objectives, together with an orientation towards meeting women's health needs, have effectively brought Bolivia's population policy extremely close to the Cairo recommendations.

Where Latin American countries have pursued full-blown state-led population programs from the outset, however, the neo-Malthusian spirit of fertility reduction is more in evidence. In Puerto Rico, for example, the government started an active campaign for female sterilization in the 1950s, and by 1997 had managed to reduce the total fertility rate from 5.2 to 2.7 (Corrêa and Reichmann, 1994: 30). Mexico, did not institute its population policy until 1973, but through major public sector provision of birth control also achieved a halving of fertility rates within a 20-year period (Table 10.2). While the Mexican government has been less coercive than Puerto Rico in its approach, its tactics have been nonetheless persuasive. Operating under the banner of slogans such as *la familia pequeña vive mejor* (the small family lives better), fertility reduction goals have been linked with the aim of bringing about improvements in people's quality of life (Sayavedra, 1997: 95). Within this framework, concerns about gender have ostensibly been paramount, with Alba (1989: 13) noting that 'The government's strategy in promoting transformations in reproductive behaviour has been to emphasize not only the benefits of family planning but also the benefits of sex education and the enhancement of women's position in the family and labour market'. New guidelines for population policy in Mexico published in 1994 highlighted family planning as a means of exercizing reproductive rights and preventing health risks for men, women and children. However, it is interesting that only international and private institutions were consulted in the process. Furthermore, the guidelines reveal more concern about men's health and sexuality, and indicate a strong preference for permanent and semi-permanent methods used by women (Sayavedra, 1997: 97–8).

Most other Latin American countries have pursued what the Southern Feminist Network DAWN has categorized as 'combined policies', which refer to mild or non-explicit population programs (Corrêa and Reichmann, 1994: 24–5). Prominent examples include Brazil and Colombia where governments themselves have not actively promoted family planning (in both instances because of resistance from the Roman Catholic Church), but have permitted non-governmental and private sector organizations to operate in their countries. In Colombia, for example, 42.4 per cent of contraceptive provision is handled by the private sector, with the rest being provided by the government and the local associate of the International Planned Parenthood Federation, Profamilia.[3] An interesting factor is that this strategy has had similar results to cases where governments have been fully involved in family planning, with Colombia having witnessed a halving of fertility rates since the 1960s.

Similar trends have occurred in Brazil, even though only 28.3 per cent of family planning services is managed by the state, a further 3.4 per cent by nongovernmental organizations, and as much as 68.3 per cent by the private sector. Until the 1970s, Brazil had followed a pro-natalist policy, not only in the anticipation that this would be positive for economic growth and in the interests of occupying its large territory, but because of religious disapproval. As Corrêa and Reichmann (1994: 38) sum up, 'Long-standing state inaction with regard to Brazilian women's reproductive needs was replaced by the invisible hands of the market, which gradually dominated the supply of contraceptives'. By the 1990s, however, the state had fended off pressure from the Church and had in place a reproductive health program that recognized the constitutional right of people to control their own fertility. By the same token, female sterilization remains officially unrecognized as a mode of family planning, and, since it is not covered by the national insurance plan, tends to be performed clandestinely by doctors when women give birth by Caesarean section (Gomes, 1994: 71). The extent of this practice was such that, by 1991, 27 per cent of married women in Brazil aged between 15 and 44 years had been sterilized (Gomes, 1994: 71–2). As revealed by a study of domestic workers in Rio de Janeiro, sterilization is preferred by women who have given birth to their ideal number of children (normally between one and three), because of negative experiences with the pill (the other main form of contraception), limited knowledge of alternatives and male reluctance to use condoms (Pitanguy and de Mello e Souza, 1997: 91). In only 2 per cent of cases where couples practice 'modern' birth control do men use condoms, and only 1 per cent of couples have opted for vasectomy (Corrêa and Reichmann, 1994: 39).

Contraceptive technologies and user groups

As might be inferred from the above, the extent to which the language of 'rights' and 'gender equality' translates into practice is in some question when women in most parts of Latin America have been the main target group for a rather restricted range of contraceptive methods. Even in Bolivia, for example, where recent state family planning initiatives have been regarded as 'woman-friendly', there is no indication of male sterilization, and, as in Brazil, male condoms are used by only 1 per cent of couples (Hardon, 1997b: 25). Next to traditional methods women in Bolivia are most likely to use the intrauterine device (IUD). Among those women who do not use contraceptives, the main reasons given are concerns about health and side effects, and lack of knowledge (Hardon, 1997b: 27). Lack of female sterilization may in part be accounted for by the fact that the consent of spouses is required (Hardon, 1997a: 17).

The latter also applies to Mexico, although a 1995 report by the Mexican Ministry of Health indicated that, as of 1992, female sterilization accounted for one-third of 'contraceptive use', followed by the IUD and the pill. Male sterilization and use of the male condom, alternatively, were negligible (Sayavedra, 1997: 98). In Colombia, a similar pattern obtains, with vasectomy applying to only 1 per cent of couples, and condoms to 3 per cent (Corrêa and Reichmann, 1994: 39). The skew towards women and limited use of barrier methods not only impacts upon women's health but also has implications for the spread of HIV/AIDS and other sexually transmitted diseases.

Fertility, education and female employment

While falling fertility is clearly facilitated by increased availability of modern contraception, we have already noted that other factors, such as 'development' and economic status, may play an equally important role in limiting births. Related influences here include education and employment, with a broad relationship noted in Latin America between family planning programs, increased educational opportunities and rising female labor-force participation (Safa, 1995b: 39). Despite cutbacks in educational spending resulting from structural adjustment programs, women's rising enrolment in educational establishments has persisted into the 1980s and 1990s, with particularly marked increases occurring in secondary education (Table 10.4). The latter is important since there is often a significant gap in fertility between those who have not completed primary school and those who have had at least one year of secondary education (UN, 1995: 92). Although the relationships between access to education and fertility are extremely complex, it would appear that education is associated with improvements in infant and maternal health, greater awareness of the advantages of limiting family size and greater receptivity to modern contraceptive techniques (Buvinic, 1995: 5–6). Certainly, detailed survey research in Mexico suggests that younger and more educated women are strongly aware of the costs of raising children and putting them through school and are therefore using contraception to limit family size (García and de Oliveira, 1997).

As for links between fertility and women's work, although female labor-force participation in Latin America appears to have been encouraged by declining birth rates (see Roberts, 1995: 93 and 128), the nature of the relationship between the two variables may be occluded by the nature of available data. As pointed out in García's and de Oliveira's (1997) study of low-income and middle-class women in the Mexican cities of Tijuana,

TABLE 10.4 Women's education and economic activity rates in Latin America

Country	Female enrolment in primary education (% of age-group)		Female enrolment in secondary education (% of age-group)		Female economic activity rate age 15+ years (index: males=100)	
	1980	1993	1980	1993	1970	1994
Haiti	70	–	13	–	79	64
Nicaragua	102[a]	105	45	44	25	37
Honduras	99	112	31	37	17	27
Cuba	–	–	–	–	24	50
Bolivia	81	–	32	–	25	31
Guatemala	65	78	17	23	15	21
Ecuador	116	122	53	56	19	24
Dominican Republic	–	99	–	43	13	19
El Salvador	75	80	23	30	26	35
Paraguay	101	110	24	38	26	26
Colombia	126	120	41	68	26	28
Peru	111	–	54	–	25	32
Costa Rica	104	105	51	49	22	29
Panama	105	–	65	–	35	40
Venezuela	104	97	25	41	26	39
Brazil	97	–	36	–	27	38
Chile	108	98	56	70	27	39
Mexico	121	110	46	58	21	37
Uruguay	107	108	62	–	35	44
Argentina	106	107	60	75	33	38

Sources: UNDP (1995, Table A2.6) and World Bank (1996, Table 7).
Note: Countries are listed in ascending order of gross national product (GNP) per capita in 1994.
– = No data.
[a] Enrolment rates in primary schools may exceed 100 per cent because some pupils fall outside the standard age range for primary education.

Mérida and Mexico City, fertility histories are usually well-documented in macrostatistics, whereas data on employment may only be collected at one point in time. This has led to a situation where (possibly inappropriately) greater primacy has accrued to the effects of fertility *on* paid work than *vice versa*. Moreover, García and de Oliveira stress the importance of taking into account the mediating effects of women's socio-economic status on interactions between fertility and paid work. For example, while women with smaller numbers of children are in some cases more likely to engage in extra-domestic income-generating activities, the relevance of fertility levels to working-class women seems less applicable, especially during periods of economic adversity (García and de Oliveira, 1997: 368). For women in this latter group, motherhood remains their primary source of identity, with employment viewed primarily as a means by which they can better fulfill their mothering roles (see also Acero, 1997: 84; McClenaghan, 1997). Among middle-class mothers, on the other hand, who are able to afford paid childcare and who have sufficient education to enter nonmanual occupations, work is more likely to be a significant aspect of their personal identity (García and de Oliveira, 1997: 381).

Fertility, marriage and household structure

Just as much as there are likely to be two-way interrelationships between fertility and female employment, fertility also appears to intersect strongly with women's age at marriage and household structure, both of which have seen changes in the past 20 or so years and constitute important demographic phenomena in their own right. Although data on women's mean age at marriage are rather patchy, and wedlock is not necessarily a prelude to having children, it appears that female age at marriage increased between 1970 and 1990 (Table 10.5). Later age at marriage, which may in part be stimulated by rising education and labor-force participation, may delay fertility and thus contribute to reducing birth rates. In Mexico, for example, the median age at marriage for the population as a whole increased from 20.8 to 22 years between 1970 and 1990,

TABLE 10.5 Women's average age at marriage, and household headship, in Latin America

Country	Women's average age at marriage (years)		Percentage of households headed by women	
	1970	1990–92	1980	1990–92
Haiti	22.4	23.8	30.0	–
Nicaragua	20.2	–	–	24.3
Honduras	20.0	–	–	20.4
Cuba	19.5	–	28.2	–
Bolivia	–	22.7	–	24.5
Guatemala	19.7	21.3	–	16.9
Ecuador	21.1	22.0	–	–
Dominican Republic	19.7	–	21.7	25.0
El Salvador	19.4	22.3	–	26.6
Paraguay	21.7	21.5	18.1	17.0
Colombia	22.4	22.6	–	22.7
Peru	21.6	–	–	17.3
Costa Rica	21.7	–	17.6	20.0
Panama	20.4	21.9	21.5	22.3
Venezuela	20.4	–	21.8	21.3
Brazil	23.0	–	14.4	20.1
Chile	23.3	23.4	21.6	21.0
Mexico	21.2	–	15.2	17.3[a]
Uruguay	–	22.9	21.0	23.0
Argentina	22.9	23.3	19.2	22.3

Sources: UN (1995, Table 33; 1997, Table 26) and UNDP (1995, Table A2.5).
Note: Countries are listed in ascending order of gross national product (GNP) per capita in 1994; data are for the years shown or the nearest available year.
– = No data.
[a] This figure is taken from the Mexican Census of 1990.

largely due to a decline in early marriages (López and Izazola, 1995: 12). In turn, the average size of family-based households (male-headed and female-headed) fell from 5.4 to 5 between 1960 and 1990 (López and Izazola, 1995: 15).

In Latin America more generally, the average number of persons in the largest urban households dropped from 5.5 to 4.9 during the 1980s, although there was little change in the smallest households (from 3.4 to 3.2 people) (ECLAC, 1994b: 76). Aside from fertility reductions, this also owes to the falling incidence of extended households (ECLAC, 1994b: 76). Between 1960 and 1990, for example, the proportion of extended households in Mexico dropped from 9.3 to 8 per cent. This is attributed partly to better health among old people and unfavorable conditions for multigenerational households in cities, which may have prompted a more independent lifestyle. Indeed, whereas 9 per cent or more of the population in the 65 year plus age-group in Mexico live alone, this is only 1 per cent across the population as a whole (López and Izazola, 1995: 56). Having said this, it is difficult to generalize this trend across different socio-economic groups, since detailed longitudinal studies of low-income communities in Mexico (and Ecuador) indicate a rise in the proportion of extended households in the crisis years (Chant, 1996; González de la Rocha, 1988; Moser, 1992).

Along with the increase in people living alone, another important factor in reducing household size is the growth of single-parent families, most of which are headed by women (ECLAC, 1994: 76). Although lone-mother households are not the only type of female-headed unit, numbers of women-headed households rose in most countries during the 1980s to reach around 20 per cent in the region as a whole (Safa, 1995b: 32; see also Table 10.5). While there are many factors involved in changes in household size, composition and headship (Chant, 1997a), migration and employment have both played a part in this, as explored in the following sections.

MIGRATION

Mirroring the dynamism in fertility, mortality and household patterns in Latin America is the fact that the population in the region is highly mobile. Gilbert (1998: 39) observes that this comprises a number of internal and international movements which stretch back centuries, the only real difference over time being the 'direction and size of the migration flows'. This said, although the progressively urban concentration of the population (now around 70 per cent) suggests that permanent migration to towns and cities has prevailed, some caution is required since many Latin American censuses fail to capture short-term movements across small distances. In Mexico, for instance, the census only documents interstate or international migration (López et al., 1993: 134). Although in Costa Rica the census records moves between smaller administrative units (namely 'cantons', the intermediate administrative entity between provinces and districts), people are only registered as having migrated where they are living in a canton different from their place of birth or to where they resided five years prior to the census enumeration (Chant, 1992: 50).[4] Although longitudinal community-level case studies can reveal much more of the complexity of migratory patterns, the problem then becomes one of extrapolation (Durand and Massey, 1992; Radcliffe, 1991a).

Bearing these problems in mind, in the following sections I attempt to identify the major characteristics of migration in Latin America in the past few decades, looking specifically at rural–urban migration, temporary movement and international flows.

These discussions are preceded by a brief review of theoretical approaches to migration and some of the underlying factors in population mobility in Latin America in general.

Approaches to migration and general factors underlying migration in Latin America

The general literature on migration and development is threaded with a wide range of theories relating to distance, direction and duration of migratory movement. In terms of overarching conceptual perspectives, the mainstream models have concentrated on what is widely taken to be the dominant type of migration, namely permanent labor migration from rural to urban areas. Three of the most significant mainstream models include:

- neo-classical/equilibrium approaches which regard migration as a rational individual response to wage-rate differentials;
- structuralist/Marxian approaches which focus on the moves impelled by spatial redistribution of economic activity, and
- structuration approaches which attempt to strike more of a balance between structural constraints and microsocial perspectives/individual agency (see Chant and Radcliffe, 1992).

Although there is value in both of the first two models in respect of explaining both short-term and long-term rural–urban mobility in Latin America, the importance of incorporating both structural and behavioral forces in the analysis of migration has increasingly been borne out by empirical studies and further emphasized with the development of the so-called 'household strategies approach' to migration. Based on the notion of the domestic unit as 'an analytical instance where the macro-structural conditions facilitating the migratory flows and family and individual determinants converge' (de Oliveira, 1991: 111), the household strategies approach views the organization of household livelihood and reproduction as central in structuring mobility, within which issues of power, ideology and identity differentiated by gender, age, marital status and so on are critical (see Radcliffe, 1991a). In adopting a more holistic view of mobility, the household strategies approach has proved of greater use in capturing the multifaceted nature of migration and in demonstrating that even if employment is a major factor, economic imperatives are strongly interwoven with social, family and cultural considerations. The household strategies approach has also been helpful in evaluating dimensions of migrant selectivity, such as age and gender.

Whatever the particular orientation of different theoretical approaches to migration, labor considerations are clearly a vital component in all of them and help to explain the rural–urban transition in Latin America in the post-war period. In a context in which most Latin American governments have promoted policies encouraging urban-based industrialization and have neglected investment in rural areas, people have moved to towns as a result of declining opportunities for livelihood in rural areas. The latter owes to the concentration of land, to the mechanization of agriculture, to unequal competition between large and small farmers and to environmental degradation. In turn, towns usually offer more employment and higher wages. In the Mexican state of Querétaro in the early 1980s, for example, minimum wage levels were 20 per cent higher in the capital of the state than in the surrounding countryside and represented an important motive for people's move to the city (Chant, 1991a: 33). Beyond this, towns and cities are usually better equipped with education and health facilities, thereby attracting people with young families (Chant, 1992; Moßrucker, 1997). As far as international migration is concerned, political factors and refugeeism figure alongside economic imperatives, again underlining the importance of holistic conceptual frameworks that recognize the variable and multidimensional nature of migration across space and through time.

Rural–urban migration

Between 1950 to 1980, 27 million people in Latin America left their farms and villages for the cities of the continent (Green, 1996: 19). While it is not always clear how much of this migration can be regarded as permanent, it was clearly a major source of urban growth during this time. Indeed, urban populations in Latin America grew at an average annual rate of 4.8 per cent between 1950 and 1980, bringing the overall urban population of the continent from a level of 40.9 per cent to 63.3 per cent (Gilbert, 1990: 44; Safa, 1995b: 32).

Although the urban population is expected to reach 75 per cent in the year 2000 (Sen, 1994: 9) and rates of urban growth in the 1980s and 1990s have continued to exceed those of national populations (compare Table 10.1 with Table 10.6), rural–urban migration has contributed less to urban expansion over time. This, in turn, marks the continuation of a process witnessed in some larger cities in Latin America in earlier decades. In Rio de Janeiro, for example, nearly 70 per cent of growth in the period 1940–50 was accounted for by migration, whereas this fell to 42 per cent between 1960 and 1970 (Roberts, 1995: 93–4). In Caracas, Venezuela, while 75 per cent of the population of the city in 1981 aged 65 years or more were migrants, this was only 49 per cent for the 25–34 year age-group, and a mere 14 per cent among persons under 15 years old (Gilbert, 1990: 49–50). Part of the reason for the decline in migration owes to higher rates of natural increase in urban areas, especially in the largest cities. This itself is a product of migratory legacies, since young people have long formed the bulk of rural–urban migrants. Indeed, from the 1970s onwards, natural increase has tended to override migration in urban growth in Latin America as a whole, and in the biggest cities, which are likely to gain most population increases *in situ*, the growth contributed by rural–urban migration is presently minimal (Roberts, 1995: 90 and 93.).

Set against this trend, there have also been diminishing levels of rural–urban migration *per se*, especially from the 1980s onwards. This is partly because the proportion of people living in rural areas is much reduced compared with the early post-war period, and partly because of lessening disparities between countryside and town. As Roberts (1995: 112) summarizes:

> the legacy of past movements and the commercialization of most rural areas mean that the distinction between rural and urban is often not a great one. Economic enterprise spans rural and urban locations. The pattern of consumption of the village may be different in scale to those of the town, but they are not different in kind.

Roberts's argument is echoed by Gilbert (1998: 50) who further observes that as a result of improvements in transport and communications, 'the city has absorbed the country', with people in places up to 50 miles or more from towns now able to make their living by selling their produce to wholesalers who supply urban markets. Beyond this, it is important to acknowledge that interactions of urban migrants with their rural relatives

TABLE 10.6 Latin America: population and urbanization

Country	Urban population as percentage of total population		Average annual growth rate of urban population (%)		Population in urban agglomerations of 1 million or more in 1990 as percentage of urban population		Growth rate of largest city 1990–95 (%)
	1980	1994	1980–90	1990–94	1980	1994	
Haiti	24	31	3.9	4.0	55	56	3.9
Nicaragua	53	62	3.9	4.2	23	28	4.3
Honduras	36	47	5.4	4.9	0	0	–
Cuba	–	–	–	–	–	–	1.1
Bolivia	46	58	4.2	3.2	14	17	3.6
Guatemala	37	41	3.4	4.0	0	0	2.3
Ecuador	47	58	4.2	3.4	14	26	2.8
Dominican Republic	50	64	4.1	3.1	25	33	3.2
El Salvador	42	45	1.9	2.7	0	0	–
Paraguay	42	52	4.8	4.4	0	0	–
Colombia	56	64	2.8	2.7	22	28	2.9
Peru	65	72	3.0	2.6	26	31	2.8
Costa Rica	43	49	3.8	3.3	0	0	2.9
Panama	50	54	2.8	2.7	0	0	2.8
Venezuela	83	92	3.5	2.9	16	27	1.3
Brazil	66	77	3.3	2.7	27	32	1.5
Chile	81	86	2.1	1.8	33	35	2.0
Mexico	66	75	2.9	2.8	27	28	0.7
Uruguay	85	90	1.0	0.9	42	42	0.6
Argentina	81	88	1.9	1.6	31	35	0.7

Sources: UNDP (1997, Table 21) and World Bank (1996, Table 9).
Note: countries are listed in ascending order of gross national product (GNP) per capita in 1994.
– = no data.

have played an important part in underpinning livelihood in rural areas. In Peru, for example, migrants moving from Santiago de Quinches to Lima, just over 100 miles north of their village, make frequent visits back to Quinches with clothing, rice and sugar (Moßbrucker, 1997: 76). In fact, given that rural relatives often mind land for their migrant kin, and, in the specific context of Quinches, make return gifts of sheep, corn and potatoes, it could also be argued that it is only by retaining rural links that *urban* survival is possible.

In addition to the above processes, there is evidence to suggest that economic restructuring in the 1980s and 1990s has reduced the pace of rural–urban migration, at least in the medium term (Becker and Morrison, 1997: 98). This is mainly because structural adjustment programs are noted to have led to a relative increase in poverty in urban versus rural areas (Stewart, 1995: 23–4). Reasons given for the growing incidence of urban poverty include the way in which structural adjustment programs (SAPs) have given rise to declining wages and employment in towns and cities through cutbacks in public expenditure and government bureaucracies (UNCHS, 1996: 117). SAPs have also promoted export agriculture which, combined with reduced subsidies on basic foodstuffs, have tended to favor rural over urban dwellers, mainly because capacity for self-provisioning from subsistence farming has given some protection against rising consumption costs (Demery et al., 1993: 3–4). Other factors which have been important in narrowing the gap between rural and urban poverty include the rising concentration of people in urban areas over time. This, amongst other things, has transferred poverty into towns and cities and exacerbated the strain on urban housing and job markets (Amis, 1995: 146). Notwithstanding the difficulties of conceptualizing urban poverty as distinct from rural poverty (Wratten, 1995), people in cities also face greater threats of poverty as a result of higher costs of living, greater dependence on cash income, a lower base of 'intangible assets' such as kinship networks, and greater exposure to environmental and health hazards arising from higher population densities, inadequate provision of drainage, sewerage and waste disposal and proximity to polluting industries (UNCHS, 1996: 111; see also Satterthwaite, 1995).

Whatever the particular constellation of events in specific countries, in Latin America in general, poverty is currently more concentrated in urban than rural areas, with 55–60 per cent of the total poor residing in cities in 1990, compared with 46 per cent in 1980 and 37 per cent in 1970 (Koonings et al., 1995: 117). While the relative incidence of poverty is still lower in cities than in the countryside, levels of urban poverty have risen more acutely in the past two decades (González de la Rocha, 1997: 5). In one or two countries, rural areas have even experienced a relative decrease in poverty. In Bolivia, for instance, the proportion of the population below the poverty line in urban areas rose from 31.1 per cent to 53 per cent between 1980 and 1989, whereas in rural areas this declined from 81.3 per cent to 76.2 per cent (Stewart, 1995: 180). In light of narrowing gaps between rural and urban poverty, Gilbert (1998: 52) notes that recession has given rise to greater interdependence between rural and urban areas, with city dwellers becoming more dependent on kin in the countryside, especially for foodstuffs.

Migration and 'secondary cities'

Notwithstanding the fall-off in rural–urban migration, it is also important to note that the growth of smaller and medium-sized cities are still likely to be fuelled by in-migration. Until the 1970s, most migrants from the countryside tended to settle in large, and particularly primate, cities, usually capitals or major ports which contained a disproportionate concentration of the national urban population. However, from the 1980s onwards, there has been greater movement by migrants from both the country and big cities to 'secondary' urban centers (Portes, 1989; Roberts, 1995: 90). In Chile, Gilbert (1994: 52) observes for Santiago that as many people are estimated to have moved out of the city as having moved in during recent decades. Thus, even if the proportion of the urban population living in cities of one million or more has risen between 1980 and 1994, this is more to do with increased numbers of large cities than the continued growth of primate centers (Table 10.6). Whereas back in 1900 there were no 'million cities' in any part of the continent, by 1990 the total with a million or more inhabitants was 38, including eight with over two million, and four with over 10 million (Ferguson and Maurer, 1996: 118; UNCHS, 1996: 48). Nonetheless, it should be noted that urban dispersion is more common in larger, more urbanized countries than in smaller ones such as Costa Rica. Still a predominantly rural-based economy, most of Costa Rica's new export industrial zones are located close to the capital San José, meaning little discernible reduction in urban primacy in recent years. Indeed, if anything, San José is currently turning into Central America's first 'mega-city' (Lungo, 1997: 57).

Government concern to promote secondary urban settlement has often taken the form of policies encouraging economic decentralization. For example, in order to spread growth more widely over the country, successive Peruvian administrations between 1958 and 1990 granted firms lower profit taxes for locating outside Lima, with the greatest reductions given to companies locating in the jungle region, the second greatest to those settling in the highlands and the third level to those operating on the coast outside Lima (Becker and Morrison, 1997: 96). Nonetheless, market-driven macro-economic policies and change have usually been more significant than specific government attempts to (re)direct migration. Secondary urbanization has been driven first and foremost by industrial relocation stemming from diseconomies of scale in metropolitan areas and/or the development and expansion of new economic activities associated with restructuring such as international tourism and export manufacturing (Gwynne, 1996a). This is certainly the case in Mexico, where Roberts (1995: 91) notes the occurrence of net out-migration from Mexico City during the 1980s to smaller cities within a 200 km radius, such as Querétaro, Toluca and Puebla, and to core cities of the *maquiladora* industry on the northern border such as Tijuana and Ciudad Juárez.

In some cases, movement to smaller towns has also been prompted by crisis-induced job losses in primate cities. In Mexico City, for example, it was estimated that crisis forced the closure of 6000 manufacturing companies between 1980 and 1988, carrying a loss of 250 000 industrial jobs (Gilbert, 1998: 51). By the same token, smaller cities have by no means escaped the ravages of recession, especially where their economies have depended heavily on the import of foreign components and technology. For example, in Querétaro, Mexico, around one-quarter of the industrial labor force lost their jobs between 1982 and 1983. Although this was partly as a result of lay-offs in construction as national and municipal authorities called a halt to public works programs, many retrenchments were in the (largely-foreign-financed) metal mechanical industry (Chant, 1991a: 61). Nonetheless, tendencies towards metropolitan deconcentration more generally in Latin America have been such that, Gilbert (1995a: 322) argues, 'The recession [has] achieved in ten years what attempts at regional development had failed to do in thirty'. Besides economic reasons for movement away from primate cities, there is also evidence to suggest that middle-class groups are moving to secondary centers as a result of concerns about the environment and quality of life (Izazola Conde and Marquette, 1995). This is hardly surprising given that although the worst environmental conditions tend to be concentrated in poor neighborhoods where there is usually inadequate sanitation and water supply, these, along with air pollution, ultimately affect all urban dwellers (UNCHS, 1996).

Having noted that the contribution of migration to urban growth has gradually been overtaken by natural increase, the latter is in many ways a product of the former, in that younger and better-educated individuals have traditionally been a disproportionate presence in rural–urban movements and have thus led to a youthful age structure in cities (Gilbert, 1990: 46). Although the depletion of migrant pools in the countryside and the progressive establishment of relatives in cities have led to some decline in this aspect of migrant selectivity over time, there are still quite pronounced differences in proportions of women and men in different types of migratory movement.

Gender selectivity in rural–urban migration

As far as long-term rural–urban migration is concerned, women have formed the bulk of population flows in post-war Latin America. In Mexico, women have been an especially significant presence in migration to large urban centers and metropolitan areas (de Oliveria, 1991: 101). This is also the case in countries such as Peru (Radcliffe, 1992), Honduras (Bradshaw, 1995b) and Costa Rica (Chant, 1992). Gender selectivity of migration has resulted in higher percentages of women among the population in major towns and cities (Gilbert, 1998: 47). Although there is some evidence of declining levels of 'feminization' over time, national averages in the late 1980s and 1990s continue to reflect the legacy of female bias in rural–urban migration flows (see Table 10.7, and compare with Chant and Radcliffe, 1992: 6, Table 1.2).

Female-selective migration in Latin America has often been explained by the fact that labor opportunities for women are considerably greater in towns than in the countryside. Indeed, whereas women's share of employment in urban areas typically ranges between 25 and 35 per cent, they are usually below 20 per cent of the workforce in rural areas. In Costa Rica, for example, where women are recorded as only 6 per cent of the rural labor force in government statistics, none of the principal agricultural activities (coffee production, banana growing, sugar cane cultivation and cattle farming) employ women in any significant number, although the coffee harvest generally recruits women and children alongside men in order to facilitate the need to pick the crop

TABLE 10.7 Urban sex ratios in selected Latin American countries

Country (year for which latest data are available)	Men per 100 women[a]	Percentage of all men in urban areas	Percentage of all women in urban areas
Nicaragua (1989)	95	58.1	61.4
Honduras (1988)	84	37.6	41.2
Cuba (1993)	97	73.0	75.8
Bolivia (1992)	94	56.6	58.5
Ecuador (1993)	97	57.9	60.4
Dominican Republic (1995)	98	59.9	63.5
El Salvador (1992)	90	49.1	51.7
Paraguay (1992)	93	48.3	52.4
Peru (1993)	97	69.4	70.8
Costa Rica (1994)	96	42.9	45.2
Panama (1995)	95	52.9	56.9
Venezuela (1990)	99	83.0	85.0
Brazil (1991)	94	74.3	76.9
Chile (1995)	95	83.1	85.8
Uruguay (1995)	91	88.1	92.0

Sources: UN (1997, Table 6).
Note: Haiti, Guatemala, Mexico, Colombia and Argentina have been omitted owing to lack of data and/or their non-inclusion in the UN table; countries are listed in ascending order of gross national product (GNP) per capita in 1994.
[a] Figures rounded up to nearest whole number.

within a three to four week period (Chant, 1997a: 131). Lest the greater availability of employment for women in towns is viewed as an unequivocally positive 'pull' factor, however, it is important to recognize that, in reality, work opportunities may be quite limited. Radcliffe (1990b: 30) notes that one of the key occupations for migrant women in Peruvian cities, domestic service, is 'only slightly better an occupation than begging and prostitution'. Not only is it low paid, but hours are often long, and women's social and recreation time is extremely limited. Bearing in mind that the interaction of demographic factors with urban economic organization makes labor markets in Latin American cities 'highly competitive and ... highly segmented' (Roberts, 1995: 127), it is unlikely that labor markets will become less differentiated or that gender selectivity will diminish in the wake of continued economic liberalization.

Over and above employment, other important factors coming into women's migration decisions include conjugal and family relationships. For example, female household heads who have been widowed or who have fallen victim to conjugal breakdown are particularly likely to move to urban areas given the social and economic difficulties of living as lone women or lone mothers in rural areas. Indeed, in Latin America in general, levels of female household headship are usually below 10 per cent in the countryside, but are often 25 per cent in towns (Browner, 1989: 467). In turn, increases in female headship are sometimes precipitated by the seasonal out-migration of men from rural areas, with short-term *de facto* female headship converting to permanent, *de jure* status where long spells of separation result in lapsed remittances, emotional stress and conjugal infidelity (Chant, 1998).

Short-term migration

The preponderance of permanent rural–urban migration in Latin America should not negate the fact that there are other important population flows involving different destinations and periodicities, even if census data make it difficult to capture the precise extent of this movement.

Short-term/circular migration has a long history in many parts of Latin America, such as the Peruvian Andes (Skeldon, 1990: 49 *et seq.*), and Honduras, Belize, El Salvador, Costa Rica, Venezuela and Ecuador (Bailey and Hane, 1995: 173 and 179). This is also true of Mexico, where temporary migration typically involves one or two members of a household at a time (Arizpe, 1982). Roberts (1985) argues that the two main goals of temporary migration are risk minimization and income maximization. Risk minimization is effected by the fact that decanting different household members to different labor markets diminishes the vulnerability attached to dependence on work in one locality. Temporary migration can be a means of maximizing income insofar as wages in higher-earning areas can often purchase more in low-wage areas. In the case of Costa Rica, wages earned in San José stretch further in Guanacaste because of cheaper costs of living and the fact that land and shelter there are more accessible and affordable (Chant, 1991b).

Temporary migration varies in duration, with Radcliffe's (1992) village-based work in the southern Peruvian Andes indicating that teenage daughters and single women in their early 20s often migrate for one or more years to work as domestic servants in urban areas. In many cases this is because they have no place in agriculture and/or are 'surplus' to household labor requirements in areas of origin. However, while women have been dominant in permanent and longer-term migration, men are more likely to be represented in short-term/seasonal mobility. For example, in rural areas of the municipality of Yuscarán in southeastern Honduras, women's higher rates of permanent migration to towns and cities are reflected in the fact that only 48.6 per cent of the population is female. Nonetheless, of the 8.1 per cent of the population who migrate on a temporary basis (with two-thirds of these returning every two weeks), as many as 66 per cent are male (Paolisso and Gammage, 1996: 25).

Similar patterns prevail in small towns in Guanacaste in northwest Costa Rica, where most low income men work in agriculture. Farming is highly seasonal in this region and most of the male populace can only be assured of finding employment during the sugar harvest between January and April. At other points in the year, therefore, many have to migrate out of the province to find work such as clearing pasture, working on banana plantations in the Osa Peninsula or the Atlantic Coast or picking coffee during the harvest period in the Central Valley (Chant, 1992).

International migration

Recognizing that until early this century Latin America was a major destination for international migrants, most international movements in the twentieth century have been *within* the continent or *from* the region to North America. The United States stands out as the most important country of destination, mainly for migrants from Mexico and other parts of Central and Meso-America. In the Mexican case, this reflects an historical legacy of annexation of half of Mexico's territory in the mid-nineteenth century, which Gledhill (1997b: 1) notes 'turned mobile workers into international migrants'. In turn, Green (1991: 59) reports that by 1990 an estimated 25 million people in the United States were 'Hispanic', with around half having come from Mexico. Although undocumented international migration makes it extremely difficult to pinpoint the exact size of migratory flows (Diaz-Briquets, 1989: 33), Dominicans are another sizeable group in the United States. Sørensen's (1985) study of Dominican migration, for example, reveals that about 15–20 per cent of the population has been involved in migration in recent years, with between 500 000 and 1 million living in the United States (mainly in New York City). As noted for the case of rural-urban migration, interactions with source and destination areas tend to remain very active, even across long distances, and can comprise substantial monetary transfers. In the Dominican case, for example, the aggregate value of migrant remittances in the mid-1980s amounted to nearly as much foreign exchange as that generated by the country's sugar industry.

As for intraregional flows of international migrants in Latin America, Argentina and Venezuela are important destination countries, with around 5 per cent of their respective populations in 1990 being foreign-born (see Thomas, 1996: 94). Whereas the origins of migrants to Argentina are quite diverse, including nationals from Bolivia and Paraguay (Green, 1991: 59), in Venezuela this is mainly restricted to Colombians, of whom half a million (around 4 per cent of the population) have been displaced by land reform, civil violence revolving around drugs and political factionalism and/or due to the desire to escape military conscription (Morris, 1995: 80–1). Costa Rica has also been an important recipient of cross-border migrants from Nicaragua, El Salvador and Panama. During the war in the region in the mid-1980s, between 4.5 and 7.5 per cent of the population was estimated to be from outside the country (Diaz-Briquets, 1989: 38). Although some of these immigrants were political refugees, many also had migrated to Costa Rica for economic reasons such as greater availability of jobs or lower costs of living (Chant, 1992). By the same token, migration *from* Costa Rica is also noted, especially among Afro-Caribbean males in Limón on the Atlantic coast. Limited occupational mobility among this group of men in a racially segmented labor market means that overseas migration (usually to work on cruise liners in Miami) is often the only route to enhancing their economic position (McIlwaine, 1997).

Conceptual frameworks for international migration have much in common with theories of rural–urban movement insofar as they emphasize (in varying degrees) wage differentials between source and destination areas, the relative risks attached to displacement, recruitment efforts and market penetration (Massey et al., 1993: 432 et seq.). Significantly, holistic approaches have again proved of most utility in revealing interconnections between different catalysts and motives underpinning international migration. This is indicated by studies of other Central American countries such as El Salvador, which in percentage terms in the region is 'probably the leading exporter of people' (Green, 1991: 58). Around one-fifth of the population is estimated to have left the country, and, when taking into account figures of temporary exiles and return migrants, this brings the level to between one-quarter and one-third (Bailey and Hane, 1995: 178). Most Salvadorean movement has been to the United States and Mexico, with large flows prompted by the lead up to civil war in 1981. However, Bailey and Hane (1995: 171) suggest that viewing Salvadoreans overseas as *either* refugees *or* economic migrants fails to capture the complexity of people's movement, especially given the wide range of migratory flows. The latter includes short-term circulation, longer-term return migration and permanent emigration, with a strong interrelatedness of economic and political forces encouraging out-migration in general (and in many cases discouraging repatriation). Noting that Salvadoreans have engaged in international migration throughout the twentieth century, Bailey and Hane (1995: 179) argue that political crisis in the late 1970s merely 'amplified Salvadorean international mobility'. More generally, economic crisis itself is also associated with a rise in international migration, especially from Mexico and other parts of Central America to the United States, not only because of declining work in home areas but because of household needs for extra income (Cornelius, 1991; Gledhill, 1995; Roberts, 1995).

As in debates on migrant selectivity in rural–urban migration, networks and cumulative causation are critically important in understanding the evolution of migration flows over time. As Durand and Massey (1992: 17) summarize, 'migration is a dynamic, developmental process in which decisions made by migrants at one point affect the course and selectivity of migration in later periods'. While first-stage migration from Mexico to the United States, for example, seems to draw preponderantly from lower-middle income groups of the population (as opposed to the poorest) reflecting the expenses attached to international movement, with time the costs and risks diminish as migrant networks are established in the United States, as well as the fact that migrant pools are reduced in source areas. Moreover, remittances may raise the class status of family back home (Durand and Massey, 1992: 17–19). As for age and gender, adult men tended to figure more prominently in earlier migration flows, with women and children following in later periods (Durand and Massay, 1992). By the same token, not all families are able to maintain cohesiveness over several years of migration, with Gledhill (1997a: 9) reminding us that '"flexible labour markets" and high international mobility of capital can have patently disruptive effects on families and communities'.

Employment

As indicated in preceding sections, patterns of employment in Latin America in the second half of the twentieth century have to be conceived within the context of a major demographic and economic transition that has given rise to a situation in which labor markets in Latin America are now predominantly urban. Related to this, employment has also been radically transformed by economic globalization in the past two decades, which according to Ward and Pyle (1995: 38) is marked by three main trends:

- a shift to export-orientated economic growth strategies under the influence of International Monetary Fund (IMF) and World Bank loan conditionalities;
- the globalization of the production and marketing operations of transnational companies;
- debt crises and recessions.

Additional factors include shifts in terms of trade and technological change, and lessening intervention of the state in economic and labor matters (Berry, 1997: 3; Sheahan, 1997: 8). In light of the multiplicity of these trends, Tokman (1989: 1071) makes the point that 'The recent economic crisis makes it even more difficult to evaluate the overall changes in the labour market in Latin America, since technological change and structural adjustment policies have occurred simultaneously with recession and declining incomes'. Although it is also hard to generalize about employment across the region, economic crisis and restructuring in the 1980s and 1990s seem to have had reasonably consistent effects on urban labor markets. The main changes include rising levels of unemployment and underemployment, deteriorating wages and working conditions, an increased supply of labor (owing to pressures on household incomes) and mounting numbers (and proportions) of informal sector workers. Given the dramatic changes which have occurred in respect of women's labor-force participation during this time, the discussion below is followed by a section which deals more specifically with gender, employment and household survival in Latin American cities.

Unemployment and underemployment

In general terms, the most troubled years for employment in most Latin American countries were the initial stages of the debt crisis in the early 1980s. Between 1980 and 1985, for example, underemployment in the region grew by 48 per cent (Safa, 1995b: 33). As for open unemployment, this rose even more precipitously, jumping to 14 per cent of the economically active population in 1984, from around 6 per cent 10 years earlier (Cubitt, 1995: 164). Moreover, levels for individual countries during this time were sometimes considerably higher. During Chile's depression of 1982–83, unemployment reached 28 per cent, compared with 16 per cent in the period 1975–81 (Sheahan, 1997: 15).

Even if rates of open unemployment by the early 1990s had declined in most parts of Latin America to between 4 and 7 per cent (ECLAC, 1994b: 28), figures for a number of countries suggest reverse tendencies. In Argentina urban unemployment in 1991 was 20.2 per cent compared with 5.6 per cent in 1986, and in Nicaragua 23.5 per cent of the labor force was unemployed in 1994 compared with 4.5 per cent in 1986 (Bulmer-Thomas, 1996b: 326). In the Nicaraguan case, the steep rise in unemployment occurred after the opening up of the economy to global financial institutions. When Violeta Chamorro became president of Nicaragua in 1990, IMF pressure resulted in the firing of 250 000 public sector employees in the next three years (Green, 1995: 56–7). Elsewhere in Latin America, the public sector also came under attack as international financial institutions pressurized governments to trim down their 'unwieldy bureaucracies' and to privatize parastatal enterprises. In Bolivia, for example, three-quarters of workers in the state mining company, COMIBOL, were dismissed following implementation of the country's 'New Economic Policy' in 1985 (Jenkins, 1997: 113). More generally, the period 1989–92 saw the share of public employment in the workforce decline from 22.5 to 19.5 per cent in Venezuela, from 21.8 to 18.7 per cent in Uruguay and from 18.3 to 15.5 per cent in Bolivia (ECLAC, 1994b: 24). Fortunately, the upturn in private waged employment in the early 1990s went some way to cushioning losses in public sector employment.

Despite the overall decline in unemployment by the 1990s, it is important to note that for certain groups in the population, current prospects are grim. Unemployment rates continue to be extremely high among the poor, and are also mounting among young people aged 15 to 24 years who have between 6 and 12 years of schooling (that is, not the least educated) (ECLAC, 1994b: 31). While open unemployment stood at 18.6 per cent in Panama in 1991, the level for 15–24 year olds was 35.1 per cent, and even in Costa Rica, which has much lower rates of open unemployment (4.2 per cent in 1992), the proportion among young persons was 9 per cent (ECLAC, 1994b: 144). These situations partly reflect the large numbers of people in this age group and partly the increased economic participation of women (ECLAC, 1994b: 31).

Wages and working conditions in the formal sector

In addition to job losses in the formal sector, the 1980s and 1990s have also seen considerable changes in working conditions, particularly in manufacturing. In general terms, this has comprised a greater incidence of subcontracting and short-term contracts, the restriction of trade union activities and the introduction of policies geared to ease processes of hiring and firing. In many cases, these changes have resulted from the pressure exerted by international financial institutions to reduce 'structural rigidities' in the workforce and to encourage greater labor 'flexibility' (Green, 1996: 109–10; Tironi and Lagos, 1991). In Bolivian manufacturing, for example, Jenkins (1997: 119) notes that there has been an increasing concentration of production in small-scale factories and workshops, and a doubling of the percentage of workers working 49 hours per week or more. In Mexico, shoe manufacturers in the city of León have farmed out increasing amounts of production to small-scale home-based workshops (Fig. 10.2) and to individual outworkers as a means of flexibilizing their operations and cutting labor costs (Chant, 1991a).

The paring-down of 'structural rigidities' has also been accompanied by labor law revisions. The New Economic Policy in Bolivia, for example, brought reductions in protection for workers and an elimination of wage indexation leaving wage levels to be bargained within individual firms (Jenkins, 1997: 113). In Peru, legislation was introduced in 1991 which gave the right to employers to hire people on 'probationary' contracts during which time employees' entitlement to fringe benefits is minimal and where no compensation is payable on their release (Thomas, 1996: 91). By the same token, it is important to note that such practices were by no means absent in large firms before the

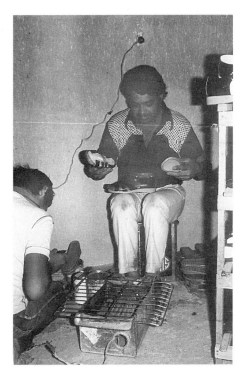

Figure 10.2 Home-based shoe workshop, León, Mexico. Photograph by Sylvia Chant.

crisis. As Roberts (1991: 118) maintains, with reference to Mexico, '"implicit deregulation" ... antedates by many years the present policy of explicit deregulation'.[5]

Wage restraints have also formed an important part of formal sector restructuring, which, coupled with inflationary costs of living, have meant negative growth rates in average real earnings in many countries in the 1980s (Table 10.8). Indeed, in general terms, the average industrial wage in Latin America fell by 17.5 per cent between 1980 and 1991, and the average minimum wage by 35 per cent (Moghadam, 1995: 122). Thomas (1996: 90–1) further notes that between 1985 and 1992, urban real minimum wages declined in all countries in Latin America except Colombia, Costa Rica, Chile, Panama and Paraguay. In Argentina, for example, the mean real wage in 1995 was only 68.5 per cent of its value 10 years previously (Bulmer-Thomas, 1996b: 325).

Wages have often been held down with the agreement of trade unions, whose bargaining strength has tended to decline with crisis and the reduction of formal sector activity in many parts of Latin America. As Koonings *et al.* (1995: 123) observe, 'Informalization and endemic poverty pose a major threat to the capacity of trade unions to organize and defend the working population'. While large-scale highly-unionized industries in key sectors of the Venezuelan economy have managed to weather the crisis triggered by falling oil revenues in the 1980s, they have been unable to halt the general decline in real wages (Méndez-Rivero, 1995: 158). Part of the problem faced by trade unions is that although the consolidation of democracy in many Latin American nations has been achieved with their active participation, the process has

TABLE 10.8 Latin America: labor force and earnings

	Size of labor force (millions)		Average annual growth rate of labor force		Annual growth rate in real earnings per employee	
	1980	1994	1980–90	1990–94	1970–80	1980–92
Haiti	3	3	1.3	1.9	–	–
Nicaragua	1	2	2.7	4.5	–2.0	–
Honduras	1	2	3.4	3.7	–	–
Cuba	–	–	–	–	–	–
Bolivia	2	3	2.6	2.6	1.7	–0.8
Guatemala	2	4	2.9	3.5	–3.2	–1.6
Ecuador	3	4	3.5	3.1	3.3	–0.7
Dominican Republic	2	3	3.1	2.7	–1.1	–
El Salvador	2	2	2.1	3.3	2.4	–
Paraguay	1	2	3.1	3.0	–	–
Colombia	9	15	4.0	2.6	–0.2	–
Peru	5	8	3.2	3.0	–	–
Costa Rica	1	1	3.8	2.8	–	–
Panama	1	1	3.1	2.6	0.2	2.0
Venezuela	5	8	3.4	3.1	4.9	–5.4
Brazil	48	71	3.2	1.9	5.0	–2.4
Chile	4	5	2.7	2.2	8.1	–0.3
Mexico	22	35	3.5	2.9	–	–
Uruguay	1	1	1.6	1.0	–	–2.3
Argentina	11	13	1.3	2.0	–2.2	–2.1

Sources: UNDP (1997, Table 16) and World Bank (1996, Table 4).
Note: countries are listed in ascending order of gross national product (GNP) per capita in 1994.
– = no data.

tended to delink them from the state, thereby depriving them of an important source of support. Chile is one exception here, where Frías and Ruiz-Tagle (1995: 141) argue that the institutional strength and stability of trade unions have increased since Aylwin came to power in 1990 marking the end of a 17-year period of military rule. In Brazil, too, the strong links of unions with powerful political factions are also argued to have given some protection (Thomas, 1996: 91). Interestingly, however, against a relative reduction of union activity in the formal sector, informal entrepreneurs have shown greater signs of organization (Koonings et al., 1995: 119). Part of this has undoubtedly been motivated by the need to survive against greater odds in a disadvantaged, but expanding, sector of the regional labor market.

The urban informal sector

Alongside the 'informalization' of labor occurring in large-scale industry and services, Latin America's 'informal sector' of employment has undergone considerable expansion during the period of crisis and neoliberal reforms. A wide variety of criteria has been used to define this 'unclear' but 'popular shorthand', first coined back in the 1970s (Gilbert, 1998: 65). These include the size of enterprises, the level of technology used in the production process, legality as a business activity, social security coverage of workers, self-employment versus waged employment and labor arrangements (Koonings *et al.*, 1995: 117 *et seq.*; Scott, 1994: 16–24; Thomas, 1995: Chapter 2). While it is true that informal enterprises are often small in scale, use rudimentary technology and are characterized by self-employment or family labor, Roberts (1994: 6) asserts that the most generally accepted definition of the informal sector is 'income-generating activities unregulated by the state in contexts where similar activities are so regulated'. Further noting that a number of labor arrangements in the 'formal sector' fit this bill and that the informal sector comprises a huge range of jobs and incomes, Roberts argues that 'The persisting interest in the idea of an informal economy lies not in its analytic precision, but because it is a useful tool in analysing the changing basis of economic regulation'. Indeed, while 75 per cent of people in micro-enterprises in 1989 were not covered by social security, this also applied to 17 per cent of workers in formal sector firms (Roberts, 1994: 16).[6]

While regulation implies legality, legality itself is a multidimensional concept, with Thomas (1997: 6) pointing out that 'being legal usually involves complying with a number of regulations often imposed by a variety of different authorities'.[7] Only 2–5 per cent of self-employed people (the biggest group of the informally employed) in Latin America as a whole have access to social security, mainly because of high costs, administrative difficulties, lack of incentives as a result of the eroding value of pensions and uncertainty in occupational prospects (Tokman, 1991: 152–3).

Growth of the informal sector

Referring to a wide and heterogeneous range of activities such as shoe shining, street selling, refuse collecting and recycling, prostitution and so on, the informal sector is primarily an employer of low-income groups. While the sector is made up in the main of low-productivity, low-profit commercial and service activities, about one-quarter of the informal labor force is engaged in manufacturing (Bromley, 1997; Grabowski and Shields, 1996: 170–1). The informal sector was already growing in many parts of Latin America during the 1960s and 1970s, and between 1970 and 1980 increased its share of the region's workforce from 16.9 to 19.3 per cent (Tokman, 1989: 1067). Although trends in informal sector employment need to be treated with caution given shifting categorizations of activity by different governments and regional organizations, growth in the pre-crisis years is usually attributed to rural–urban migration and the consequent creation of a labor surplus in cities (Portes and Schauffler, 1993). From 1980 onwards, however, rates of increase in informal employment seem to have been greater, especially in urban areas, with labor surpluses occurring more as a result of changes in urban economies *per se* rather than demographic processes alone. Indeed, between 1980 and 1990 informal employment in Latin American cities increased from an overall average of 25.6 per cent to 30 per cent (Gilbert, 1995b). Much of this expansion occurred in the period between 1980 and 1985, when the number of people employed in the informal sector in Latin America increased by 39 per cent (Tokman, 1989: 1067). Over the decade as a whole, it also vastly outstripped growth in the formal sector. Between 1980 and 1990, for example, the growth in own-account workers and workers in small firms in seven countries (Argentina, Chile, Colombia, Costa Rica, Brazil, Venezuela and Mexico) was approximately four times that of the combined numbers employed in large firms and in the public sector (Thomas, 1996: 88–9). On the basis of self-employment, unpaid family labor and owners of firms with fewer than five workers it was estimated that by 1989 informal employment occupied 33 per cent of workers in Guatemala City, 30 per cent in Tegucigalpa and 28 per cent in San Salvador (Roberts, 1995: 123). Higher levels still apply to countries such as Cuba, where liberalization of the economy during its 'Special Period' dating from the collapse of communism in the Soviet Union has resulted in job losses and the cessation of full-employment policies (Molyneux, 1996). By January 1996, there were over 160 000 people registered as self-employed and a proliferation of small-scale economic activities. While traditionally suspicious of informal sector work, the informal sector is now recognized and accepted by the state as an important part of Cuban people's survival (Molyneux, 1996: 33). In Nicaragua, too, the effects of economic embargo during the Sandinista administration combined with a war-ravaged countryside and weak industrial base have meant that many people have had to create their own sources of employment. By 1990, the informal sector occupied nearly half the labor force in Managua, excluding domestic servants (Roberts, 1995: 124). Only in Costa Rica, where state commitment to providing social welfare and to holding down unemployment has remained strong, has there been little increase in the informal sector. This stabilized at around 22 per cent of the economically active population during the 'lost decade', although there was also evidence of rises in the numbers of unstable and lower-paid jobs (Lungo, 1997: 58; Roberts, 1995: 124).

The main reasons underpinning the general upsurge in informal employment in Latin America during the 1980s were cutbacks in public employment, the closure of formal sector firms in the wake of increased competition provoked by lowered tariff barriers, declining labor demand in the formal sector and increased growth of the labor force (as a result of demographic growth *per se* and of pressure on household incomes during the economic crisis) (Alba, 1989: 18–21; Gilbert, 1995a: 327). Many family firms have also been pushed into informality because of declining abilities to pay registration, tax and labor overheads, bearing out the argument that 'informality for the self-employed is basically a household survival strategy in the face of unemployment and declining real wages' (Roberts, 1995: 124). As Thomas (1996: 99) summarizes, the 'top-down' informalization promoted by governments and employers has been matched by a 'bottom-up' informalization stemming from the need for retrenched formal sector workers and newcomers to the labor market to create their own sources of earnings.

Activities and working conditions in the informal sector

As the above discussion will suggest, it is hardly surprising that the informal sector has become increasingly competitive during the years of crisis and restructuring. As Miraftab (1994: 468) argues:

> Poor people have had to concentrate their daily activities with much greater intensity around the issue of survival. This has implied not only longer hours of work, but also the need to be extremely innovative to earn a living at the edges of the urban economy (Escobar Latapí and González de la Rocha, 1995).

Yet, although ever more creative strategies to generate income can be witnessed both in the streets and houses of Latin American cities (Fig. 10.3), competition is such that, according to International Labor Organization (ILO) figures, there was a 42 per cent

FIGURE 10.3 Home-based production of Christmas lanterns out of Coca-Cola© cans, Querétaro, Mexico. Here, imaginative use is made of a special Santa edition of the Coca-Cola© can. Tiny holes are punched on the outside of the can, and a metal ring attached to the bottom of the inside provides support for a candle. The product is usually sold in the street after dark to maximize sales. Photograph by Sylvia Chant.

drop in income in the informal sector between 1980 and 1989 (Moghadam, 1995: 122–3).

Limits to the expansion, if not the viability, of informal sector employment are presented by lower purchasing power among the population in general and greater numbers of people needing to work (Roberts, 1991: 135). The latter is in part the legacy of high fertility in the 1960s and 1970s and in part a result of the increased participation of women in the workforce. Indeed, the saturation of the informal sector is often argued to have hit women the hardest given their disproportionate concentration in the sector and the fact that their limited resource base confines them to the lowest productivity ventures within it (Bromley, 1997: 135; Moser, 1997; Scott, 1994). In Costa Rica, where women constitute 41 per cent of informal workers, low-income women I interviewed in towns in Guanacaste province in 1989 complained that their resources were so limited that all they could do was to sell small quantities of snacks such as home-made sweets, flavored ices and pastries outside local schools or on the streets. Since most of their neighbors were forced to do the same, some felt it was not worth it at all, thereby contributing to the so-called 'discouraged worker' effect (Baden, 1993: 13). In order to get around the problems of making ends meet, many low-income women in Costa Rica and in Mexico may resort to taking a variety of jobs, such as combining part-time domestic service with home-based or community-based activities such as tortilla making and selling (Fig. 10.4) or personal services such as hairdressing.

Beyond this, it should also be noted that the informal sector is unlikely to thrive as long as the formal sector remains in a state of stagnation. Detailed empirical studies in

Figure 10.4 Home-based tortilla making for door-to-door sale, Querétaro, Mexico. Tortillas, made from maize or wheat flour, are a primary Mexican staple. Although it is possible to buy tortillas from tortilla factories or from supermarkets, many people prefer the hand-made variety. Photograph by Sylvia Chant.

Latin America have revealed that the informal sector is linked to the formal sector in numerous (and subordinate) ways, that enterprises with the fewest direct links with the formal sector are likely to be the least dynamic economically and that, over time, the informal sector is increasingly likely to lose its independent basis for subsistence (see Roberts, 1991: 132).[8] Taking into consideration the point that 'the informal sector in Latin America and elsewhere is an adjunct to the large-scale sector of the economy, producing those goods for which the market is so reduced and so risky that large-scale enterprises are not interested in entering' (Roberts, 1995: 121), links include the formal sector's use of informal sector enterprises for production, distribution, marketing and retail (Thomas, 1996: 56-9). In times of crisis, declining fortunes in the formal sector are likely to cut off valuable sources of contracts and supplies to the informal sector. Thus, although this sector has continued to expand during the years of crisis and restructuring (and recovery in some countries in the early 1990s), it has not been able to absorb all the job losses in the formal sector, which undoubtedly accounts for the fact that open unemployment has risen in an unprecedented manner during the past two decades.

Theoretical and policy perspectives on the informal sector

Leading on from the above, the behavior of labor markets in Latin America over the past 20 years has also borne out the importance of eschewing notions of labor-market dualism whereby the formal and informal sectors are conceived as discrete or autonomous entities. Instead, and along the lines of Moser's ground-breaking neo-Marxian paper on 'petty commodity production', the labor market is more appropriately conceptualized as a continuum of productive activities. This entails recognition of complex gradations of formality and linkages between different enterprises (which may well be more exploitative than benign) (Moser, 1978). Over and above definitions and conceptualizations of the informal sector, other debates on urban labor markets in developing countries over the past 30 years have included interrelated issues of the role of the informal sector in creating jobs and incomes, and questions of policy interventions.

Until relatively recently, Tokman (1989: 1072) notes, there was no explicit policy towards the informal sector in Latin America because of the belief that labor surpluses would eventually be absorbed by the modern sector. In some respects this was borne out empirically by the shrinkage of the informal sector in countries such as Argentina, Chile, Colombia, Peru, Mexico and Venezuela between 1950 and 1980 (Gilbert, 1998: 71). On top of this, the view of many economists and civil servants that informal activities were 'parasitic', 'unproductive' and tantamount to 'disguised unemployment' meant that there was little will to promote expansion of the sector (Bromley, 1997: 124). In effect, the informal sector has been seen as an employer of 'last resort' or a fragile means of basic subsistence in a situation in which social welfare provision for those excluded from the formal labor force is minimal or nonexistent (Cubitt, 1995: 163; Gilbert, 1998: 67). As such, informal sector entrepreneurs have routinely been forced to operate on the margins of the labor market through the prohibitive costs attached to becoming legal and through lack of government support, often facing harassment or victimization in the process (Thomas, 1996: 56–7). Indeed, while the process of 'becoming legal' varies across countries and occupations, it is usually extremely time-consuming and in some instances can cost firms their entire annual profits (Tokman, 1991: 147). With reference to street occupations in Cali, Colombia, Bromley (1997: 133) echoes that 'regulations are excessively complex, little known, and ineffectively administered, resulting in widespread evasion, confusion and corruption'. In short, the productive potential of what some describe as the 'true capitalist entrepreneurs' of developing countries is 'stunted by excessive and inappropriate interventions on the part of the state via redundant regulation and thwarting red tape' (Tokman, 1989: 1068).

In the past 10–15 years, however, government authorities, along with planners and social scientists, have come round to the idea that the informal sector is a sector of entrepreneurship rather than a 'poverty trap' (Cubitt, 1995: 175). This has led to arguments for greater priority to be given to the design of policies for the informal sector, including state support for credit, raw material supplies, management training, skills and market information (Szirmai, 1997: 208). This change in attitude has in part been motivated by a more general shift towards deregulation which Tokman (1989: 1068) notes has undermined the association between informality and stagnation/marginalization and placed 'greater emphasis on the entrepreneurial potential of informal sector workers'. More benign attitudes to the informal sector also owe to recognition that even if the sector is not necessarily an 'engine of growth', it is undoubtedly a creator of employment.

Hernando de Soto, together with his colleagues at the Instituto de Libertad y Democracia in Peru, has been the principal protagonist of moves to unleash the potential of informal economic activities. De Soto's ground-breaking book, *The other path*, published in 1989, championed the informal sector as 'a grass-roots uprising against unjust and excessive regulations' created by governments in the interests of the society's powerful and dominant groups (Bromley, 1997: 127; see also de Soto, 1989). De Soto's basic argument is that in a context where the legal system is designed to accord with the interests of the economic elite, illegality is the poor's only, and justifiable, alternative. Thus, although the informal sector is technically 'illegal', it is not criminally so, but is more to do with 'nonconformity' with bureaucratic rules and regulations (de Soto, 1989). This argument is grounded within the distinction made by Age of Enlightenment thinkers such as Adam Smith and Thomas Paine, between 'natural laws' and 'formal laws'. Whereas the former are based on moral principles that are socially necessary, right and just, the latter are the artificial creations of governments (Bromley, 1997: 127). Many informal activities may not conform to formal laws, but in being orientated to gaining a livelihood and fulfilling social responsibilities (providing for dependents, for example), are clearly in the best interests of society. Emphasizing the fundamental ways in which the existence of the informal sector relieves unemployment, provides a gainful alternative to crime and harnesses the entrepreneurial talent of the disaffected masses, de Soto suggests that the state would be best advised to stimulate and protect informal entrepreneurs and to grant them more freedom.

Although de Soto's work has gained support from both left and right of the political spectrum, it is important to recognize the wider implications of his arguments, one important one being that advocating decontrol of economic enterprise sets a precedent for greater deregulation in the formal sector. This, in turn, mirrors the broader objectives of the IMF, the World Bank and other international agencies to liberalize production and markets in developing regions (Portes and Schauffler, 1993: 55). Greater tolerance of poor working conditions in the informal sector also helps to depress unemployment figures and can therefore be politically expedient. For these kinds of reasons, many people have stressed the importance of remembering that the informal sector basically exists because of poverty and in this way is not really a solution to economic and labor-market disadvantage. As summed up by Thomas (1995: 130), the informal sector 'is a picture of survival rather than a sector full of entrepreneurial talent to be celebrated for its potential to create an economic miracle'.

Nonetheless, given that the informal sector is a large and probably persistent feature of urban labor markets in Latin America and other developing regions, it is clear that, in the short term at least, measures are needed to help it operate more efficiently and with better conditions for its workers. Possible options here include (1) a repeal of regulations and policies which obstruct entrepreneurship without serving any legitimate public regulatory purpose and (2) a cessation of policies which assist 'favored' sectors such as large enterprises (Chickering and Salahdine, 1991: 6). The fact that most governments in developing countries have supported policies of subsidizing the large-scale capital-intensive sector via credit, direct and indirect market protection (tariffs, quotas and so on) has effectively discriminated against the informal sector (Grabowski and Shields, 1996: 172). As for point (1), even if governments cannot waive regulations to any great extent, they might consider simpler and diminished requirements and/or allow for progressive implementation (Tokman, 1991: 155).

On the supply side, Rodgers (1989) suggests policies geared to education and training to promote the diversification of the informal sector, together with subsidies, investment and enhanced access to credit. Moves of this nature would need to be planned carefully since education is often perceived as irrelevant to informal sector activities, besides which people may lack the basic literacy and numeracy skills to benefit from training programs (Thomas, 1995: 112–4). Tokman (1989: 1071) also reminds us that since 'the informal sector is both complex and multi-dimensional ... no single prescription will improve the standard of living of those employed within it'. As such, policies might at least be decentralized in order to accord with needs and skills in different localities (Portes and Schauffler, 1993: 56).

GENDER AND THE URBAN LABOR MARKET

Aside from the general changes noted in Latin American urban labor markets over the past few decades, the configuration of the labor force has also been marked by important shifts in gender composition, with an overall decline in men's share of employment, before and during crisis and restructuring. Bearing in mind that women's economic activities often fall outside the net of registration in official figures because of their informal and part-time nature, between 1950 and 1980, the size of the female labor force purportedly tripled in Latin America, meaning that by 1980, women's share of the overall workforce (including agriculture) was 26 per cent (compared with 18 per cent 30

TABLE 10.9 Women's share in labor force and occupational categories

	Female share of labor force (%)		Female share in sector, 1990 (%)			
	1980	1994	Administrative and managerial	Professional, technical and related work	Clerical and sales	Services
Haiti	45	43	32.6	39.3	88.3	65.2
Nicaragua	28	36	–	–	–	–
Honduras	25	28	27.8	50.0	59.6	72.4
Cuba	–	–	–	–	–	–
Bolivia	33	37	16.8	41.9	64.7	72.5
Guatemala	22	25	32.4	45.2	39.3	51.7
Ecuador	20	26	26.0	44.2	40.9	63.5
Dominican Republic	25	29	–	–	–	–
El Salvador	27	33	17.7	43.3	59.7	72.3
Paraguay	26	28	16.1	51.2	46.2	71.8
Colombia	25	35	27.2	41.8	45.5	69.6
Peru	24	28	22.1	40.9	52.1	37.6
Costa Rica	21	29	23.1	44.8	40.4	59.3
Panama	30	33	28.9	50.7	57.5	55.8
Venezuela	27	33	18.6	55.2	45.7	57.5
Brazil	28	34	–	–	–	–
Chile	26	31	19.4	51.9	46.3	72.5
Mexico	27	32	19.4	43.2	41.7	45.0
Uruguay	31	40	20.6	61.1	45.9	67.7
Argentina	28	30	–	–	–	–

Sources: UNDP (1995, Table A2.7) and World Bank (1996, Table 4).
Note: countries are listed in ascending order of gross national product (GNP) per capita in 1994.
– = no data.

years earlier) (Safa, 1995b: 32). According to estimates from the Inter-American Development Bank, women's overall share of the labor force looks set to rise to 27.5 per cent by the year 2000 (Thomas, 1996: 95), although United Nations Development Programme (UNDP) figures for 1994 for many countries suggest that this overall projection may well be exceeded (Table 10.9).

Part of the general rise in female employment over the past four decades can be attributed to the decline in agricultural employment relative to the growth of industry and services. In the Dominican Republic and Puerto Rico, Safa (1995b) notes that the disintegration of the sugar economy has led to employment losses for men, with the shift to urban-based labor-intensive manufacturing and services having favored female workers. Indeed, in Puerto Rico, unemployment rates have been higher for men since the 1950s. Over and above this, there has been an increasing supply of women to the labor force as a result of rural–urban migration, rising education and lower fertility (Safa, 1995a: 169).

While women's share in the labor force was already growing before the recession (Gilbert, 1998: 74), the fact that this has continued in the wake of tightening labour–market conditions suggests that economic restructuring has also played an important part in the process. Indeed, Humphrey (1997: 171) notes that even during the 'lost decade' in Brazil, women's share of employment in the São Paulo Metropolitan Area grew from 33 to 38 per cent between the beginning and end of the 1980s. A number of factors pertaining to both the demand and supply side of the labor market are relevant to these tendencies, as discussed below.

Women's employment and economic restructuring

On the supply side of the labor market, numerous studies of poor urban communities show that pressures on household income have been the main impetus behind women's increased involvement in remunerated work (Benería, 1991; Chant, 1996; González de la Rocha, 1988; Moser, 1997). In many cases, this is because of the declining purchasing power of male breadwinners' wages, and, in others, because men have lost their jobs altogether. In Uruguay, Nash (1995: 155) notes that the participation of women in the workforce increased from 38.7 to 44.2 per cent between 1981 and 1984 as a result of rising levels of male unemployment. In Cisne Dos, a low-income barrio in Guayaquil, Ecuador, detailed survey work by Moser (1997: 36–7) reveals that the labor-force participation of female spouses in male-headed households increased from 31.5 per cent in 1978, to 45.5 per cent in 1993, mainly as a response to declining incomes and the casualization of men's jobs.

On the demand side of the labor market, the recruitment of female workers has persisted during the past 15 or so years of economic restructuring as Latin American countries have placed greater emphasis on strategies of export promotion (Safa, 1995b: 33). This has increased competition among firms, which have often resorted to increasing their female workforce, either directly by recruitment in industrial plants (especially common in Mexico, Brazil, Venezuela, Puerto Rico and the Dominican Republic) or on a subcontracted piecework basis. Although on the one hand this may have increased income-generating opportunities for women, there are decidedly more benefits for employers – reducing production costs, saving on social security contributions, fragmenting the workforce and taking advantage of the low 'aspiration wages' of married women (Benería and Roldan, 1987; Miraftab, 1994: 469; Peña Saint Martin, 1996). In addition, it is also argued that sectors and occupations which were already marked by a high degree of feminization (such as the labor-intensive *maquiladora* industry, domestic service and other low-skilled tertiary occupations) were less hard hit by recession than sectors in which men were prevalent such as heavy industry and construction (de Barbieri and de Oliveira, 1989: 23; Ward and Pyle, 1995). In Brazil, for example, whereas 59.5 per cent of economically active women were employed in services in 1977, this had risen to 72.6 per cent by 1988 (Pitanguy and de Mello e Souza, 1997: 77). Many of these jobs are in domestic service, which is a major employer of unskilled women.

Upward trends in women's employment are not confined to Latin America, and it is significant that a worldwide increase in female labor-force participation has occurred during a time in which the conditions of employment have deteriorated considerably. As Moghadam (1995: 115–16) notes, 'The global spread of flexible labour practices and the supply-side structural adjustment economic package coincide with a decline in labour standards, employment insecurity, increased joblessness, and a rise in atypical or precarious forms of employment'. Moreover, as Acero (1997: 72) observes, the casualization of work in the formal sector and the growth of the informal sector have affected women to a greater extent than men. Indeed, despite the fact that unemployed women are more likely to be classified as economically inactive 'housewives', it is interesting that in general terms, female unemployment rates are higher than those of their male counterparts (Moghadam, 1995: 111; Monteón, 1995: 51).

Although women still tend to be restricted to a narrower range of occupations than men, Safa's study of Puerto Rico, Cuba and the Dominican Republic reveals some diminution of occupational segregation as women's participation increases in the professions, clerical work and the public sector (Safa, 1995b: 39). Another interesting development in the years of the crisis, also noted as having led to some blurring of the boundaries between men's and women's activities, is the increased use of the home as a site of production (Miraftab, 1994: 487–8). Yet in Brazil, Humphrey (1997: 171) argues with reference to industrial employment that 'The continuing entry of women into the labour force has not in any way undermined the gender division of labour. Gender segregation and gender inequalities remain as great as ever'. Part of this pattern is attributed by Humphrey not only to the way that labor-market divisions mirror gender divisions in wider society, but because 'gendered occupations and work structures are constructed within the factory and then institutionalized and legitimated through segmented labour markets'. Indeed, for Brazil, more generally, women are still confined predominantly to low-income occupations, and their average wage is only 54 per cent of men's (Pitanguy and de Mello e Souza, 1997: 73). Evidence from Colombia also suggests that income differentials may be greater in the informal than the formal sector, with women's average incomes being 86 per cent of men's in the latter compared with 74 per cent in the former (Tokman, 1989: 1071). Given that, within certain industries, such as textiles and electronics, women seem to be locked into the least skilled jobs, and that there are tendencies towards the 'masculinization' of export manufacturing employment as automation increases (Acero, 1997; Roberts, 1991: 31), it is possible that more women will be further cornered into the informal sector over time.

Women's work and urban households

While women are clearly still a very vulnerable group in the labor force, with the limited jobs available to them usually paid at lower rates than their male counterparts,

Safa (1995b: 33) notes that the crisis has increased the 'importance and visibility of women's contribution to the household economy as additional women enter the labor-force to meet the rising cost of living and the decreased wage-earning capacity of men'. Indeed, much of this increase in female labor force participation has been among wives and mothers (see González de la Rocha, 1988: 214–15; Selby *et al.*, 1990: 174). A study of Mexico based on official national statistics indicated that women aged 20–49 years, increased their labor-force participation rate from 31 to 37 per cent between 1981 and 1987 (cited in González de la Rocha, 1991: 117). Over one-quarter of married women in Mexico were recorded as working in 1991 compared with only 10 per cent in 1970 (CEPAL, 1994: 15), and the highest levels of economic activity are now in the 35–39 year age cohort (43 per cent). In Costa Rica, while there was only one female worker for every three men in the 20–39 year age cohort in 1980, by 1990, the gap had narrowed to one in two (Dierckxsens, 1992: 22).

These developments have an important series of ramifications. Nash (1995: 162) argues that 'the rising rates of marital instability, often caused by unemployment or the forced migration of men to find new sources of work puts even greater responsibility on women for the care and welfare of dependents'. In a similar vein, González de la Rocha (1994: 141–2) argues with reference to her research on Guadalajara, Mexico, that, in addition to rising demands on their time, women's employment has brought few changes in domestic power relations, because women's earnings are usually so much lower than men's, because women cannot necessarily control their own earnings and because women are still very much tied to childcare and domestic responsibilities. Indeed, given that there has been such little movement by men into reproductive tasks, it is not surprising that women's labor burdens have tended to grow in the past 15–20 years (Chant, 1996: 298; Langer *et al.*, 1991: 197; Moser, 1997).

By the same token, some research suggests that employment can be a source of power and negotiation for women, with positive outcomes for conjugal relations. In my own studies of low-income households in the Mexican cities of León, Querétaro and Puerto Vallarta, for example, there seems to be greater collective negotiation and scrutiny over financial affairs where households have diversified their income strategies (Chant, 1991a). With reference to Puerto Rico, Cuba and the Dominican Republic, Safa (1995a: 58) also notes that although the cultural norm of the male breadwinner remains decidedly embedded in the workplace and the state, 'women's consciousness and increased bargaining power in the household have increased as a result of the massive incorporation of women into the labor force in recent decades'. This said, McClenaghan (1997: 29) notes for the Dominican Republic, that even where women are the primary providers, men are usually still acknowledged as *el jefe*.

Although there is clearly debate on the subject, the 'myth of the male breadwinner' has undoubtedly been challenged through women's declining dependence on male incomes and growing economic participation in their own right (Safa, 1995b: 33). Moreover, alongside gender-selective migration, it is highly likely that these tendencies have been important in accounting for growing numbers of female-headed households in recent decades (see Chant, 1997a: Chapters 3 and 4).[9] Indeed, many case studies in different parts of Latin America have indicated that levels of female household headship tend to be greater in areas where women's rates of employment are high (Bradshaw, 1995a; Chant, 1991a; Fernández-Kelly, 1983; Safa, 1995a).

Gender relations, employment and household structure

While women's increased labor-force participation has arguably expanded their potential autonomy and led in some cases to the formation of independent households, the latter may also have been precipitated by an intensification of domestic strife during the crisis period. Aside from the general insecurity and exhaustion associated with increased poverty, unequal burdens of labor and male consternation at their wives going out to work have been associated with rising levels of intrahousehold conflict (Benería, 1991; Gledhill, 1995: 137; González de la Rocha et al., 1990). As Selby et al. (1990: 176) argue:

> Male dignity has been so assaulted by unemployment and the necessity of relying on women for the subsistence that men formerly provided, that men have taken it out on their wives and domestic violence has increased ... the families which have been riven by fighting and brutality can easily be said to be the true victims of economic crisis.

Certainly, loss of employment and dependence on women's earnings can strike an extremely discordant note at the core of masculine identities, as noted for Mexico City by Gutmann (1996). This, in turn, may have destabilizing effects on families, as appears to be the case in northwest Costa Rica (Chant, 1997b). By the same token, men who have reacted violently towards women can and do change, often as a result of women's instigation (Gutmann, 1997).

CONCLUSIONS

Reflecting on recent trends in population, migration, employment and gender relations in Latin America, it is clear that globalizing forces have exerted increasingly significant influences on their evolution in the late twentieth century. In many cases, these forces have taken the shape of market-driven macro-economic shifts, with global institutions playing an important role in the formation of policy. In general terms neoliberalism does not seem to have benefited Latin American societies, and, in particular, seems to have impacted extremely negatively on lower-income groups.

While there are signs in some countries in Latin America that the debt crisis is now behind them and that real growth is possible, the prospects for sustained development in the region as a whole are doubtful (Rey de Marulanda, 1996: 29; Thomas, 1995: 105). Indeed the deficits engendered by the 'lost decade' of the 1980s require that economies in the 1990s and beyond must be extra dynamic to compensate for stagnation of the debt crisis years (Cordera Campos and González Tiburcio, 1991: 29). As Jenkins (1997: 108) identifies with reference to Bolivia, for example, the New Economic Policy has not been successful in stimulating a dynamic industrial sector and in improving the longer-term chances of economic performance. As with other economies with low levels of infrastructure and human resources there are difficulties of insertion into mainstream economic activity and international markets (Jenkins, 1997: 126). More worrying still, perhaps, is Batley's (1997: 338) observation that as growth has resumed in some liberalized economies in the region income disparities have increased.

The scope for redressing these tendencies is rather limited; first, because of current demographic and labor-market trends and, second, because of the diminishing resources of people and governments to counter the effects of growing polarization. The latter are interrelated insofar as struggles to combat economic deterioration have become progressively privatized within households as a result of the erosion of government services and subsidies (Benería, 1991: 171 and 176). Since social expenditure remains small in most parts of the region, and has become increasingly targeted and compensatory during the 1990s, the possibilities for effective social policy-making are now considerably reduced (Batley, 1997: 339). While low-income households have traditionally been able to manage some self-defence by mobilizing their own (albeit limited) resources, González de la Rocha (1997: 2) cautions that persistent poverty over nearly two decades has greatly weakened the survival mechanisms of urban households. In many cases, too, the cumulative depletion of time, energy and income has led to a situation in which women in particular are carrying burdens within their households that are close to unsustainable and which are seriously affecting their physical and psychological health. There are clear needs for interventions that ensure that the poor are prevented from engaging in protracted self-exploitation that jeopardizes their own chances of survival, and those of their children.

With respect to the relevance to these points of the specific themes covered in the chapter, we have seen that the 1980s and 1990s witnessed a slowdown in population growth in Latin America, mainly as a result of increased women's access to, and use of, contraception. This in turn has been influenced directly and indirectly by global initiatives for fertility reduction, although, at the end of the day, market-driven poverty has been a much more decisive factor in limiting births among low-income groups than government or agency propaganda (and indeed has also displaced the resistance to family planning traditionally presented by Roman Catholicism). Regrettably, falling fertility does not seem to have been accompanied by substantial improvements in maternal and child health, with signs in many parts of the region that these are deteriorating. If the spirit of the Cairo consensus is to be carried through to the letter, it would seem imperative for governments to implement the full range of services itemized in the Plan of Action. Here, international funding could play a major role in integrating basic medical entitlements and sanitary infrastructure within reproductive health initiatives.

Another critical issue facing Latin American governments is the need for population policy to address the issue of labor-force absorption (Alba, 1989: 29). The present age structure of the population means that demographic growth is still high, and in the immediate future (or at least into the first decade of the twenty-first century) the numbers of people seeking work will continue mounting. Moreover, even if in the longer run decreasing fertility rates may lead to a fall in new entrants to the labor force, this may be offset by reductions in infant mortality and rising life expectancies which may lead to greater numbers of older people having to provide for themselves (Thomas, 1995: 108). The proliferation of households via the formation of female-headed units and the decline in extended households may also lead to greater individualization of survival strategies at the grass roots, which in turn will put greater pressure on labor markets.

While rural–urban migration has traditionally answered some problems of survival in Latin America, it is clear that many urban economies in the region are losing their powers of absorption over time. In a situation in which urban deterioration rather than rural dynamism is stemming the flow of rural–urban migrants, there is likely to be greater pressure for external migration, particularly to high-wage countries such as the United States (Massey et al., 1993: 463). Given that Latin American governments are unlikely to be able to develop policies which will effectively reduce flows of potential migrants (Becker and Morrison, 1997: 103), securing access to overseas labor markets would seem to be paramount, at least as a short-term safety mechanism. Such prospects are remote on any grand scale, however, and for the bulk of the population who will remain in the region economic prospects look bleak, especially for those lacking skills and assets. As Batley (1997: 338) sums up, 'Poor people may seek survival within the introverted economies of the dispossessed, trading between themselves on the basis of weak inflows from the "upper circuit" of the money economy'. Indeed, there is little doubt that there will be a growing incidence of 'flexible' labor contracts in the formal sector and declining opportunity in the public sector. Set against a growing economically active population in cities, the informal sector will be under further pressure to absorb new people (Thomas, 1995: 111). Many of the latter will be women, who, despite the competitive conditions of informal employment and the rigors of their dual labor burden, are unlikely to want (or to be able to afford) to retreat into the home on a full-time basis. Indeed, for those women who migrate, and for the growing number that are heading their own households, full-time domesticity is an unrealistic scenario.

Given that women's rising level of education over time has played a part in stimulating greater female economic activity, and that this is one sector which has broadly held up during the 'lost decade', one of the more practicable ways for Latin American governments to brace themselves for the twenty-first century might be to devote greater funding to education as and when recovery permits. In particular, increased vocational education in expanding spheres such as information technology could bolster economic prospects in a number of ways. On one hand, it could enhance indigenous economic activity insofar as the more skilled and educated the labor force, the greater its likely productivity, which in turn could provide an important market for goods and services and reduce the need for external orientation (Szirmai, 1997: 90–1). Alternatively, and in view of the likely powerlessness of Latin America to resist the forces of globalization, it could attract greater foreign investment into the region. As Mitter (1997: 26–7) points out, the youth of developing country populations set against demands for new technological skills are likely to motivate Northern companies to spread information-intensive aspects of production to the South in the next few decades.

The dedication of resources to education could be complemented by other (relatively modest) investments in human capital such as primary health care, and housing and neighborhood improvements which could help to enhance the prospects of 'development from below'. Indeed, while recognizing that Latin American governments will be restricted in their range of options by the aggressive market-orientated strategies of global economic institutions, without due attention to the individuals and households who make up their societies, attempts to shape a more positive and equitable future for the continent are likely to be doomed to failure.

Acknowledgements

My thanks go to Dr Cathy McIlwaine, and to the editors of this volume, for their helpful comments on an earlier draft of this chapter.

Endnotes

1 My research in Mexico began in 1982 and has concentrated on three cities in the country – Querétaro, León and Puerto Vallarta. My principal research projects have focused upon low-income housing, family structure, household survival strategies, female employment and women-headed households, with funding over the years provided by the UK Economic and Social Research Council (ESRC), the Leverhulme Trust and the Nuffield Foundation. My research in Costa Rica dates back to 1987 and has been carried out in various localities in the northwest province of Guanacaste. Previous projects have included analyses of migration and women-headed households, with my most recent research during the summer of 1997 involving a pilot study of men, households and poverty, funded with support from the ESRC (award R000222205) and the Nuffield Foundation [SOC/100 (1554)].

2 Smyth (1994: 2) points out that whereas 'birth control' refers to the rights of people freely to regulate the timing, spacing and number of their children, 'population control' refers to policies aimed at controlling the same variables for demographic ends.

3 In order to avoid confrontation with the Roman Catholic Church, national affiliates of international organizations such as the International Planned Parenthood Federation (IPPF), have often incorporated the term 'family' in their nomenclature. Aside from 'Profamilia' in Colombia, 'Benfam' in Brazil is another example (Corrêa and Reichmann, 1994: 17).

4 This observation refers to the most recent census of Costa Rica which was taken in 1984.

5 Explicit deregulation refers to the formal abandonment or erosion of legislation, whereas implicit deregulation relates to the 'inadequate implementation or systematic bypassing' of regulations (Standing, 1989: 1077).

6 The micro-enterprises referred to by Roberts (1994) were legally registered with the federal and local authorities, but, since social security contributions are by far the most costly aspect of legality, there is greater likelihood that employers will not pay (see also Note 7).

7 More specifically, Tokman (1991: 143) identifies three types of legality with which informal sector enterprises may not comply:
- legal recognition as a business activity, which involves registration and possible subjection to health and security inspections;
- legality in respect of payment of taxes;
- legality in respect of labor matters such as compliance with official guidelines on working hours, social security contributions and fringe benefits.

See also Roberts (1991: 115).

8 It should also be borne in mind that the formal sector has depended very much on the informal sector for its own dynamism (Gilbert, 1998: 67–9). As Becker and Morrison (1997: 93) sum up, 'In reality, the informal sector enjoys a largely symbiotic relationship with the modern manufacturing and service sectors'.

9 This is not to say that female-headed households have been absent from Latin American societies in previous historical periods.

Further Reading

Cubitt, T. 1995 *Latin American society*, 2nd edn. Longman, Harlow, Essex. A very accessible and useful introduction to development in Latin America, which relates a number of the issues discussed to Third World development more generally. The book covers population, migration and employment, along with topics such as shelter, social relations, values and institutions, and the state.

Furedy, F. 1997 *Population and development*. Polity Press, Cambridge. Although this book does not concentrate specifically on Latin America, it provides an excellent historical review of

North–South perspectives in debates over population and development, and questions the relationships between actual demographic trends and their ideological representations. Essential reading for the global context of approaches to fertility management in Latin America.

Gilbert, A. 1998 *The Latin American city*, revised edn. Latin America Bureau, London. A lively, up-to-date account of urbanization in Latin America, which, like the present chapter, attempts to consider urban trends from the perspective of low-income groups. Amply illustrated with maps, photographs, tables and diagrams, the book comprises 8 chapters, one of which (Chapter 3, 'The move to the city') focuses on rural–urban migration, and another (Chapter 4, 'The world of work') analyses urban employment in the region.

Roberts, B. 1995 *The making of citizens: cities of peasants revisited*. Edward Arnold, London. This book is a fully updated and substantially revised version of Roberts's classic text *Cities of peasants*, published by Arnold in 1978. The main focus is Latin America, but Roberts's discussions of urbanization and underdevelopment in the early part of book take a wider global perspective. Of particular relevance to the present chapter are his discussions in Chapter 4 ('Migration and the agrarian structure'), which examines long-term trends in rural–urban migration in Latin America and their implications for urban growth, and Chapter 5 ('The urban economy and the organization of the labour market'), which looks at employment in cities. Also of considerable relevance for gender dimensions of employment and survival is Chapter 7 ('Urban poverty, the household and coping with urban life').

Thomas, J.J. 1995 *Surviving in the city: the urban informal sector in Latin America*. Pluto Press, London. This is possibly the most thorough account of the nature and behavior of the informal sector of employment in Latin American cities in recent years. It traces the history of conceptualizations of informal sector activity and its wide-ranging characteristics. Particular attention is paid to the evolution of informal employment in the wake of urban growth, the debt crisis and economic restructuring in the late twentieth century. Attention is also given to longer-term prospects for the urban informal sector in Latin America, and to the implications of different policy interventions.

United Nations Centre for Human Settlements (HABITAT) 1996 *An urbanizing world: global report on human settlements 1996*. Oxford University Press, Oxford. This comprehensive and impressively up-to-date report was prepared for the second United Nations World Conference on Shelter and Human Settlements held in Istanbul in June 1996. The report is nearly 600 pages long and covers a wide range of issues relating to cities and urbanization at a global scale, including shelter, infrastructure, urban services, environmental conditions, urban finance, and settlement management and planning. Aside from reference to Latin American examples in the thematic discussions, there is also a separate section dealing with regional aspects of population and urbanization in Latin America and the Caribbean, which addresses demographic growth, rural–urban mobility and international migration.

Part 7

Rural and urban transformations

11

RURAL DEVELOPMENT: FROM AGRARIAN REFORM TO NEOLIBERALISM AND BEYOND

CRISTOBAL KAY

This chapter argues that the agrarian reform programs of the 1960s to the 1980s largely failed to incorporate the peasantry into the development process and that the subsequent neoliberal policies followed by most Latin American countries since the 1980s have further deepened the exclusionary character of the region's modernization.[1] New capitalist groups have emerged and prospered in the rural areas while traditional landlords have further dwindled. The peasant economy, although still an important provider of employment and staple foods, is a relatively declining sector and many peasants have been marginalized as producers, being condemned to bare subsistence and forced to seek wage employment. A more complex and heterogeneous agrarian structure exists today in comparison with the centuries-old bimodal *latifundia–minifundia* system.

The rural economy and society have experienced profound changes in recent decades. While in 1960 over half the Latin American population was rural, today it is only one quarter. Furthermore, agriculture's share in the value of total Latin American exports declined from approximately half to one-fifth, and agriculture's contribution to the gross domestic product (GDP) fell from 20 per cent to under 10 per cent. Paradoxically, although agriculture's importance has declined, the rural areas have gained a new prominence with the emergence of ethnic and environmental movements (Chapter 9).

The chapter begins by discussing the causes, objectives and outcomes of Latin America's agrarian reforms in terms of their impact on production, income distribution, employment, poverty, gender relations and socio-political integration. It then examines the changes in the technical and social relations of production in the countryside ushered in by government policies supportive of the modernization of the *haciendas* (large landed estates) and capitalist farms within the context of globalization and the shift from inward to outward orientation. The influence of these transformations on the peasantry and their future prospects under neoliberalism are subsequently analysed. Finally, the emergent social movements, which are challenging the imposition of neoliberal policies in the countryside, are examined.

THE LOST PROMISE OF AGRARIAN REFORM

From the 1850s to the 1930s the haciendas expanded and achieved a dominant position within Latin America's agrarian structure. This expansion was often achieved by displacing the rural indigenous population to marginal areas. This was the golden age of the hacienda system, landlords being at the height of their economic power, political influence and social prestige. Only in Mexico was this dominance successfully challenged by the revolutionary upheavals of 1910–17. However, it was not until the populist government of Cárdenas during 1934 to 1940 that the hacienda system finally lost its predominant influence in Mexico. The Bolivian revolution of the early 1950s also dealt a major blow to the landlord system with the implementation of an extensive agrarian reform program.

The Cuban revolution of 1959 signalled the final demise of the hacienda system in most Latin American countries. Fearful of the spread of revolution to other countries in the region and the spectre of socialism, the US government launched the Alliance for Progress initiative. This encouraged governments throughout the region to implement agrarian reform programs by providing economic aid. Consequently, from the 1960s to the 1970s a spate of agrarian reforms took place in Latin America, among them in Chile, Peru, Ecuador and Colombia. In the late 1970s and 1980s following the Sandinista revolution in Nicaragua and the civil war in El Salvador agrarian reforms were also carried out in those countries. Only in Argentina has agrarian reform been completely absent. In Brazil strong opposition from landlords stalled any significant agrarian reform but there has been some minor land redistribution since the restoration of democratic rule in the mid-1980s.

Prior to agrarian reform Latin American governments had adopted policies encouraging the modernization of the hacienda system. The introduction of import substitution industrialization (ISI) policies in the post-war period had already begun to transform the traditional agrarian system. Such government measures as subsidized credits for the purchase of agricultural machinery and equipment, improved quality of livestock, fertilizers, high-yielding-variety seeds and technical assistance programs aimed to stimulate the technological modernization of large landed estates. The social relations of production had also begun to change. Labor-service tenancies and to some extent sharecropping began to give way to wage labor (Goodman and Redclift, 1981). Some landlords sold part of their estates to finance improvements on the remainder of their property, thereby advancing a landlord process of 'transformation from above' (Kay, 1988a).

Despite the pre-agrarian reform attempts at modernization the agricultural growth record of Latin America in the post-war period remained poor, especially with respect to domestic food production. Agricultural production grew at an annual rate of 3.2 per cent between 1950 and 1964 but by only 0.3 per cent in per capita terms (ECLA, 1968: 314). Although agricultural exports grew faster than production of domestic food crops, Latin America's position in the world agricultural market deteriorated. Furthermore, food imports increased by 44 per cent between 1950 and 1965 while agricultural exports increased by only 26 per cent, and agriculture's net contribution to foreign exchange earnings began to deteriorate, placing an additional strain on Latin America's balance-of-payment problems.

Latin America's agriculture was inefficient and wasteful of resources. A highly unequal land tenure system was largely to blame for this state of affairs but inadequate government support for the sector was also a factor. Most of agriculture's growth stemmed from an increase in the area cultivated rather than an increase in yields. Extensive growth without major technical and social transformations clearly predominated over intensification of agriculture. While the crop area increased by 51 per cent yields grew by only 24 per cent, or at an annual rate of 1.6 per cent, between 1950 and 1964 (ECLA, 1968: 315).

In explaining Latin America's poor agricultural performance structuralists emphasized the high degree of land concentration while neoclassical and monetarist interpretations stressed government policy, in particular price and trade policy which allegedly discriminated against agriculture (Valdés and Siamwalla, 1988). Government price controls on some essential food commodities and an exchange rate policy which overvalued the local currency and thus made food imports cheaper and agricultural exports less profitable acted as a disincentive to agricultural production (Valdés et al., 1990). While it is generally accepted today that ISI policies adopted by most governments in Latin America discriminated against agriculture, the fact that large agricultural producers were often compensated by countervailing policies is generally ignored (Kay, 1977). Landlords received highly subsidized credits and benefited from cheap imports of agricultural machinery and inputs as a consequence of the trade policy and from special technical assistance programs. Thus government policy was biased not just against agriculture but within agriculture against peasants and rural workers (Kay, 1981).

While landlords no longer dominated the political system in the post-war period in many Latin American countries, they still exerted a major influence on government policy and could swing the power of the state in their favor with regard to peasants and rural workers (Huber and Safford, 1995). Tenants had to pay high rents (either in money, kind or labor services) and agricultural workers were paid low wages and had to endure poor working conditions. Rural labor was largely unorganized and confronted a series of legal obstacles to unionization. Working conditions throughout rural Latin America were exploitative and repressive (Duncan and Rutledge, 1977).

Latin America's bimodal agrarian structure was seen by structuralists as inefficient and unjust as well as having detrimental social and political consequences. It was argued that agrarian reform, by modifying the uneven income distribution, would widen the domestic market for industrial commodities, strengthen the industrialization effort by increasing the supply of agricultural commodities and have a beneficial impact upon foreign exchange. Last, but not least, reformists blamed land concentration for the social inequality, marginalization and poor living conditions of the majority of the rural population in Latin America (Feder, 1971).

The bimodal land tenure system

Latin America had one of the most unequal agrarian structures in the world. At one extreme were the *minifundistas* who owned very small landholdings, and, at the other, were the *latifundistas* who owned very large landholdings in the form of plantations, haciendas and *estancias* (cattle ranches). By 1960 *latifundios* constituted roughly 5 per cent of farm units and about 80 per cent of the land; *minifundios* constituted 80 per cent of farm units but only 5 per cent of the land (Barraclough, 1973: 16).² The middle-sized farm sector was relatively insignificant. Subsequent studies have shown this bimodal characterization to be overexaggerated; tenants had a significant degree of control over resources within the estates and medium farmers had access to better quality land and were more capitalized, thereby contributing more to agricultural output than originally estimated. Despite this evidence of greater heterogeneity, Latin America still had one of the most polarized agrarian systems in the world.

Peasant holdings were the main providers of employment, accounting for about half of the agricultural labor force, four-fifths of whom were unpaid family workers. Large estates employed less than one-fifth of the agricultural labor force (Barraclough, 1973: 22). In 1960 an estimated one-third of the total agricultural labor force was landless and a variety of tenancy arrangements were widespread, an estimated one-quarter (or more) of agricultural workers being tenants or squatters.

This agrarian system was clearly inefficient. On the one hand *latifundios* underutilized land by farming it in an extensive manner and leaving a significant proportion uncultivated. On the other hand, *minifundios* were wasteful of labor, using too much labor on too little land. Not surprisingly, while labor productivity was much higher on *latifundios* than on *minifundios*, the reverse was the case regarding land productivity. Average production per agricultural worker was about 5–10 times higher on *latifundios* than on *minifundios*, while production per hectare of agricultural land was roughly 3–5 times higher on *minifundios* relative to *latifundios* (Barraclough, 1973: 25–27).³ Given that much rural labor was unemployed or underemployed and land was relatively scarce, it was more important from a developmental perspective to raise land productivity than to increase labor productivity.

Causes and objectives of agrarian reform

The most far-reaching agrarian reforms have tended to be the outcome of social revolutions. Such was the case in Mexico (1917), Bolivia (1952), Cuba (1959) and Nicaragua (1979). However, radical agrarian reforms were also undertaken by elected governments, as in Chile during the Frei (1964–79) and Allende (1970–73) presidencies, or even by military regimes, as in Peru during the government of General Velasco (1969–75). Less wide-ranging agrarian reforms in terms of the amount of land expropriated and the number of peasant beneficiaries were carried out largely by civilian governments in the remainder of Latin America. The major exception is Argentina where to date no agrarian reform has taken place and agrarian reform has not formed part of the political agenda. The uniqueness of the Argentinian case is explained in part by the relative importance of family farming and middle-sized capitalist farms as well as by the low proportion of the rural population. Paraguay and Uruguay had colonization programs but in neither country has a significant agrarian reform taken place.

Agrarian reforms have generally been the outcome of political changes from above. Although in some instances these were responding to social pressures from below, few agrarian reforms in Latin America were the direct result of peasant uprisings. Urban social forces and even international forces, as in the case of the Alliance for Progress, played an important role in bringing about agrarian reform. While the peasantry was not an important social force behind agrarian reform legislation, it did significantly influence the process itself. Thus those areas where rural protest was strongest tended to receive the most attention from agrarian reform agencies.

In espousing agrarian reform, governments were pursuing a variety of objectives. A major objective, and the primary one of the more technocratic type of agrarian reform, was a higher rate of agricultural growth. Thus only inefficient estates were to be expropriated and more entrepreneurial-minded estates were encouraged to modernize further. It was expected that less land would be left idle and that land would be cultivated more intensely thereby increasing agricultural output. Another economic (and social) objective was equity. A fairer distribution of income was regarded as facilitating the ISI process by widening the domestic market for industrial goods. A more dynamic agricultural sector would lower food prices, generate more foreign exchange and create more demand for industrial commodities. Thus the underlying economic objective was to speed up the country's industrialization process.

Agrarian reforms also had social and political objectives. By distributing land to peasants, governments hoped to ease social conflicts in the countryside and to gain the peasantry's political support. By means of land redistribution and measures assisting the creation or strengthening of peasant organizations, governments aimed to incorporate the peasantry into the social and political system. Giving peasants a stake in society would strengthen civil society and the democratic system. More radical types of agrarian reforms were particularly keen to organize and mobilize the peasantry in order to weaken landlord opposition to expropriation.

Governments also aimed to increase their support among the industrial bourgeoisie whose economic interests could be furthered by agrarian reform. However, this was more problematic as industrialists often had close ties with the landed class and were fearful that social mobilization in the countryside could spill over into urban areas. Political links between landlords and the urban bourgeoisie were far closer than commonly thought and the bourgeoisie generally placed their political interests before short-term economic gains. They were well aware that agrarian reforms could gain a momentum of their own and spill over into urban unrest. This would intensify workers' demands for higher wages, better working conditions and even lead to demands for the expropriation of urban enterprises.

Although agrarian reforms were largely instituted from above, once expropriation was under way conflicts in the countryside often escalated. Peasants demanded a widening and deepening of the agrarian reform process; landlords opposed such demands and pressurized the government to suppress the increasingly bold actions of the peasants. This was particularly the case in countries where political parties and other organizations used the reformist opening in the country's political system to strengthen peasant organizations and assist their social mobilization. Support, or lack of it, from urban-based political parties and urban social groups was often crucial in determining the outcome of the reform process.

Reach of agrarian reforms

The scope of agrarian reform in Latin America varied greatly both with regard to the amount of land expropriated and with regard to the number of peasant beneficiaries. The agrarian reforms in Bolivia and Cuba were the most extensive with respect to the amount of land expropriated, about four-fifths of the country's agricultural land being expropriated. In Mexico, Chile, Peru and Nicaragua almost half the country's agricultural land, and in Colombia, Panama, El Salvador and the Dominican Republic between one-sixth and one-quarter, of the agricultural land was expropriated (Cardoso and Helwege, 1992: 261). A smaller proportion of agricultural land was affected by agrarian reform in Ecuador, Costa Rica, Honduras and Uruguay (CEPAL/FAO, 1986: 22). In Venezuela about one-fifth of the land was affected by agrarian reform, but most of this had previously belonged to the state and was largely in areas to be colonized. Thus Venezuela's agrarian reform was mainly a colonization programme.

Cuba, Bolivia and Mexico had the highest proportion of peasants and rural workers who became beneficiaries of agrarian reform. In Cuba and Bolivia about three-quarters of agricultural households were incorporated into the reformed sector, while in Mexico it was less than half. In Nicaragua, Peru and Venezuela the proportion of beneficiaries was about one-third, in El Salvador one-quarter and in Chile one-fifth. In Panama, Colombia, Ecuador, Honduras and Costa Rica agricultural families who benefited from land redistribution constituted about 10 per cent of the total number of agricultural families (Cardoso and Helwege, 1992: 261; Dorner, 1992: 34). In other countries the proportion was even lower.

The fact that agrarian reforms with the widest scope were the outcome of revolutions indicates the importance of the question of political power. Where landlords were defeated and displaced from power, the wider was the scope of the agrarian reform. In some instances, landlords have been able to reverse some or all of the gains of agrarian reform. This has followed a major political upheaval – such as a counter-revolution or military *coup d'état* as in Guatemala and Chile with the overthrow of the Arbenz government in 1954 (Brockett, 1988) and Allende in 1973 (Kay, 1978), respectively.

Peasants have sometimes been able to push the agrarian reform process further than intended by governments and redirect it according to their interests. Peasant communities in Peru who had largely been excluded from land redistribution later gained direct access to land from the reformed sector, often after violent clashes between community members (*comuneros*) and the police. In Nicaragua peasants succeeded in pressurizing the Sandinista government to adopt a less state-centered agrarian reform policy than the one which had privileged state farms since 1979. After 1984 some reformed enterprises were transferred directly to peasant beneficiaries in either cooperative or individual ownership. This shift in policy was also provoked by the desire to reduce the influence of the contras among the peasantry and to stimulate food production (Utting, 1992). Following this policy change, the amount of expropriated land redistributed to peasant beneficiaries in individual ownership trebled from 8 per cent in the period 1981–84 to 24 per cent in the period 1985–88 (Enríquez, 1991: 91–2). Peasant beneficiaries also gained more favorable access to scarce inputs, modifying the earlier advantageous treatment given to state farms. However, civil war and the resulting economic deterioration of the country meant that peasants still faced a difficult situation. In Colombia, Ecuador and currently in Brazil peasants have also resorted to land invasions, which have resulted in expropriation and improved access to land (Petras, 1997; Veltmeyer, 1997). Nevertheless these land invasions lacked (or lack) the scope and significance of those in Mexico, Chile and Peru.

In many Latin American countries, however, peasants were not able to extend the expropriation process or prevent landlords from blocking or reversing the process. In most countries agrarian reform remained limited in scope in terms of land expropriated and peasant beneficiaries. Despite an explicit commitment to agrarian reform and peasant farming, a large majority of Latin American governments implemented timid agrarian reforms and failed to support peasant farming to any significant extent. Rhetoric prevailed as governments were either too weak to implement a substantial agrarian reform or had the underlying intention of promoting capitalist farming (Thiesenhusen, 1995).

Collectivist character of the reformed sector

Collective and cooperative forms of organization within the reformed sector were surprisingly far more common than the capitalist context of Latin America, with the exception of Cuba, would lead us to expect. In Mexico, particularly since the Cárdenas government of the 1930s, the *ejido* has dominated in the reformed sector. *Ejidos* are a collective type of organization, although farming is largely carried out on a household basis. Until recently it was illegal to sell *ejido* land. In Cuba state farms had predominated since the early days of the revolution and by the mid-1980s most individual peasant farmers had joined production cooperatives. Production cooperatives and state farms were the dominant farm organization in Chile's reformed sector during the governments of Frei and Allende (1964 to 1973). This was also the case in Peru, during Velasco's agrarian reform after 1969, in Nicaragua during the Sandinista revolution (1979–90) and in El Salvador during the Christian Democrat regime of 1980–89. Only a small proportion of expropriated land was distributed directly as private peasant family farms.

An important explanation for the statist and collectivist character of Latin America's most important agrarian reforms lies in their inherited agrarian structure. Prior to reform large-scale farming prevailed in the form of plantations, haciendas and *estancias*. Governments feared that subdividing these large landed estates into peasant family farms might lead to a loss of economies of scale, reduce foreign exchange earnings as peasant farmers would switch from export-crop to food-crop production, impair technological improvements, limit the number of beneficiaries and reproduce the problems of the *minifundia*. Furthermore, a collective reformed sector reduced subdivision costs, allowed more direct government control over production and, in some instances, marketing and could foster internal solidarity. In those countries pursuing a socialist path of development as in Cuba, Allende's Chile and Nicaragua under the Sandinistas a collectivist emphasis was also underpinned by political and ideological factors. In some cases collective forms of organization were regarded as transitory as in Chile and El Salvador. As beneficiaries gained entrepreneurial and technical experience a gradual process of decollectivization was envisaged.

Agrarian reform policy-makers throughout Latin America greatly underestimated the relative importance of sharecropping and labor-service tenancies within large landed estates. National census data generally failed to record, or to record accurately, the number of peasant tenant enterprises within the hacienda system. This led them to underestimate the difficulties of organizing collective farming and the pressure which beneficiaries would exercise within the collective enterprise for the expansion of their own family enterprise. The new managers of the collectives, generally appointed by the state, had far less authority over the beneficiaries than landlords had and were unable to prevent its gradual erosion from within.

The enduring influence of the pre-reform landed estates on the post-reform situation is startling. In this sense the collectivist character of the reformed sector should not be overstated as it was often more apparent than real. In Peru about half the agricultural land of the reformed sector (collective and state farms) was cultivated on an individual basis. In Chile and El Salvador the figure was about a fifth and only in Cuba was it insignificant. This reflects the varying degrees of capitalist development and proletarianization of the agricultural labor force in each of these countries before the agrarian reform. The differences between types of estates, such as plantations and haciendas, were also reflected in the character of post-reform enterprises and any subsequent process of decollectivization.

One feature of Cuba's agrarian reforms (1959 and 1963), which is not often mentioned, is the fact that Castro's government greatly extended peasant proprietorship, giving ownership titles to an estimated 160 000 tenants, sharecroppers and squatters. Before the revolution peasant farmers had numbered only about 40 000 (Ghai *et al.*, 1988: 10 and 14). Cuban agriculture was dominated by sugar, grown on large plantations, and the agricultural labor force was largely proletarian. A large proportion of seasonal sugar-cane cutters (Fig. 11.1) came from urban areas. The plantation sector was taken over by the state without much difficulty. Over time state farms were amalgamated into even larger units, becoming giant agro-industrial complexes under the direct control of either the Ministry of Agriculture or the Ministry of Sugar. Cuban policy-makers were great believers in 'large is beautiful'. It was not until almost two decades after the revolution that the Cuban leadership launched a campaign for the cooperativization of peasant farmers. They were encouraged to form Agricultural and Livestock Production Cooperatives (CPAs), having resisted joining state farms, and within a decade over two-thirds of all peasant farmers had done so. CPAs were clearly

FIGURE 11.1 A group of Cuban manual cane cutters having their lunch break, 1985. They form a *brigada de macheteros* belonging to a state farm. By 1985 only a small percentage of cane was cut manually. The quality of the cane is superior than mechanically cut cane and there is less damage to the soil as the cane cutting machines are very heavy and compact the soil. Photograph by C. Kay.

outperforming state farms (Kay, 1988b) and they became an example to state farms and led to the transformation of state farms into a new sort of producer cooperative, a process which is still ongoing.

Assessment of agrarian reforms

Agrarian reforms can be assessed in narrow economistic terms or in broader systemic and institutional terms. They can be evaluated in terms of their impact on growth, employment, income distribution, poverty and socio-political participation as well as on the wider development context. More recent evaluations have included the impact of agrarian reforms on gender divisions and on the environment.

While agrarian reform may be a precondition for sustainable development, it is not a sufficient condition. Agrarian reform should not be regarded as a panacea for all the ills afflicting Latin American rural economies and societies, yet the enthusiasm of the initial campaigns and proposals for agrarian reform were often seen in this light. Agrarian reforms were perceived as a way of liberating the peasantry from landlordism with its associated feudal and exploitative conditions. They were seen as a way of achieving equitable rural development which would reduce rural poverty. They were also considered important for facilitating Latin America's struggling industrialization process by expanding the domestic market and easing foreign exchange constraints.

Given that agrarian reforms were seen as a panacea, it is paradoxical that governments failed to provide the financial, technical, organizational and other institutional support needed to ensure their success. In many instances the continuation of ISI policies and the persistent discrimination against agriculture in terms of price, trade and credit policy made the task of creating a viable agrarian reform sector impossible. Clearly, mistakes in design and implementation of agrarian reforms also contributed to their eventual unravelling.

Most agrarian reforms failed to fulfil expectations for a variety of reasons. In some instances, agrarian reform was implemented in a half-hearted fashion by governments which paid lip service to agrarian reform for domestic or foreign political purposes, be it to gain votes from the peasantry or aid from international agencies. In other instances fierce political opposition from landlords, sometimes with the support of sectors of the bourgeoisie, restricted reforms.

Agricultural production

The impact of agrarian reform on agricultural production has been mixed. Most analysts agree that results fall well below expectations. In Mexico agricultural production increased by 325 per cent from 1934 to 1965, the highest rate in Latin America during this period, which was the result of the impetus given to agrarian reform by the Cárdenas government and the supportive measures for agricultural development. Thereafter, Mexican agricultural performance has been poor (Thiesenhusen, 1995: 41). In Chile, agrarian reform initially had a very favorable impact on agricultural production. This increased by an annual average rate of 4.6 per cent between 1965 and 1968, but deteriorated thereafter (Kay, 1978). It is estimated that much of the initial increase in agricultural output came from the commercial farm sector. This is not surprising given that landlords were often allowed to retain a part of the estate, referred to as a *reserva*. They thus kept the best land and farm equipment enabling them to intensify production. The reformed sector performed reasonably well at first, receiving much government support in the form of credits, technical assistance, marketing facilities, mechanization and so on. However, as the expropriation process escalated and strained the administrative and economic resources of the Chilean state, the reformed sector faced increasing problems. Internal organizational difficulties began to arise as beneficiaries devoted more time to their individual plots than to the collective enterprise.

In Peru, agrarian reform failed to increase agricultural production as the average yearly growth rate of 1.8 per cent from 1970 to 1976 was similar to the pre-reform rate of the 1960s (Kay, 1982: 161). During the period 1970 to 1980 the average annual growth rate of agriculture was negative (ECLAC, 1993: 76). But in the 1980s agriculture recovered, growing by 2 per cent a year, although this was still below the 2.2 per cent population growth rate (IADB, 1993a: 261 and 267). The reformed sector, plagued with internal conflicts between government-appointed managers and beneficiaries, was partly responsible for this poor performance. The state exacerbated matters by its failure to provide resources or adequate technical training to beneficiaries and by its continued adherence to a cheap food policy, which reduced the reformed sector's profitability.

In Nicaragua a series of factors conspired against the economic success of the 1979 agrarian reform. In the decade before the agrarian reform agriculture had been stagnant. After agrarian reform in the 1980s agricultural output declined on average by 0.9 per cent a year (IADB, 1993a: 267). The armed conflict between the contras and the government severely disrupted production. Other contributing factors were the insecurity of tenure which inhibited investment by private farmers, the mass slaughter of livestock by farmers fearful of being expropriated, shortages of labor, disruption of the marketing system and, last but not least, mismanagement of the reformed enterprises (Enríquez, 1991). In El Salvador the 1980 agrarian reform was implemented during a period of civil war which came to an end in 1992 (Seligson, 1995; Paige, 1996). GDP declined by 0.4 per cent a year while agriculture fell by 0.7 per cent a year in the 1980s (IADB, 1993: 263 and 267). The commonly held view that individual farming is superior to collective farming is not borne out in El Salvador. Yields achieved on the collective land of the producer cooperatives of the reformed sector were often higher than those obtained on family plots either within or outside the reformed sector (Pelupessy, 1995: 148).

Income distribution and poverty

The gains in income distribution deriving from agrarian reforms were also less than anticipated. In Peru, it is estimated that Velasco's agrarian reform redistributed only 1–2 per cent of national income through land transfers to about a third of peasant families (Figueroa, 1977: 160). Sugar workers on the coast, already the best-paid rural workers, benefited most whilst *comuneros*, the largest and poorest group amongst the peasantry, benefited least (Kay, 1983: 231–2). The initial positive redistributivist impact of many agrarian reforms in Latin America was often cancelled out by the subsequent poor performance of the reformed sector and by macro-economic factors such as unfavorable internal terms of trade and foreign exchange policy. In addition, by excluding the poorest segments of the rural population – members of peasant communities, *minifundista* smallholders and seasonal wage laborers – from land redistribution many agrarian reforms increased socio-economic differentiation among the peasantry and failed to reduce significantly rural poverty. Tenant laborers and the permanent wage workers, who generally became full members of the reformed sector, sometimes continued the landlord practice of employing outside seasonal labor for a low wage or renting out pastures or other resources of the reformed sector to *minifundistas* and *comuneros*. This was particularly the case in Peru, El Salvador and Nicaragua.

Gender relations

In terms of reducing gender inequalities the assessment is rather negative. Most land reform legislation ignored the position of women, failing to include them explicitly as beneficiaries, to give them land titles or to incorporate them into key administrative and decision-making processes in the cooperatives, state farms and other organizations emanating from the reform process. Even in Cuba, women made up only one-quarter of production cooperative members and were even fewer on state farms (Deere, 1987: 171). In Mexico women constituted 15 per cent of *ejido* members, and in Nicaragua and Peru women were only 6 per cent and 5 per cent of cooperative members, respectively. Women were excluded as beneficiaries because of legal, structural and ideological factors. The stipulation that only one household member could become an official beneficiary (that is, a member of the cooperative or receiver of a land title) tended to discriminate against women given the assumption that men were head of the household (Deere, 1985). The agrarian reform in Chile reinforced the role of men as main breadwinners and gave only limited opportunities for women to participate in the running of the reformed sector, despite some legislation to the contrary during the Allende government (Tinsman, 1996).

Socio-political integration and participation

The greatest contribution of agrarian reforms may lie in the stimulus given to institution building in the countryside. Governments facilitated the organization of the peasantry into trade unions and cooperatives of various kinds, such as producer, marketing and credit associations. This brought about a considerable degree of integration of the peasantry into the national economy, society and polity. Prior to reform, insurmountable obstacles lay in the way of peasants creating their own organizations. Political parties began to vie for the peasant vote and extended their networks to rural areas where in the past reformist and left-wing political parties in particular had often been excluded by the landed oligarchy. With agrarian reform peasant participation in civil society was much enhanced. Many peasants, when granted a land title, felt that only then had they become citizens of the country. By weakening the power of landlords and other dominant groups in the countryside, agrarian reforms encouraged the emergence of a greater voice for the peasantry in local and national affairs.

Governments often sought to establish peasant organizations that would extend and consolidate their influence in the countryside. Governments were more successful in gaining the allegiance of peasants from the reformed sector. However, they were not always able to keep their allegiance. Some peasant organizations came to regard government patronage as a hindrance to the pursuance of their aims and sought a degree of autonomy by breaking free from the government's co-optation. The agrarian reform in Chile brought about a major organizational effort. While in 1965 only 2100 rural waged workers were affiliated to agricultural trade unions, this figure increased to 282 000 by the end of 1972 (Kay, 1978: 125). This meant that about 80 per cent of rural waged workers were members of trade unions, an unusually high figure within the Latin American context. Strikes and land seizures by farm workers escalated as they became organized, gained in self-confidence and had less to fear from repression. Landlords could no longer easily dismiss striking farm workers nor count on swift retribution from the state against a peasant movement which was demanding an acceleration and extension of the expropriation process. The intensified conflicts in the countryside contributed to the military *coup d'état* which led to the violent overthrow of the Allende government.

In Nicaragua the Sandinista agrarian reform also provoked a major organizational effort of the peasantry (Enríquez, 1997) The government helped to set up the *Unión Nacional de Agricultores y Ganaderos* (UNAG) in 1981 and by 1987 one-fifth of all agricultural workers had joined (Blokland, 1992: 154). UNAG also managed to wrench a greater degree of autonomy from the state over time and has remained the most important peasant and farmer organization in the countryside to this day.

In short, agrarian reforms were often restricted in scope and thwarted in their aims by opposition forces or by government mismanagement. However, in those countries where agrarian transformation went deeper and where social exclusion was significantly reduced, social stability and political integration has facilitated economic development. Hence it is possible to argue that, from a longer-term perspective, agrarian reforms have made a major contribution to the democratization of society. Whilst agrarian reforms marked a watershed in the history of rural society in many Latin American countries, the root causes of social and political instability will remain as long as relatively high levels of rural poverty and peasant marginalization persist.

The neoliberal unravelling of the agrarian reform

The inauguration of neoliberal governments led to counter-reforms, the privatization of the reform sector and the ending of agrarian reforms. Neoliberal land policies have shifted priorities away from expropriation of estates, which typified the populist agrarian reform period, towards privatization, decollectivization, land registration, titling and land-tax issues. Legislation has also been introduced in some countries which has facilitated the break-up of indigenous communities and the sale of their land. Chile was the first to initiate this process in late 1973, Peru followed in a more gradual manner after 1980, Nicaragua since 1990 and Mexico and El Salvador since 1992. Some expropriated land was returned to former owners (particularly in Chile) but most was subdivided into family farms known as *parcelas* and sold to members of the reformed sector who henceforth were referred to as *parceleros* (Jarvis, 1992). In some instances a sizeable number of estate peasants were unable to secure a parcel, both for political reasons and because of financial circumstances, and joined the ranks of the rural proletariat. Nevertheless, this process of land subdivision significantly increased the area under the individual control of peasant farmers. However, after some years a fair proportion of *parceleros* were unable to keep up their land payments or finance their farm operations and had to sell part or all of their *parcela* to the capitalist farm sector.

The most significant symbol of the neoliberal winds sweeping through Latin America was the change in 1992 of Article 27 of Mexico's Constitution of 1917. This had opened the road to Latin America's first agrarian reform and enshrined the demand for 'land and liberty' by the peasant insurgents during the Mexican revolution. No government had dared to modify this key principle of Mexico's Constitution but the forces of globalization and neoliberalism proved too strong to resist and the government took the risk of tackling this hitherto sacred cow (Randall, 1996). This new agrarian law marks the end of Mexico's agrarian reform, allows the sale of land of the reformed sector and the establishment of joint ventures with private capital including foreign capitalists (DeWalt et al., 1994).

Agrarian reform and the subsequent neoliberal unravelling of the reformed sector have thus given rise to a more complex agrarian structure. It has reduced and transformed the *latifundia* system and has enlarged the peasant sector and the commercial middle-to-large-farm sector.

Decollectivization also increased heterogeneity amongst the peasantry as the levelling tendencies of collectivist agriculture were removed. Capitalist farmers have been the ones to benefit from the liberalization of land, labor and financial markets, the further opening of the economy to international competition, the new drive to exports and the withdrawal of supportive measures for the peasant sector. Their greater land, capital and technical resources, their superior links with national and especially international markets and their greater influence on agricultural policy ensure that they have been more able to exploit the new market opportunities than have peasant farmers.

The continuing search for agrarian reform

Poverty, exclusion and landlessness are still far too common in Latin America. The land issue has not yet been resolved. The contemporary struggle for a piece of land by the mass of landless peasants in Brazil, spearheaded by the Movimento (dos Trabalhadores Rurais) Sem Terra (MST; or the Landless Rural Workers Movement), clearly illustrates this with its campaign of selective seizure of estates and massive demonstrations (Box 11.1). The era of radical agrarian reforms, however, is over. There has been a shift from state-led and interventionist agrarian reform programs to market-orientated land policies. Paradoxically, such land policies have turned out to be much driven from above by the state and international agencies. Thus future state interventions in the land tenure system are likely to be confined to a land policy which focuses not on expropriation but on progressive land taxes, land colonization, land markets, registration, titling and secure property rights. A variety of studies are indicating that such land policies have not turned out to be the promised panacea. While the potential benefits of clearly defined property rights may be substantial given that about half of rural households lack land titles (Vogelgesang, 1996) the economic and socio-political context under which small farmers are operating conspire against them. The result is 'modernizing insecurity'. Peasants in the end turn out to be the losers from these land-titling projects because of their weak position in the market as well as in the political system, which is unable to protect their land rights.[4]

As the search for agrarian reform continues (Thiesenhusen, 1989) issues such as prices, markets, credit, technical assistance, wages, regionalization and globalization currently exercise a major influence on agriculture's performance and the peasants' well-being. While major agrarian reforms, especially of a collectivist kind, are unlikely to recur it is certainly premature to argue that current land policies and neoliberal measures are settling the agrarian problem in Latin America whose resolution will still require changes in the unequal and exclusionary land-tenure system.

It can be concluded from the above that agrarian reforms provide a framework for growth, equity and sustainable development in rural society only when accompanied by complementary policies and appropriate macro-economic measures. Whilst clearly facilitated by a favorable external environment, internal transformations remain critical

> **BOX 11.1 The battle for land in Brazil: land for the landless**
> (reproduced from Serrill, 1996: 35)
>
> **The Landless Peasant Movement: the Movimento Sem Terra (MST)** Some 1500 peasants from northern Pará state in the Brazilian Amazonia wanted land, and they were hungry enough, desperate enough, to take bold action to get it. On April 17 they blocked a highway in Eldorado de Carajás to draw attention to their demand for the right to settle on idle farmland nearby. To their consternation the state government responded with busloads of heavily armed military police. After the cops fired a volley of tear gas, the peasants charged, waving machetes, hoes, scythes and a few pistols. The police opened fire with automatic weapons.
>
> The result was the bloodiest confrontation in the 30-year history of Brazil's land reform movement. Nineteen demonstrators died and 40 more were wounded by the police fusillade. The scene, filmed by a local television newsman and broadcast repeatedly in the following day, stunned all of Brazil. . . .
>
> In the vanguard of land reform is the Landless Peasant Movement, a combative and well-organized group whose strategy is to illegally occupy uncultivated farmland. Brazil has more than 100 such squatter camps – home to 37,000 defiant families. The death toll from these rich vs. poor confrontations is high, and mounting. Since 1979 more than 200 peasants have been killed by police or by hired guns in land conflicts around the country; rarely has anyone been prosecuted for these crimes.

for determining the outcome of the agrarian process. Rather than regarding agrarian reform as a panacea, it is best seen as an instrument of transformation, albeit an important one, for the achievement of these objectives. The main legacy of agrarian reform is that it has hastened the demise of the landed oligarchy and cleared away the institutional debris which prevented the development of markets and the full commercialization of agriculture, albeit after the unravelling of the reformed sector. Thus the main winners have been the capitalist farmers. Although a minority of *campesinos* (peasants and landless laborers) gained some benefits, for the majority the promise of agrarian reform remains unfulfilled.

GLOBALIZATION, NEOLIBERALISM AND LATIN AMERICAN AGRICULTURE

In the post-agrarian reform era the major force shaping the rural economy and society has been the shift over the past one or two decades to an outward-orientated development strategy largely based on primary exports. This has further integrated Latin America's agricultural sector into the world economy. The debt crisis of the 1980s and the adoption by most Latin American countries of 'structural adjustment programs' (SAPs) stimulated agricultural exports in the hope that these would alleviate the region's foreign exchange problems. As a result of the export drive, agricultural exports have been growing much faster than production for the domestic market.

These shifting production patterns have modified the rural social structure in Latin America. It has largely been the capitalist farmers who have been able to take advantage of, and benefit from, the new opportunities: the financial, organizational and technological requirements of the export products being beyond the reach of the peasant economy. Nevertheless, through agribusiness contract farming, some smallholders have been able to participate in the production of agro-industrial products for export or for high-income domestic urban consumers. This integration of some sections of the peasantry as producers into the agro-food complex has accentuated the socio-economic differentiation process. Some peasants have been able to prosper through capital accumulation, thereby evolving into 'capitalized family farmers' (Lehmann, 1982) or 'capitalist peasant farmers' (Llambí, 1988). Others have become 'proletarians in disguise' – formal owners of a smallholding but tied to, and dependent on, agribusiness and earning an income similar to the average rural wage. Another category is that of 'semi-proletarians' whose principal source of income is no longer derived from the household plot but the sale of their labor power for a wage. Furthermore, a significant number of peasants have been 'openly' and fully proletarianized.

Latin America's agricultural modernization and performance

Agriculture continues to provide a major share of Latin American foreign exchange earnings although its contribution declined substantially in the 1970s and 1980s. Agricultural exports accounted for 44 per cent of total exports in 1970 but only 24 per cent in 1990 (ECLAC, 1993: 81). In only exceptional cases, such as in Chile, has the share of agriculture in total export earnings risen. Since the 1960s subsistence crops, which are produced mainly by the peasant sector, grew at a much lower rate than export crops, produced largely by the medium and large commercial farm sector. This reverses the trend of the 1950s and early 1960s in which agricultural production for the domestic market grew faster than production for export.

The modernization and liberalization of agriculture based on the growth of an export sector followed upon earlier modernization strategies. During the 1960s and 1970s a shift towards the intensification of Latin American agriculture took place. Many Latin American governments encouraged the modernization of the hacienda system through such measures as subsidized credits for the purchase of agricultural machinery and equipment, better quality livestock, fertilizers and improved seed varieties as well as the delivery of technical assistance programs. Consequently, large commercial farmers began to shift to higher value added crops which were in increasing demand by urban consumers and to capitalize their enterprises through land improvements, increased irrigation, upgrading infrastructure and mechanization. This process of modernization can be characterized as the 'landlord road' to agrarian capitalism as landlords themselves transformed their large landed estates into commercial profit-orientated farms (Kay, 1974). Also, green-revolution-type technologies, involving improved seeds, were increasingly adopted. In 1970 about 10 per cent of Latin America's wheat area was sown with high-yield varieties but today this has risen to 90 per cent. The spread of the green revolution, a technological package much favored by transnational agribusinesses, also contributed to the increased use of fertilizers and pesticides.

This intensification of agriculture meant that growth in output was achieved by an increase in the productivity of the various factors of production. Up to the 1980s the expansion of agriculture's land area still accounted for 60 per cent of output growth;

thereafter the intensive margin predominated as a source of agricultural growth (Ortega, 1992: 123). However, this process of capitalization has proceeded unevenly in different Latin American countries. In Brazil, agriculture continues to expand to an important extent via the extensive margin as a result of the colonization of the Amazonian frontier. Furthermore, within agriculture capitalization has been largely confined to the commercial farm sector, leaving peasant agriculture relatively unaffected.

New relations of production and changes in the rural labor structure

The technological transformation of agriculture discussed earlier has largely been confined to what has become known in Latin America as *agricultura empresarial* (entrepreneurial agriculture) which refers to the modernizing capitalist farms. Macro-economic policy, favoring the development and diffusion of capital-intensive technologies and the bias of extension services in favor of commercial farmers, has widened the technological gap between capitalist agriculture and the peasant economy, reinforcing a bimodal agrarian structure. It is difficult, if not impossible, for peasant farmers to adopt new technology. Not only is it too risky and expensive, but also it is inappropriate for small-scale agriculture and the inferior soils of peasant farming. In addition, the harmful environmental consequences of fossil-fuel-based technology is increasingly being called into question. The capital-intensive (and often import-intensive) nature of this technology is also inappropriate for Latin American economies as it requires too many scarce capital resources (such as foreign exchange) and too few workers from the abundant labor supply.

The modernization of the *latifundio* has been accompanied by a structural shift in the composition of the agricultural labor force. Compared with the traditional personalistic and clientelistic relations which existed between landlords and peasants, the relations between capitalist farmers and peasants are increasingly mediated by impersonal market forces and characterized by new forms of exploitation and subordination.

Four major changes in the composition of the labor force can be highlighted:

- the replacement of tenant labor by wage labor;
- within wage labor, the growth of temporary and seasonal labor;
- the increasing feminization of the agricultural labor force;
- the 'urbanization' of rural workers.

The decline of tenant labor

Tenant labor used to supply most of the *latifundio*'s permanent and temporary labor needs. Tenant labor became more expensive than wage labor for landlords, as the rent income received from tenants (sharecroppers, labor-service tenants, or others) became lower than the profits landlords could earn by working the land directly with wage labor. Mechanization, which was attractive because of the availability of government-subsidized credits, turned direct cultivation by landlords into a more profitable activity than tenancy. Thus the higher opportunity costs of tenancies and tenant laborers resulted in their being replaced by wage laborers, leading to an 'internal proletarianization' process (Kay, 1974). Landlords also employed fewer tenants for political reasons because of the changing political climate of the 1950s and 1960s so as to reduce the internal pressure for land reform. Already in 1973 the proportion of wage labor within the economically active agricultural population varied between 30 per cent and 40 per cent in most Latin American countries, and in a few cases it was over 50 per cent (Ibáñez, 1990: 54–6).

The growth of temporary and seasonal wage labor

Within the shift to wage labor, there has been a marked increase in the proportion of temporary, often seasonal, wage employment. In many countries permanent wage labor has declined, even in absolute terms, while in almost all countries temporary labor has greatly increased. In Brazil it is estimated that in 1985 permanent wage labor had fallen to a third of rural wage laborers; the remaining two thirds being employed on a temporary basis (Grzybowski, 1990: 21). In Chile the shift from permanent to temporary labor has also been dramatic. While in the early 1970s approximately two thirds of wage labor was permanent and a third temporary, by the late 1980s these proportions had been reversed (Falabella, 1991).

This growth of temporary labor is partly connected to the expansion of agro-industries that export seasonal fruit and vegetables and is particularly evident in those Latin American countries that export these products. This has led to the increasingly 'casualized' or precarious nature of rural wage labor. Temporary workers are generally paid by piece rates, are not usually entitled to social security benefits and have no employment protection. These changes in employment practices towards more casual and flexible labor enable employers to increase their control over labor by reducing workers' rights and bargaining power. Their introduction has been facilitated by regressive changes in labor legislation, introduced often by the military governments but continued by their neoliberal civilian successors. The expansion of temporary wage labor therefore represents a deterioration in the conditions of employment.

This casualization of rural labor has contributed to the fracturing of the peasant movement. Although seasonal laborers can be highly militant they are notoriously difficult to organize because of their diverse composition and shifting residence. Thus the shift from permanent to seasonal labor in the countryside has generally weakened peasant organizations making it difficult for them to negotiate improvements in their working conditions either directly with their employers or indirectly by pressurizing the state.

The feminization of rural seasonal wage labor

Associated with the expansion of temporary and/or seasonal wage labor is the marked increase in the participation of women in the labor force. In the past, rural women worked as day laborers, milkmaids, cooks or domestic servants on the landlord's estate. They also found seasonal wage employment during the labor-intensive harvests on coffee, cotton and tobacco farms. With the increasing commercialization of agriculture and the crisis of peasant agriculture an increasing proportion of rural women have joined the labor market (Lara Flores, 1995). This has resulted in a renegotiation of gender relations within the household as can be ascertained from Box 11.2.

The rapid expansion of new export crops such as fruits, vegetables and flowers, however, has opened up employment opportunities for women. Agro-industries largely employ female labor since women are held to be more readily available, more willing to work on a seasonal basis, accept lower wages, are less organized and, according to employers, are better workers for activities which require careful handling. Although

> **BOX 11.2 Household gender relations and non-traditional agricultural exports** (reproduced from Barrientos et al., 1999)
>
> **Women go out to work: Chilean 'temporeras'** In the neighboring community, both Rachel and Ximena had grown up working in the grape packing plants during the summer and they wanted to continue once they had married and moved to Tomé Alto. They discovered that in Tomé Alto there was no precedent for women to go out to work.
>
> > We were the first to work in the packing plants. We arrived in 1979 and no women went out to work. The men here were very machista and didn't let them. When our husbands met us they knew that we worked in the grapes but we still had to convince them to let us continue. We heard that Don Jaime was looking for workers and so we went together without our husbands' permission to look for work. We were the only ones from El Tomé to look for work ... then one year they were short of workers and they asked us if we knew anyone. We took some women who wanted to work and now everyone works.
>
> They effectively acted as pioneers, introducing the idea that women could go out to work in the harvest.

they are employed in generally low-skilled and low-paid jobs, for many young women these jobs provide an opportunity to earn an independent income and to escape (at least partially and temporarily) from the constraints of a patriarchal peasant-family household. Even though the terms of their incorporation are unfavorable, this does not necessarily imply that gender relations have remained unchanged. Furthermore, with the rural women's rising incorporation into the formal labor market they have begun to exercise increasing influence in the affairs of peasant organizations and, in some instances, have even established their own organizations (Stephen, 1993).

In Mexico, about 25 per cent of the economically active rural population are employed in fruit and vegetable production and half of them are women. In Colombia over 70 per cent of the labor employed in the cultivation of flowers for export and about 40 per cent of coffee harvesters are women (ECLAC, 1992: 103). In Chile about 70 per cent of temporary workers in the fruit export sector are women, employed mainly in the fruit packing plants (Fig. 6.3, p. 149). It is estimated that in Ecuador in 1991 69 per cent of workers in non-traditional agro-export production were women (Thrupp, 1996: 69). The rapid expansion of nontraditional agricultural exports (NTAEs), in which female labor is a key ingredient, has occasionally extracted high environmental and health costs, as these activities are accompanied by an intensive use of chemical fertilizers, pesticides and herbicides which pollute the environment and create dangerous health hazards (Box 11.3).

The 'urbanization' of rural labor

An additional dimension to the growth of temporary wage labor concerns the geographical origins of the workers so employed. An increasing number of temporary workers come from urban areas. In Brazil about half of the temporary workers employed in agricultural activities are of urban origin. They are known as 'bóias frias' ('cold luncheons'), as they go to work with their lunch box containing cold food, and

> **BOX 11.3 Environmental and health costs of 'successful' nontraditional agricultural exports** (reproduced from Stewart, 1996: 132)
>
> **The price of a perfect flower** Carmenza bends tenderly over her baby son, adjusting the shawl in which he is carefully swaddled. He is so young that he does not even have a name, but he's been admitted to hospital with bronchial problems. 'The doctor said it was because of the pesticides I breathed in while I was pregnant' Carmenza says. She is one of the 72,000 workers, most of them women, who work in Colombia's flower industry. For the price of a small bunch of carnations in a typical London florist, she works eight to twelve hours a day – perhaps more during the peak season before Valentine's Day when 14 million roses are jetted off to the United States alone. She is proud of the beautiful blossoms she grows. But she is worried about her health and that of her child.
>
> Fainting, dizziness, skin irritations, respiratory ailments, neurological problems, premature births – these are some of the symptoms which flower employees and doctors say are due to pesticides at work. At one regional hospital, up to five people a day arrive, acutely poisoned, with another twenty showing signs of chronic effects.

'volantes' ('fliers', or floating workers), who reside on the periphery of cities or towns and fluctuate between rural and urban employment. About three-quarters of female volantes are employed in the coffee-growing industry and when there is no agricultural work they tend to look for employment in the urban areas largely as domestics (ECLAC, 1992: 98). The growing presence of labor contractors (*contratistas*) who hire gangs of laborers from small towns and cities for work in the fields means that the direct employer is not always even the farm owner or manager. Increasingly, rural residents have to compete with urban laborers for agricultural work, and vice versa, leading to more uniform labor markets and wage levels.

PEASANT FUTURES: A PERMANENT SEMI-PROLETARIAT?

The internationalization of Latin America's agriculture, the demise of the hacienda system and the increasing dominance of entrepreneurial agriculture are having a profound impact on the peasantry. How are these major transformations affecting the development of the peasant economy, especially in the wake of the increasingly widespread and entrenched neoliberal policies pursued by most governments throughout Latin America? Can the peasant economy provide adequate productive employment and rising incomes? Will peasant producers be able to increase productivity thereby stemming the erosion of their past role as a major supplier of cheap food or will they become a mere supplier of cheap labor to the capitalist entrepreneurial farm sector? Or, will they become fully proletarianized?

The contemporary significance of the peasant economy

The peasant household farm sector is still significant within Latin America's rural economy and society. The peasant economy has not faced a unilinear decline. In particular, land subdivision in the reformed sector in Chile and Peru and, more recently, Nicaragua has significantly expanded the peasant sector. But these are exceptions that might also prove to be temporary. It is estimated that peasant agriculture in the 1980s in Latin America constituted four-fifths of farm units, possessed a fifth of total agricultural land, over a third of the cultivated land and over two-fifths of the harvested area (López Cordovez, 1982: 26). The peasant economy accounted for almost two-thirds of the total agricultural labor force, the remaining third being employed by entrepreneurial or capitalist farms. Furthermore, peasant agriculture supplied two-fifths of production for the domestic market and a third of the production for export. Their contribution to food products for domestic consumption is particularly important. At the beginning of the 1980s, the peasant economy provided an estimated 77 per cent of the total production of beans, 61 per cent of potatoes and 51 per cent of maize. In addition, the peasant economy owned an estimated 24 per cent of the total number of cattle and 78 per cent of pigs.

While the peasantry is far from disappearing, it is not thriving either since their relative importance as agricultural producers continues to decline. According to de Janvry et al. (1989b) the Latin American peasantry is experiencing a double squeeze. First, they face a land squeeze. By failing to acquire additional land to match their increased numbers, the average size of peasant farms has decreased. This decline of the peasant sector mainly concerns the small peasantry (*minifundistas*) which accounts for about two thirds of peasant farm households. Their average farm size decreased from 2.1 ha in 1950 to 1.9 ha in 1980. The remainder of the peasant sector retained an average farm size of 17 ha, partly through the implementation of redistributive land reforms (de Janvry et al., 1989a: 74).[5] Second, peasants face an employment squeeze as employment opportunities have not kept pace with the growth of the peasant population and as they face increased competition from urban-based workers for rural employment.

This double squeeze on the peasant economy has led many peasants to migrate, feeding the continuing and high rate of rural out-migration (see Chapter 10). Peasants have also responded by seeking alternative off-farm sources of income (such as seasonal wage labor in agriculture) and/or nonfarm sources of income. In many Latin American countries over a quarter of the economically active agricultural population currently reside in urban areas. The proportion of the economically active rural population which is engaged in nonagricultural activities is rising, reaching over 40 per cent in Mexico and averaging about 25 per cent in others (Ortega, 1992: 129). Thus nonfarm employment is expanding faster than farm employment in rural Latin America. This trend means that an increasing proportion of total peasant household income originates from wages, whereas income from their own-farm activities often comes to less than half the total (de Janvry et al., 1989a: 141).

This process, which can be called semi-proletarianization, is the main tendency unfolding among the Latin American peasantry. It is the small peasantry who can be more accurately characterized as semi-proletarian as between two-fifths and three-fifths of their household income is derived from off-farm sources, principally from seasonal agricultural wage employment on large commercial farms and estates. As the small peasantry is the most numerous, it can be argued that this process of semi-proletarianization is dominant. However, it is less marked in those few Latin American countries where land reforms significantly increased peasant access to land.

The Latin American peasant sector has increasingly become a refuge for those rural laborers who are unable or unwilling to migrate to the urban areas and who cannot find permanent employment in the capitalist farm sector. Thus, while the peasant economy increased its share of employment by 41 per cent between 1960 and 1980, employment in capitalist agriculture increased by only 16 per cent (de Janvry et al., 1989a: 59). Furthermore, rapid technological improvements in the capitalist farm sector and the insufficient land and capital resources of the peasant farm sector and its technological stagnation make inevitable a decline in the peasants' role as agricultural commodity producers unless corrective measures are taken by the state.

In short, Latin America's peasantry appears to be trapped in a permanent process of semi-proletarianization and of structural poverty. Their access to off-farm sources of income, generally seasonal wage labor, enables them to cling to the land, thereby blocking their full proletarianization. This process favors rural capitalists as it eliminates small peasants as competitors in agricultural production and transforms them into cheap labor. Semi-proletarianization is the only option open to those peasants who wish to retain access to land for reasons of security and survival or because they cannot find alternative permanent productive employment, either in the rural or the urban sector.

Structural adjustment and rural poverty

Agricultural modernization in Latin America, with its emphasis on capital intensive farming and the squeeze on the peasant economy, means that rural poverty remains a persistent and intractable problem. Structural adjustment programs and stabilization policies of the 1980s had a detrimental impact on poverty, although more in the urban than the rural sector (Altimir, 1994). Adjustment policies exacerbated poverty as government expenditure on social welfare, subsidies to basic foods and other essential services were cut back (see Chapter 10). However, some governments reduced this negative impact by targeting welfare payments more closely and by introducing poverty alleviation programs.

Estimates of poverty vary because of the inadequacy of the data and the different methodologies and definitions employed. Despite differing estimates, all authors agree that poverty is far too high and shifting to the urban areas as a result of the continuing high rates of rural out-migration. However, the proportion of people in poverty still remains higher in rural areas. The incidence of rural poverty is particularly high in Haiti, Bolivia, Guatemala, Honduras, Nicaragua, El Salvador, the Dominican Republic, Brazil, Mexico, Peru, Ecuador, Paraguay and Venezuela, in each of these countries over half of the rural population live in poverty (Ibáñez, 1990: 20; Mejías and Vos, 1997). Within countries, indigenous communities and rural women are particularly vulnerable to poverty.

The main cause of rural poverty is structural, being related to the unequal land distribution and the increasing proportion of semi-proletarian and landless peasants. Tackling the root causes of poverty will require major land redistribution and rural investments, raised employment opportunities and improved agricultural productivity, particularly of smallholders. Only by such an assault on various fronts will it be possible to alleviate rural poverty significantly. Latin America's poverty is directly related to the unresolved agrarian question. How long government neglect of the rural poor is sustainable remains an open issue.

State, market and civil organizations: what future for the peasantry?

Neither the state-driven ISI development strategy (1950s to 1970s) nor the market-driven policies of the 1980s and 1990s have been able to resolve the peasant question. Rural poverty and unequal rural development are still with us. It was only during the brief land reform interlude, which brought in its wake major peasant organizations and mobilizations, that sections of the peasantry began to emerge from their marginalized situation only to see their hopes for a better future vanish with the counter-reform and neoliberal project. However, these past upheavals have created new opportunities as well as constraints. Calls for new thinking for new policies for rural development practices are multiplying. Such voices are seeking to find new ways of combining state action with market forces and civil organizations so as to make a fresh attempt to resolve the agrarian question (de Janvry et al., 1995). One of the key actors in this process has to be the peasantry and the rural proletariat.

What then are the prospects for a peasant path to rural development? It is well known that access to capital, technology and domestic and foreign markets, as well as to knowledge and information systems, are becoming increasingly important relative to access to land in determining the success of an agricultural enterprise. Even though in recent decades some peasants managed to gain access to land through agrarian reforms this by no means secured their future survival. The widening technological gap between the capitalist and peasant farm sectors have prompted those involved with the peasants' well-being to urge international agencies, governments and nongovernmental organizations (NGOs) to adapt existing modern technologies to the needs of the peasant sector and to create more 'peasant-friendly', appropriate and sustainable technologies. Such a policy, however, runs the danger of relying exclusively on technological fix, while the sustainability of peasant agriculture depends on wider social and political issues and particularly a favorable macro-economic context. In short, a viable peasant road to rural development raises questions about development strategy and ultimately about the political power of the peasantry and their allies.

In recent years, concerned scholars and institutions have become increasingly vociferous in pointing out the adverse impact of Latin America's 'selective' agricultural modernization on the peasantry. As opposed to the 'concentrating and exclusionary' character of this process, they call for a strategy that includes the peasantry in the modernization process (Murmis, 1994). This would be part of the democratization of rural society and some authors speak of 'democratic modernization' to highlight this link (Chiriboga, 1992). Currently, suggestions are being made with a view to 'changing production patterns with social equity' in Latin America and for the 'productive reconversion' of agricultural producers so as to meet the challenges of an increasingly global world economy in the new millennium (ECLAC, 1990). To forward these aims, special government policies in favor of the peasantry (a form of positive discrimination) are proposed, to reverse the past bias in favor of landlords and rural capitalists. The achievement of broad-based growth requires proactive state policies so as to overcome market failures and biases against the poor while at the same time harnessing the creative and dynamic forces of markets in favor of the rural poor.

Reconversion and nontraditional agricultural exports (NTAEs)

The key for the development of peasant farmers and their transition to 'capitalized peasant farms', especially in these days of privatization, liberalization and globalization is to enhance their market competitiveness. For this purpose some governments in Latin America are beginning to design policies for the 'reconversion' of peasant farming which has been referred to in a variety of ways such as 'readaptation to more profitable options' and 'new productive and market options'. In a broad sense reconversion aims at enabling and improving peasant agriculture's ability to adapt to its increasing exposure to global competition and to enter into the more dynamic world market. This is to be achieved through a series of specific peasant programs with the purpose of raising productivity, enhancing efficiency and shifting traditional production and land-use patterns to new and more profitable products thereby increasing the peasants' competitiveness (Figures 11.2 and 11.3) (Kay, 1997).

Governments and NGOs concerned with promoting the development of peasant farmers proposed a series of measures for facilitating their participation in the lucrative agricultural export boom. It was almost exclusively capitalist farmers who initially reaped the benefits of the thriving 'nontraditional agricultural export' (NTAE) business as they had the resources to respond relatively quickly to the new outward-looking development strategy of the neoliberal trade and macro-economic policy reforms. In view of the dynamism of the NTAE sector it was thought that a shift in the production pattern of peasant farmers to these products would spread the benefits of NTAE growth more widely and ensure their survival. However, experience has been rather mixed.

Figure 11.2 Field of lilies, Chile, 1994. A rural extension officer provides technical assistance to Mapuche peasant smallholders in the south of Chile. This indicates both the wide reach of globalization and the efforts of the Concertación government to assist smallholders to find alternative sources of income. The bulbs are provided by a private firm from Holland which sends them to Chile so as to take advantage of the counter-season and allow the bulbs to multiply. The bulbs are grown not for the flower but for reproduction. The government's extension agency, INDAP, provides technical assistance and credit for smallholders and supported the construction of a packing and storage plant. Bulbs are exported back to Holland and the firm then sells them worldwide. Photograph by C. Kay.

FIGURE 11.3 Inside a smallholder's greenhouse, Peumo, Chile, 1996. Smallholders in Chile are becoming more specialized and technical: here carnations are grown for national markets but only family labor is used. The smallholder benefits from being a member of a local cooperative both for technical and marketing assistance. Photograph by C. Kay.

To analyze the impact of NTAE growth on smallholders and rural laborers Carter *et al.* (1996: 37–8) argue that this depends on three factors: first,

> whether small-scale units participate directly in producing the export crop and enjoy the higher incomes generated from it (which we call the 'small-farm adoption effect'); second, whether the export crop induces a pattern of structural change that systematically improves or worsens the access of the rural poor to land (the 'land access effect'); and third, whether agricultural exports absorb more or less of the labour of landless and part-time farming households (the 'labour-absorption effect').

They examine the cases of agro-export growth in Paraguay (based on soybeans and wheat), in Chile (based on fruit) and in Guatemala (based on vegetables). Their findings reveal that only in the case of Guatemala was there a broad-based growth as both the land access and net employment effects were positive, while the opposite happened in Paraguay, resulting in exclusionary growth. The Chilean case had elements of both as the net employment effect was positive but the land access effect was negative. Thus in Chile the fruit-export boom has been partly exclusionary, as many peasant farmers (in this case largely *parceleros*) have sold part or all of their land as they were squeezed by the export boom, and partly inclusive, as the shift from traditional crops to fruit-growing increased labor demand.

Even if a larger proportion of peasant farmers were to adopt the new export crops it is far from certain that this will ensure their survival. Thus the much fancied NTAE rural development policy of many Latin American governments cannot be considered as a panacea, especially if no complementary measures are taken to create 'level playing fields' (Carter and Barham, 1996). The Chilean experience is in this regard illustrative (see Chapter 6 and Gwynne and Kay, 1997).

First, there has been a very low adoption rate of NTAEs by small-scale farmers because of financial, technical, risk and other factors. Second, even those who did switch to NTAEs were far more likely to fail as compared with capitalist farmers as they were less able to withstand competitive pressures because of their disadvantaged position in marketing, credit, technology and other markets. According to Murray (1996) three stages can be distinguished in the transition of peasant farmers (largely *parceleros*) to fruit production for global markets. In the first stage only a small percentage undertake a limited production of fruit for local and national markets. In a second stage a larger proportion switches to fruit growing as well as to the expanding fruit export economy. However, in the third stage, which in Chile began in the late 1980s and is continuing today, peasant farmers begin to get squeezed as a result of the increasing competitive nature of the export market. As a consequence of rising debts, among other factors, many are forced to sell all or part of their land thereby contributing further to the ongoing process of land concentration. Such an ongoing process of land concentration is also happening in other Latin American areas in which NTAEs are taking hold.

Food import substitution (FIS) and sustainable development

An almost forgotten alternative or additional possibility to NTAEs for peasant farmers is to enhance their comparative advantage in staple food production. This can be achieved through a program of 'food import substitution' (FIS) as suggested by de Janvry (1994). More radical proposals call for the redevelopment of the peasant economy through an 'autonomous development' strategy that is seen as the key for sustainable development in rural areas. However, for an autonomous development strategy to succeed major supportive policies by the state are required, such as specifically targeted protectionist measures to counteract the distortions in the world food market arising from subsidies to farmers in the developed countries.

Policies aimed directly at strengthening the position of the peasantry in local and global food markets would entail the creation of 'level playing fields'. At present these market fields are heavily biased against peasant farmers and rural laborers. Import substitution in staple foods has the advantage not only of saving valuable foreign exchange but also of enhancing food security, employment and possibly a more equitable income distribution, especially if it is peasant farmers who undertake this FIS. The expansion of peasant food output has also the advantage of being more ecologically friendly as they use less chemical inputs as compared with capitalist farmers and also relative to NTAEs. Instead of viewing NTAEs and FIS as being in conflict or as alternatives, they can be seen as complementary. In Schejtman's (1994) view it is possible to envisage a positive correlation as those peasants who are able to go into lucrative agro-exports can use their increased incomes, knowledge and market experience derived from NTAEs to invest in raising the productivity of their traditional food crops.

Nongovernmental organizations (NGOs)

A new relationship has to develop between the state and civil society by which the state devolves some of its powers, initiatives, financing and activities to local governments and civil organizations such as NGOs, producer and consumer organizations, trade unions, women and ecological associations and political parties. These should play an increasing role in policy formulation and implementation. NGOs are known to be particularly able to establish close working relationships with grass-roots organizations and their constituency. Such increased participation of individuals and civil organizations in economic, social and political affairs is likely to strengthen the democratic processes. By creating a more participatory framework it might be possible to establish mechanisms for regulating and governing the market for the benefit of the majority in society.

In some instances governments in Latin America have already begun to subcontract certain activities such as technical assistance for peasant farmers to NGOs, as well as giving greater powers and resources to local administrative agencies. It is too early yet to assess the significance and impact of such initiatives. However, NGOs face a dilemma when they become to depend too closely on government resources and appear to be implementing government policy, as they might lose the support from the grass-roots and thus their legitimacy. But if NGOs are in turn able to influence government policy by making it more sensitive and friendly towards peasant, gender, indigenous and ecological issues then this closer relationship is only to be welcomed. Generally, NGOs have limited resources and this constrains the coverage of their activities to a limited number of beneficiaries. In those countries where the state has been drastically downsized NGOs have often been used as a palliative to overcome the abdication of social responsibility by the state. Thus the closer links between state and NGOs can be a mixed blessing (Bebbington and Thiele, 1993).

The key agrarian question: assets and power

The increasing competitive gap between peasant and capitalist farming due to agriculture's unequal modernization limits the survival of the peasant producers and perpetuates rural poverty. A few enlightened neoliberals accept that rural markets in Latin America are distorted and biased against the peasantry and hinder the pursuit of efficiency and maximization of welfare (Binswanger et al., 1995). The slogan of 'getting prices right' is certainly not a panacea for rural development and its proper achievement entails structural reforms. A major step in tackling the agrarian question requires a redistribution of assets as well as the empowerment of peasants and rural workers. While land titling programs for peasants, which became fashionable in the 1980s, may give greater security of tenure and thereby encourage investment they are restricted in scope. Although agrarian reforms are no longer on the political agenda, except in Brazil, the problem of land concentration remains. Land policy reforms are far from dead as a broad-based and sustainable development strategy requires a fairer distribution of land assets.

However, access to finance and knowledge are increasingly important assets in today's globalized world. This calls for government policies that facilitate peasant access to these other two crucial assets through market reforms, human resource development and special credit and technical assistance programs. Some of these projects can be implemented by NGOs and the private sector. Governments have to give greater priority to rural education and to infrastructural works, such as irrigation and road projects, that are targeted at smallholder communities.

Such policy reforms have little chance of succeeding unless peasants and rural workers develop their own organizations such as producer associations, cooperatives and trade unions. It is only through the creation of a countervailing power by peasants and rural workers and by exercising constant pressure that they will be able to shape the future to their advantage rather than having continually to accept the disadvantages of the past and present. While undoubtedly the state, political parties and NGOs can provide the necessary supportive role, the development of such organizations depends on the determination of peasants and workers themselves. Although it is difficult to develop such organizations it is also true that the removal of structural constraints of the kind mentioned earlier is surely going to facilitate the empowerment of peasants and rural workers.

Whether or not these proposals will be adopted is an open question, but there are grounds for some optimism as new opportunities have emerged for going beyond the debt crisis. Real exchange rate devaluations should favor peasant farmers, as they make more intensive use of labor and less use of chemical inputs, compared with capitalist farmers whose costs of capital and tradable inputs increase. Meanwhile trade liberalization has removed some biases against agriculture, although it is important to remember that 'urban bias' was not the main cause of all rural ills. These changes provide incentives for import substitution in staple foods that should benefit peasant farming. New technological advances in agro-ecology and social forestry, although still limited in their application, tend to favor peasant farmers. Last, but not least, the rapid expansion of NGOs has certainly made governments more sensitive to issues of poverty, equity, gender and ecology. The extent to which these new opportunities are resulting in meaningful changes in favor of the peasantry remains to be seen.

The new peasant movement: indigenous and environmental issues

The neoliberal project has certainly not gone unchallenged by peasants. The peasant rebellion in Chiapas, the most southern and indigenous region of Mexico, at the beginning of 1994, was fuelled by the exclusionary impact of Mexico's agricultural modernization on the peasantry and by fears that Mexico's integration into the North American Free Trade Agreement (NAFTA) will marginalize them further (Harvey, 1994; de Janvry et al., 1997). Undoubtedly, Mexico's peasant economy cannot compete with the large-scale mechanized maize and cereal farmers from North America unless special protective and developmental measures are adopted in their favor (Collier, 1994).

Indeed the Chiapas rebellion has come to symbolize the new character of social movements in the countryside in Latin America which are at the forefront of the struggle against neoliberalism (Veltmeyer et al., 1997: Chapter 10). The peasantry is striking back and it would be a serious mistake to dismiss these new peasant and indigenous movements in Latin America as the last gasp of rebellion. These new movements are shaping new class and ethnic identities in which the protagonists are affirming their own history and capacity to make history. The confident assertions, from opposite political spectrums, on 'the end of history' (Fukuyama, 1992) and the death of the peasantry (Hobsbawm, 1994: 289) are proving to be premature.

Over the past decade the peasantry has re-emerged as a significant force for social change not only in Mexico but also in Brazil, Ecuador, Bolivia, Paraguay, Colombia and El Salvador, among other Latin American countries. In Brazil the principal protagonist in the countryside has been the landless rural workers movement of the MST which has spearheaded over a thousand land invasions or take-overs of estates, demanding their expropriation. This comes as no surprise as land inequality is particularly acute in

Brazil, where 4 per cent of farm owners control 79 per cent of the country's arable land (Veltmeyer *et al.*, 1997: 181). In these land occupations a variety of peasants were involved, mainly rural semi-proletarians or proletarians, such as waged workers, squatters, sharecroppers and tenants. Through their direct actions, which also involve blocking highways and sit-ins at local offices of the state's agrarian reform institute, they had by 1994 pressurized the government to settle over 120 000 families since the beginning of their actions 10 years earlier. In this struggle there have been many casualties as *fazendeiros* (landlords) and their hired guns took the law into their own hands, generally with absolute impunity. Many protestors also died and were wounded in clashes between the militarized police and the landless peasants (Box 11.1).

Environmental and ethnic questions have become increasingly important political issues as the fate of the tropical forests and the fate of indigenous peoples have become more intertwined. Environmental movements have become struggles for social justice as native groups have been displaced and their livelihoods threatened through the actions of companies exploiting the natural resources through logging, mining, oil extraction, the building of dams for hydroelectric power stations and deforestation for pastureland and cattle raising. The ensuing conflicts between these companies, cattle ranchers and the local population, which have often led to casualties, activated human rights groups in defence of the victims. This coalition of indigenous, peasant, environmental, human rights and other organizations became one of the major forces in the fight for social justice (Kaimovitz, 1996).

In the case of Chiapas, with its large Mayan Indian population, the linkage of the indigenous question to environmental concerns gave the uprising wider support and strength. In Brazil the building of the Transamazon Highway in the 1970s led to large-scale deforestation and expansion of pastureland as corporate capital was lured by tax rebates, subsidies and cheap credit to Amazonia. This led to large-scale migration of settlers, largely from impoverished northeastern Brazil, to tropical forest areas and contributed to the environmental deterioration of these areas. This expansion of grazing and mining encroached on lands used by indigenous groups and rubber tappers in what Dore (1995: 262) has called the most extensive enclosure movement in history. This sparked off the rubber tappers movement as well as the actions of native indigenous groups in defence of their livelihoods, which brought the Amazon environmental issue to world attention. The assassination of 'Chico' Mendes in 1988 did provoke an international outcry (Box 11.4). His murder led to strong national and international pressure which prompted the government into action by acceding to some of the rubber tappers demands with the establishment of extractive reserves which attempt to reconcile the conservation of the forest with its sustainable use in supporting local livelihoods. The first extractive reserve was created in 1990 and many others followed thereafter (Hall, 1996).

Over the past decade there has been a resurgence of Indian ethnic identity in many Latin American countries which has revitalized and changed the character of the social movement in the countryside (see Chapter 9). In Ecuador, major social mobilization took place in 1990 and 1994 both of which were organized by the Confederation of Indigenous Nationalities of Ecuador (CONAIE). This organization derived its strength and following through its ability to incorporate organizations of Indians from the highlands as well as from native ethnic groups of the Amazon. During an entire week in 1990, tens of thousands of Indian peasants blocked highways, organized marches in various cities and seized government offices (Zamosc, 1994). Their protest was brought

> **BOX 11.4 Chico Mendes and the Brazilian rubber tappers' environmental movement** (reproduced from Hall, 1996: 93)
>
> **Chronicle of a death foretold** Several years have passed since the fateful day of 22 December 1988 when Francisco Alves Mendes was struck down by an assassin's bullet. As leader of the Xapuri Rural Workers' Union, located in the western Amazonian state of Acre, 'Chico' Mendes had helped spearhead one of the most significant social–environmental movements in Latin America, that of the rubber tappers or *seringueiros*. His murderers, recently arrived cattle ranchers or *fazendeiros* from the south of Brazil who resented the tappers' growing power to resist land-grabbing and the destruction of their rubber stands, were eventually caught, tried and convicted. That the assassins were brought to justice at all was due largely, however, to the massive international campaign which immediately followed the murder. The perpetrators' subsequent escape from goal came as no real surprise in a country where 99 per cent of homicides arising from rural land conflicts go unpunished.
>
> Chico Mendes had lived in the shadow of death for a decade. At his fortieth birthday party on 15 December 1988, one week before he was killed, he stoically predicted, 'I don't think I am going to live until Christmas.' Mendes was the ninetieth rural activist to have been murdered in Brazil that year. The previous 89 had produced little or no impact, but his death provoked a massive national and global reaction which has had lasting repercussions for Amazonia. Over 4000 people accompanied the funeral cortege in Rio Branco on Christmas day, while the world's news media reverberated with headlines and leading articles examining events leading up to the assassination and highlighting the rubber tappers' struggle to defend their livelihoods.

about by the economic recession resulting from the structural adjustment package. In the second mobilization of 1994 the protest was directed specifically against the introduction of neoliberal policies, especially the new so-called 'Agrarian Development Law'. This law threatened the communal lands of indigenous groups, facilitating their privatization and ultimately favoring their transfer to capitalist farmers through the market mechanism. Thousands of indigenous communities, representing all of the country's ethnic nationalities, participated in this major 1994 mobilization. Peasants, small farmers, trade unions and popular organizations also joined this protest and international environmental and human rights organizations offered their support (Picari, 1996). In Bolivia, an historic 'march for territory and dignity' took place in 1990 in which hundreds of people from lowland indigenous groups 'trekked from the Amazon rainforest through the snow-capped Andes en route to the capital city to protest logging on indigenous lands and to demand legal rights to these lands' (Albó, 1996: 15).

This new peasant movement differs from past social movements in the countryside for various reasons. First, ethnic groups have a greater presence than in the past. There is also a greater degree of ethnic consciousness and, in some cases, even demands for national autonomy expressed in self-government and territorial sovereignty (Box 11.5). While governments have not yielded to the claim for national autonomy, countries such as Bolivia, Ecuador, Colombia and Brazil have changed their respective constitutions recognizing the multi-ethnic character of the nation and including provisions which recognize the linguistic, cultural, social and territorial rights of the various indigenous groups.

> **BOX 11.5 Zapatistas and indigenous struggles for land and human rights** (reproduced from Ross, 1997: 35)
>
> The Zapatista National Liberation Army (EZLN) has been in the forefront of the struggle for indigenous autonomy, which incorporates concepts of control over the exploitation of natural resources found on or under Indian lands. This position directly challenges the Mexican Constitution, which says that the mineral wealth of the subsoil belongs to the entire nation. It is an even greater challenge to the neoliberal idea that the lands belongs to those who buy them.
>
> As the process of globalization facilitates access to Mexico's natural resources, farmers and indigenous peoples, for whom land, water, forests and mineral wealth are linked to culture and community, are resisting such penetration with mounting militancy. If these struggles against the depredations that globalized resource exploitation inflicts upon rural lands and communities are ecological ones, then Mexico's environmental movement is among the most dynamic in the Americas. But regardless of definition, this emerging constellation of struggles over land, the environment and human rights is now a powerful engine for social change in Mexico.

Second, owing to the social transformations of the peasantry, the movement has acquired a more urbane and international dimension. This has been helped by factors such as the more fluid relationships between the urban and rural sectors, the greater mobility of the rural population, the improvements in rural education and the more pervasive influence of the media. Thus the new leadership has become more adept at promoting the movement's objective by making skilful use of the media thereby reaching a wider audience both nationally and internationally.[6] To the 'globalization from above' they have been able to generate a 'globalization from below', by-passing national governments and creating international pressure on them as well as on other international organizations by appealing to the foreign constituency (Kearney and Varese, 1995). For example, the World Bank stopped financing large-scale hydroelectric dams in Brazil as a result of such actions.

Third, the peasant movement has achieved a greater degree of autonomy from political parties and the government. While this is partly a result of the greater maturity achieved by the movement through past struggles it has also been made possible by the political vacuum resulting from the crisis of the left-wing parties. The world crisis of socialism could not fail to weaken socialist organizations and often led to left-wing parties adopting elements of the neoliberal agenda. The weakening of the landlords' former dominance over the peasantry in countries with radical agrarian reforms has also opened up new political spaces for the peasantry as well as new social actors. There is more willingness to resort to direct action such as land invasions, road blocks, occupation of government offices and even towns, sometimes armed with weapons.

Fourth, peasant movements have far closer links to NGOs that have made important contributions in creating or strengthening grass-roots organizations in the countryside. International NGOs have also provided a useful vehicle for raising worldwide support for the new movements' causes, especially if these concern ecological, social justice and human rights issues. Women have a greater presence in these new peasant movements, although still less than their relative importance would warrant. Women have figured especially prominently in some of the indigenous and human rights movements.

This new character of the social movement in the countryside does not mean that traditional concerns have vanished. Thus demands for better wages and working conditions, land, improved prices for peasant products, greater and cheaper access to credit and technical assistance continue to be made and in some instances with even greater urgency than in the past. But new issues have emerged, such as the environment, and some issues have become either less prominent or have acquired a different meaning. The land question has acquired a new connotation with the conflicting territorial claims made by capitalists, small settlers and indigenous groups as well as by the new ecological concerns.

The new peasant and indigenous movements are not struggling for a mythical past and utopia. They do, however, reject contemporary modernity with its current neoliberal and globalization processes as this is exclusionary and often threatens their survival, whether physically, socially or culturally. Such modernity is regarded as reckless, hypocritical and bigoted (Zamosc, 1994). Instead they are struggling for a different modernity which rests on their own emancipatory project which includes greater control over their lives, more security and a better standard of living. Thus the challenge for the new movements is to use the current processes of modernization for their own interests by controlling them, taking them further and, where possible, developing their own alternatives. Thus peasants have to increase their ability to master their own environment as well as to participate in more favorable terms in the global environment (Bebbington, 1996).

Conclusions

This chapter shows how Latin America's rural economy and society have been transformed in recent decades as a consequence of the widening and deepening of capitalist relations in the countryside and its further integration into the world economy. Latin America's agriculture is now an inherent part of the new world-food regime. Agroindustrial modernization and globalization have profoundly changed the technical and social relations of production in the countryside. Furthermore, the recent shift to economic liberalism is intensifying these changes and bringing about new structural transformation.

This form of modernization has benefited only a minority of the rural population and has excluded the vast majority of the peasantry. The beneficiaries are a heterogeneous group, including agro-industrialists, capitalist farmers and some capitalized peasant households. The losers are the semi-proletarianized and fully proletarianized peasantry, the majority of rural laborers whose employment conditions have become temporary, precarious and 'flexible'. Some landlords have also lost out, especially in countries where more radical agrarian reforms were implemented or where they succumbed to competition following trade liberalization.

With the increasing integration of Latin America's rural sector into the urban sector, the boundaries between rural and urban have become ambiguous. The massive rural out-migration has partly 'ruralized' the urban areas, and the countryside is becoming increasingly urbanized. Urban and rural labor markets have become more closely interlinked. The land market has become more open and competitive, enabling urban investors and international capital to gain greater access to agricultural land. Competition among agricultural producers has intensified as a result of the more fluid situation in the land, capital and labor markets.

While the rural economy and society are less important today than in the past, it still retains critical significance in most Latin American countries. The 'lost decade' of the 1980s, when structural adjustment programs proliferated throughout Latin America, reveals the strength of the rural economy in confronting the debt crisis and responding to changed circumstances such as the new impetus to export agriculture. To ignore the agrarian question of unequal access to land, rural poverty and exclusionary modernization is ill-advised. In Brazil and Guatemala, the land problem has not yet been properly addressed whilst in many others it remains unresolved. Rural poverty remains widespread, and discrimination against indigenous communities is still pervasive. The continuing promotion of agro-exports further depletes natural resources, and societal forces are still not strong enough to prevent ecological deterioration. Nevertheless, the environmental movement has emerged as a major social force in recent years, forcing governments to introduce environmental legislation, but the practical outcome is still unclear.

Although the shift from a state-centered inward-directed development process to a neoliberal market-orientated and export-orientated model has weakened the power of traditional peasant organizations through the fractioning of rural labor, new social conflicts have erupted in the countryside. A new peasant and indigenous movement has emerged in the countryside and it will be politically difficult to continue to impose the neoliberal model upon the peasantry regardless of its consequences. It is possible that rural conflicts might even become more violent than in the past because of the fact that the state has been weakened in its mediating and incorporating capacity being unable to deal with the effects of the current unequal and excluding pattern of rural modernization.

A radical shift to a post-liberal development strategy is required. This change has to be shaped by the creative interaction between civil society and an activist but democratic state, in which the new peasant and indigenous movement must have a crucial role so as to ensure that market forces are harnessed for a participatory, inclusionary and egalitarian development process.

Endnotes

1 Many ideas for this text were first presented in my chapter in S. Halebsky and R.L. Harris (eds) *Capital, power, and inequality in Latin America* (Boulder, CO, Westview Press, 1995).
2 These data are derived from the so-called CIDA land tenure studies. For a résumé, see Barraclough (1973). The Alliance for Progress prompted a comprehensive study of Latin America's agrarian structure during the first half of the 1960s. The CIDA (Comité Interamericano de Desarrollo Agrícola, or the Inter-American Committee for Agricultural Development) studies represent the most ambitious collective research to date of Latin America's land tenure. In the mid-1960s reports on seven countries were published: Argentina (1965), Brazil (1966), Colombia (1966), Chile (1966), Ecuador (1965), Guatemala (1965) and Peru (1966), followed subsequently by two or three other country reports. The CIDA studies had a major influence on shaping a certain view of the Latin American agrarian question as well as on the design of agrarian reform policies. They conveyed a bimodal view of Latin America's land tenure system and were used by governments to lend scientific weight to the case for agrarian reform legislation.

3 The data reflect the situation during the 1950s and very early 1960s.
4 For useful evaluations of recent neoliberal land policies in Latin America, see Shearer *et al.* (1990), Stanfield (1992), Carter and Mesbah (1993) and Jansen and Roquas (1998).
5 The precariousness of smallholders is underlined by the fact that about 40 per cent of *minifundistas* lack property titles to the land they farm (Jordán *et al.*, 1989: 224).
6 The Chiapas case comes again to mind. The Zapatistas have even their own website, and a large number of solidarity organizations promote their cause through the internet.

Further Reading

Bebbington, A., Thiele, G. 1993 *Non-governmental organizations and the state in Latin-America: rethinking roles in sustainable agricultural development.* Routledge, London. The most comprehensive analysis on NGOs active in the rural sector in Latin America. It takes a sympathetic approach and yet is able to analyse the shortcomings of this important new actor.

Deere, C.D., León, M. (eds) 1987 *Rural women and state policy: feminist perspectives on Latin American agricultural development.* Westview Press, Boulder, CO. A pioneering study on gender issues in the Latin American countryside.

de Janvry, A. 1981 *The agrarian question and reformism in Latin America.* The Johns Hopkins University Press, Baltimore, MD. A classic, if slightly dated, text which provides the best analysis of Latin American agrarian developments from a dependency theory perspective. The author is one of the most prolific and well-known writers on agrarian development issues.

Goodman, D., Redclift, M. 1982 *From peasant to proletarian: capitalist development and agrarian transitions.* Basil Blackwell, Oxford. Discusses whether there is an inevitable process which transforms the peasantry into a rural proletariat and analyses the various paths of transition to capitalism in the countryside. Examines in particular the development of agrarian capitalism in Brazil and Mexico.

Kearney, M. 1996 *Reconceptualizing the peasantry: anthropology in global perspective.* Westview Press, Boulder, CO. A post-peasant perspective which emphasizes the manifold identities of the 'peasantry' in the new global era, analyses the way in which peasants define themselves and considers the distinction between rural and urban as problematic. Draws on the experience of indigenous peoples in Mexico or originating from Mexico but living (sometimes temporarily) in the United States.

Thiesenhusen, W.H. (ed.) 1989 *Searching for agrarian reform in Latin America.* Unwin Hyman, Winchester, MA. The most comprehensive study to date of agrarian reforms in Latin America, written by experts in the field and masterfully edited by the doyen on this topic.

Weeks, J. (ed.) 1995 *Structural adjustment and the agricultural sector in Latin America and the Caribbean.* Macmillan, London. A thorough evaluation of the profound shift from an import substitution to an export-promotion development strategy and the accompanying structural adjustment policies and their impact on the region's agriculture.

12

CITIES, CAPITALISM AND NEOLIBERAL REGIMES

COLIN CLARKE, DAVID HOWARD

Cities have played an important historical role in the economic development of Latin America and the Caribbean: first, as relay posts for the extraction of primary products (both mineral and agricultural) *en route* to overseas metropoles, and for the receipt of imported manufactured goods for local consumption and forwarding to rural areas; second, as commercial centers articulating the post-colonial national economy and its regional components; and, third, as places where national society has matured into a class stratification with certain citizen rights. At the end of the twentieth century, the majority of Latin American and Caribbean citizens are urban dwellers (Table 12.1), and city-systems, taken country-by-country, perform, in variable proportions, all the tasks enumerated above.

Cities in both the developed and the underdeveloped (once colonial) world were linked, historically, first to mercantilism and later to capitalism. But this chapter shows that late-twentieth-century dependent capitalism, one of the facets of which was import substitution industrialization (ISI), failed to absorb surplus labour *in toto* in Latin American and Caribbean cities, thereby allowing a small-scale informal economy to exist outside the formal, large-scale industrial and commercial sector which the state had helped to construct. Urban population growth, as a result both of high natural increase and of cityward migration, was engaged, in part, in the informal sector. In Latin America and the Caribbean, women outnumbered men in the migratory movements to the cities, and rapidly concentrated, when they were employed, in urban informal occupations.

Since the 1980s, under the influence of neoliberal policies of a markedly neocolonial kind emanating from the World Bank, formerly state-protected urban economies have been opened, noncompetitive heavy industry has been stripped out and public expenditure has been cut, resulting in some white-collar unemployment and a massive increase in self-employment. In general, the middle class has survived this process better than the urban poor, who have found their life chances seriously curtailed. The informal sector, which in urban Latin America was squeezed despite population growth

TABLE 12.1 Urbanization in Latin America, 1965–95

Country	Urban population as a percentage of the total population			Average annual growth rate of urban population (%)		Urban population living in the largest city (%)		
	1965	1985	1995	1965–1980	1980–1995	1960	1980	1995
Argentina	76	84	88	2.2	1.8	46	45	41
Bolivia	40	44	58	2.9	3.9	47	44	34
Brazil	50	73	78	4.5	3.0	14	15	–
Chile	72	83	86	2.6	2.0	38	38	36
Colombia	54	67	73	2.8	2.7	17	26	21
Ecuador	37	52	58	5.1	3.9	31	29	21
Paraguay	36	41	54	3.2	4.7	44	44	48
Peru	52	68	72	4.1	2.9	38	39	41
Uruguay	81	85	90	0.7	1.0	56	52	45
Venezuela	72	85	93	4.5	3.3	26	26	25
Costa Rica	38	45	50	3.7	3.7	67	64	72
El Salvador	39	43	45	3.5	2.0	31	29	26
Guatemala	34	41	42	3.6	3.6	41	36	23
Honduras	26	39	48	5.5	5.2	31	33	35
Nicaragua	43	56	62	4.6	3.9	41	47	46
Panama	36	41	56	3.4	2.8	61	66	37
Dominican Republic	35	56	65	5.3	3.8	50	54	52
Mexico	55	69	75	4.5	3.1	28	32	34

Source: World Bank (1997) and Thomas (1995: 2).

between 1950 and 1970, has subsequently expanded (sometimes in the form of workshops to which slimmed-down factories subcontract) to mop up surplus labor expelled from the formal sector. In many cities, it is now the largest employer of labor, so much so that, in the Caribbean, informality has come to permeate the urban – and rural – scene.

This chapter looks at the implications of these neoliberal changes for urban social stratification, for poverty, household organization and housing tenure and for social inequality, citizenship and identity. Before embarking on these issues, however, it is important to examine the previous failure of industrialization to underpin urbanization in Latin America and the Caribbean after World War Two and to explore the nature of the informal sector which has not become the transitory feature that Marxists and non-Marxists had anticipated.

INDUSTRIALIZATION AND THE INFORMAL SECTOR

Two key factors have affected urban industrialization in Latin America: first, the historical role of a strong, centralized state; second, the process of industrialization, which has relied on large-scale enterprises and relatively high levels of capital and imported technology (Roberts, 1995). Post-war dependent capitalism failed to absorb surplus labor in the large-scale, formal sector of Latin American and Caribbean cities, as the combined impact of natural increase and migration increased the urban labor force by an annual rate of 5 per cent between 1940 and 1970 (Roberts, 1995). Rather, there was a heavy reliance on petty services and industries of a small scale and haphazard kind, such as those involved in domestic service, shoe and clothes making, traditional and modern handicrafts, metalwork and machinery repair, trade (especially the market and street selling of food and clothing), shoe shining, car watching, and casual labor of all kinds.

This small-scale or informal sector has been described as entailing 'the circumventing of regulations, benefits, payments of taxes, and so on by employers and the unequal and selective application of such by the state' (Rakowski, 1994: 4). It is often deemed to be synonymous with extreme urban poverty and is characterized by piecemeal strategies for survival among low-income groups living in rented slums or squatter settlements. It is, however, highly heterogeneous and fragmented and incorporates a wide range of conditions and flexible contingencies.

The origin of the idea of the formal–informal dichotomy was rooted in the traditional-versus-modern duality, which had previously been used in urban analysis to differentiate labor markets (Kay, 1989). The modern sector was characterized, in general, by the influence of foreign companies or interests, capitalist penetration and capital-intensive technologies and professional and government activities. Geertz (1963), researching in South East Asia, had used the idea of economic dualism to distinguish between the bazaar and firm-centered economy. Actors within the bazaar economy were treated as economically irrational since their aim was to minimize business risk rather than to maximize profit. Given the dominance of the capitalist system, logic dictated that the bazaar economy would eventually be replaced by the profit-driven, firm-centered economy. Hart (1973), working on West Africa, has been credited with the elaboration of the formal–informal sector concept, outlining an economic duality in Third World labor markets which distinguished between wage earners and the self-employed. The dual division was based to a large extent on the rationalization of work practices.

Hart's definition of informal and formal sectors was taken up by the International Labor Organization (ILO) in the early 1970s. The former sector was characterized by enterprises which tend to exhibit ease of entry, reliance on indigenous resources, family ownership, small-scale operations, labor-intensive and labor-adapted technology and skills acquired outside formal education and to operate in unregulated and competitive markets (Hart, 1973). Conversely, enterprises in the formal sector had the opposite attributes of difficult entry, a frequent reliance on nonlocal resources, individual ownership, large-scale operations, capital-intensive technology and formally acquired skills and protected markets.

Santos (1979), using the terms 'upper circuit' and 'lower circuit' (to show that both are merely facets of dependent capitalism) for the Third World, and MacEwan Scott (1994) for Lima, have argued that the independence of the two sectors, formal and informal, is illusory. A plethora of business linkages and complex socio-economic networks join informal and formal activities when it is to the advantage of the upper circuit that they should do so. Moreover, the exploitative nature of formal–informal linkages has been emphasized in Birkbeck's study of waste paper collection in Cali, Colombia (Birkbeck, 1979). Garbage pickers worked for the local paper manufacturer by collecting waste paper, which made up a third of the raw material requirement in the paper industry, yet the informal and flexible nature of their activities dictated that they never gained access to basic workers' rights and remained wholly subordinate to the exploitative terms set by the factory owner. Similar sector linkages have been revealed to be widespread in retailing and manufacturing; in most instances large, highly capitalized enterprises have used outworking to cheapen labor costs by isolating, disadvantaging and exploiting the self-employed in the informal sector.

Mazumdar (1976) considered the key urban dichotomy to be that between unprotected and protected labor markets. The potential of the unprotected informal sector to expand at a more rapid pace than the formal sector was emphasized; its greater economic efficiency and its capacity to create more employment opportunities for urban dwellers were key features. But, while the relative dynamism of informal enterprises was lauded, these small-scale activities still remained subordinate and dependent on linkages with the formal sector. This dynamism of the informal sector has been championed relative to the inefficiency of large-scale, bureaucratic enterprises by de Soto. Researching conditions in Peru, he noted the ability of small-scale and medium-scale enterprises to generate wealth and employment in Latin American urban economies. The informal sector, he argued, is an efficient and spontaneous response to the incapacity of the state to provide for the basic needs of its population (de Soto, 1989). Small-scale industries thus survive and even profit in the interstices of cumbersome state regulation, which should be removed to give full play to capitalist forces. A similar perception had already been voiced by Kowarick (1979), who argued that the marginal sector was one of the pivots on which the economy turns.

The concept of the urban informal sector is now generally acknowledged as a useful focus of attention, but in analytical and theoretical terms is judged to be somewhat blunt-edged. In an attempt to divert the discussion away from overly neat dichotomies, the urban economy has been treated as a continuum of productive activities. Moser refers to petty commodity production – emphasizing the mutual articulation of different modes of production – as 'a form of production that exists at the margins of the capitalist mode of production, but is integrated into and subordinate to it' (1994: 20). Petty commodity production challenges former notions that the informal sector would eventually be subsumed by the formal development of the economy. Instead, it is argued that such production will remain inexorably subordinate to, and dependent on, the dominant capitalist mode.

Informal enterprises have been equated with small-scale activities, either in terms of their actual size or with respect to the level of fixed capital or financial turnover involved. 'The urban low-income population represents a market of low profitability to large-scale enterprises, thus creating a space for the small-scale entrepreneur to offer low-cost services or goods' (Roberts, 1995: 117). The large-scale sector has generated less than half of all urban employment opportunities in Latin America since World War Two. Much contemporary industrialization is based on capital-intensive technology, or requires a skill base which the general level of education cannot adequately provide. Wage differentials expand, creating a widening gap between skilled and semi-skilled or unskilled labor. 'Capital-intensive industrialisation thus produces significant income inequalities in the urban sphere, first by differences in income between the large modern sector firms and small enterprises and, second, by the wide income differences found *within* modern sector enterprises' (Roberts, 1995: 120, emphasis in original).

The informal and formal sectors thus coexist in a dynamic relationship which may be benign or exploitative; Tokman eschews this dichotomy by using the term 'heterogeneous subordination' to describe intersectoral relations (1978). In Latin America, the informal sector has been viewed as an adjunct to the large-scale sector of the economy (Roberts, 1995; Bromley, 1978; Portes and Schauffer, 1993). It produces cheap components and semi-finished products for the formal sector and goods of low profitability with limited and risky markets which large-scale enterprises are not interested in entering. These small-scale markets, given their low profitability and restricted capacity for capital accumulation, ultimately limit the expansion of informal activities, except under circumstances of constant population growth.

Between the 1940s and 1970s the informal sector in Latin America declined in relative importance. Oliveira and Roberts (1994) suggest that self-employment, an approximate guide to informal sector trends, declined from 29 to 20 per cent of total urban employment over those three-to-four decades. During the 1980s and 1990s, in contrast, the informal sector grew noticeably in absolute and relative terms (Table 12.2), and Portes and Schauffer (1993) estimated informal employment in Latin America at 31 per cent in the late 1980s. Uncertain economic conditions, resulting in a decline in the ability of the formal sector to generate jobs, have led to a corresponding increase in informal or precarious employment as people seek wider work opportunities.

However, the ideas that guided Latin American industrial development after World War Two were the opposite of those later rehearsed by de Soto. This program was relatively successful in Mexico, largely because ISI policies and joint ventures between US and Mexican capital made full use of the country's cheap labor. The informal sector nevertheless remained a vital part of the larger economy. State policies enhanced the commitment to industrial development, but around 37 per cent of Mexico City's working population is engaged in the informal sector. The proportion oscillates around this figure, depending on macro-economic conditions. The growth in informality from 34 per cent in 1981 to 40 per cent in 1987 responded to the contraction in formal jobs in manufacturing and the public sector (Ward, 1998), as unserviceable oil-loans at the start of the decade had to be covered by borrowing from the International Monetary Fund (IMF), which subsequently introduced a restructuring package.

Structural adjustment required massive cuts in public expenditure and the opening up of the economy to foreign imports. Import tariffs were removed and the manufacturing sector reorientated from inelastic domestic consumption towards manufacturing for export. A prime example of the latter are the *maquiladoras*, which numbered 1400 in 1990 and are heavily concentrated in, but are not exclusive to, such border cities as Tijuana and Ciudad Juárez. Over a million people are employed in the *maquiladora* industries, many of which are owned by US firms, sell their products in US markets, but draw on cheap, mostly female, labor. Outworking from factories to small workshops often characterizes these developments.

The urban economies of Latin America and the Caribbean have changed markedly under contemporary policies of trade liberalization and export-orientated industrialization. Governments are less willing, or less able, to protect internal markets as they had done under the previous policy of import substitution, so that many large firms have contracted or disappeared, and employment in small firms in the informal sector has expanded (Table 12.2). Structural adjustment packages also involve cuts in public sector funding and investment, coupled to privatization, and thus dictate that future expansion of employment is unlikely to occur within the public sector (Table 12.2). The changing nature of urban economies has increased the reliance of the urban labor force on precarious employment and has limited access to formal employment opportunities, particularly for workers with low-skills levels. Some jobs may offer full-time, year-round employment, but on a short-term contract basis. Employers thus avoid many of the costs associated with welfare provision for their labor force, and ever

TABLE 12.2 The structure of nonagricultural employment in some Latin American countries, 1980–92

Country and Year	Informal Sector[1]				Formal Sector		
	Own-account workers	Domestic service	Small firms	Total	Public sector	Large private firms	Total
Latin America							
1980	19.2	6.4	14.6	40.2	15.7	44.1	59.8
1985	22.6	7.8	16.6	47.0	16.6	36.5	53.1
1990	24.0	6.9	21.8	52.7	15.6	31.7	47.3
1992	25.0	6.9	22.5	54.4	14.9	30.8	45.7
Argentina							
1980	20.4	6.0	13.0	39.4	18.9	41.8	60.7
1985	22.9	6.5	13.3	42.7	19.1	38.2	57.3
1990	24.7	7.9	14.9	47.5	19.3	33.2	52.5
1992	25.9	7.8	15.9	49.6	17.7	32.7	50.4
Brazil							
1980	17.3	6.7	9.7	33.7	11.1	55.2	66.3
1985	21.1	9.1	14.5	44.7	12.0	43.4	55.4
1990	21.0	7.7	23.3	52.0	11.0	36.9	47.9
1992	22.5	7.8	22.5	54.1	10.4	35.4	45.8
Chile							
1980	27.8	8.3	14.3	50.4	11.9	37.7	49.6
1985	24.4	9.8	19.1	53.3	9.9	36.8	46.7
1990	23.6	8.1	18.3	50.0	7.0	43.0	50.0
1992	23.0	7.5	19.0	49.5	8.1	42.3	50.4
Colombia							
1980	25.3	6.7	20.5	52.5	13.8	33.7	47.5
1985	28.0	7.0	20.7	55.7	12.4	31.8	44.2
1990	25.1	6.2	27.8	59.1	10.6	30.2	40.8
1992	25.4	5.9	29.0	60.3	9.9	29.6	39.5
Costa Rica							
1980	16.3	6.1	14.0	36.4	26.7	36.9	63.6
1985	17.2	6.2	17.1	40.5	26.3	33.1	59.4
1990	17.6	5.6	22.0	45.2	23.0	31.7	54.7
1992	20.9	5.8	23.0	49.7	20.9	29.4	50.3
Mexico							
1980	18.0	6.2	24.9	49.1	21.8	29.1	50.9
1985	23.5	6.4	21.4	51.3	25.5	23.2	48.7
1990	30.4	5.6	19.5	55.5	25.0	19.6	44.6
1992	30.5	5.5	20.0	56.0	24.5	19.5	44.0
Venezuela							
1980	21.2	4.5	8.8	34.5	25.6	39.8	65.4
1985	21.3	4.9	13.7	39.9	24.5	35.6	60.1
1990	21.4	5.0	22.1	48.5	22.6	28.9	51.5
1992	22.3	5.0	23.6	50.9	20.0	29.1	49.1

Source: PREALC (1993; 4, Table 2) in Thomas (1995: 46–47).

larger numbers of urban workers are forced to depend on insecure jobs with low wages.

THE STATE AND THE CHANGING SOCIAL STRATIFICATION

Andre Gunder Frank remarked that development and underdevelopment were two sides of the same coin: they were dynamically but asymmetrically interrelated – power lay with the advanced capitalist countries, and the passing of colonialism made no difference (1971). Political colonialism merely gave way to neocolonialism or economic dependence. But state-led attempts to escape from the colonial pattern had already been made, especially in independent Latin America, several decades before anyone talked about dependency theory. These projects were industrial in nature and urban in location.

The major long-term state-endorsed strategy for capitalist development in Latin America and the Caribbean prior to the late 1970s had been the substitution of imports of consumer goods by identical commodities manufactured locally. This policy was introduced in Argentina and Uruguay during and after World War One and spread throughout the continent as the principal development strategy. It was taken up in Mexico, where tariff walls were raised to disadvantage foreign imports; and capital was often mixed – US and Mexican. This suggests that Latin American countries are not totally 'trapped' by dependency; they are, however, severely constrained by it.

In the Caribbean after 1945 a policy was developed to give tax incentives to local and foreign manufacturers to produce goods for the local market behind protective tariff barriers. This strategy was initiated in Puerto Rico under Operation Bootstrap and taken up elsewhere in the Caribbean, notably in Jamaica via the Pioneer Industry Incentive Law (1949) and the Industrial Incentives Law (1956). Also introduced (following experimentation in Puerto Rico) was a program for creating industry that would operate solely for export, the Export Industry Incentive Law (1956), whereby machinery and raw materials could be assembled free of import duties, but the finished products had to be exported (mostly to the United States and Canada). Additionally, tax incentives were given – depreciation on machinery, for example, could be written off over the first five years, leaving little to pay by way of local taxes.

These incentives were given academic respectability by the Caribbean economist, Professor Arthur Lewis (1954), who wrote about 'economic development with unlimited supplies of labour'. The focus of this strategy was to mop up surplus urban labor; the concept of 'marginality' and the 'informal sector' had yet to be 'discovered'. In Jamaica between 1960 and 1970, about 10 000 jobs (most of them in the capital city, Kingston) were created in manufacturing plant stimulated by these programs, though this was less than the number of jobs required to absorb the annual addition to Kingston's labor force at this time.

A major problem with tax-exemption industrialization was that it existed to absorb surplus urban labor, though modern industry is capital, not labor intensive. Moreover, industrialization by import substitution caters to a restricted middle-class market; the demand for consumer goods is inelastic and restricted by the existence of the nonconsuming masses and the small size of the middle class. In Brazil, for example, the military government created the 'economic miracle' of the late 1960s and early 1970s by the perverse policy of concentrating income in middle-class hands and encouraging foreign companies to establish their high-tech, consumer-orientated factories there (Furtado, 1970). When the established policy of ISI came to be revised in the 1970s, under the influence of the World Bank (with its free-trade mission), it involved the abolition of state protectionism, trade liberalization (as prefigured in the export industry incentive laws) and the turning of urban economies inside out.

The 1980s and 1990s were characterized throughout Latin America by the reduced direct role of the state in domestic economies, insisted on by the World Bank as part of its structural adjustment packages to rescue economies encumbered by debt. These debts had been run up after the OPEC (Organization of Petroleum Exporting Countries) oil-price rise of 1973 – either by the oil rich borrowing against anticipated income (such as Mexico or Venezuela) or by the oil deficient (such as Brazil and Jamaica), who needed to borrow to balance their exchequer. Fiscal austerity measures to limit government spending and to combat inflation, linked with the emphasis on the privatization of state enterprises, has noticeably cut back public sector employment and altered the nature of urban labor markets in much of Latin America and the Caribbean during the past decade. Chile was the first to pursue this path of development under the military dictatorship of Pinochet in the 1970s. During the 1980s and 1990s many civilian governments, for example in Mexico, Argentina, Peru and Jamaica, followed similar policies of economic neoliberalism, the key principles being trade liberalization and comparative advantage.

As Latin American and Caribbean governments have pursued similar neoliberal policies, so urban social stratifications have become more akin than previously. Educational attainment is an increasingly necessary attribute for obtaining superior occupational status and higher wages, though the impact of family status and social connections remain significant influences among the affluent classes. More students are entering and completing secondary and tertiary courses, though this increase serves to illustrate the growing polarization in many Latin American and Caribbean societies in terms of occupational differentiation and income inequality.

Traditional business elites maintain their dominance in many instances, particularly through their control of large-scale enterprises in the service (including tourism) and manufacturing sectors. The growing influence of international finance and the importance of operating within globalized markets has required urban elites to seek wider alliances and economic networks. While extensive privatization of state enterprises throughout Latin American has, to an extent, side-lined the state's direct role in the urban economy, structural adjustment policies maintain the state's importance but via the *indirect application* of externally devised economic strategies.

While the state has been of great significance for the development of business contacts, subsidies, labor regulation and import licences, large-scale companies are nonetheless focusing more on outward-orientated business strategies that reduce their dependence on the government and domestic class alliances. Somewhat in contradiction to this shift has been the increased importance of bureaucratic–technological groups, who, with their high levels of educational attainment, have obtained significant managerial and administrative positions. During the 1980s this class accounted for around 16 per cent of the Latin American urban population (Oliveira and Roberts, 1994).

Urban workers' changing occupational status relative to their position in informal or formal labor markets is an important aspect of the current discussion. The variable extent of social welfare provision by public or private employers, growing income differentials and unequal access to educational facilities impinge critically on the life chances of urban household members. Employees covered by their employers' contractual obligations towards proper conditions of work and social welfare provision have been termed the formal working class (Portes, 1985). Those without such cover, usually including the self-employed and family workers, are considered as informal workers – in a reworking of Hart's dichotomy.

Roberts, however, has emphasised the special class position of the informal petty bourgeoisie, owners of small-scale enterprises characterised by low levels of technology and precarious market conditions. This class, it is argued, makes up a larger proportion of the social structure in smaller cities compared with larger ones, where the formal working class assumes relatively greater importance (Roberts, 1995: 141). Portes (1985) estimated that the informal petty bourgeoisie made up 10 per cent of the urban economically active population for Latin America as a whole, the formal working class represented 22 per cent, while the informal working class accounted for 60 per cent.

The basic idea behind structural adjustment policies advocated by the World Bank since the late 1970s has been to cut public expenditure – on the civil service, health and education, to cut taxes to stimulate the private sector, and to encourage the development of industrial zones manufacturing for export to a hemispheric or global market. Kingston, Jamaica, for example, had an export manufacturing zone of approximately 10,000 workers in the late 1980s, mostly women and non-unionised, producing clothing and other similar goods for the US market. Much of the investment has come from Pacific Rim countries, such as South Korea and Japan; and more recently (*The Economist*, 1998a) Asian supervisors have been introduced to extract higher-quality work from the locals. A closer look at what has been happening in the labor market in Kingston illustrates the difficulties of securing formal employment.

Anderson (1987) recognized a formal sector in Kingston in which goods and services are supplied to a standardized format. It consists of a primary sector involving industry, professional services, businesses and the government sector and a secondary sector (upper and lower) with less protection and lower incomes. The remainder of the occupations, domestic service, informal manufacturing, (street) selling, petty services and agriculture were grouped in the informal sector. In 1977, 1983 and 1989 the decline in the government sector can be detected, accompanied by a shift into the more weakly protected secondary inferior sector (for example, nonunionized women's labor in the garment industry) (Table 12.3). If the lower secondary sector is added to the informal sector, then the informal sector for men declined from 60.4 per cent in 1977 to 59.4 per cent in 1983 and to 53.3 per cent in 1989; while the same pattern for women was 52.6 per cent to 50 per cent to 45 per cent respectively. In mitigation of these rather favorable figures, one has to bear in mind the decline in the purchasing power of Kingstonians caused by the devaluation of the Jamaican dollar, the dissolution of the State Trading Corporation (which subsidized basic foods) and the increased costs of unregulated transport. Many people in downtown Kingston receive remittances from abroad, and these are essential for purchases of food and clothing.

Further data have been compiled by labor-force surveys carried out by the Jamaican government, focusing on all those aged 15–65 years old (Table 12.4). In 1977, 1983 and 1989 unemployment for men rose from 17.5 per cent to 21 per cent and then dropped to 11.4 per cent; for women the figures followed the same pattern but were much higher – 29.9 per cent, 35.3 per cent and 21.8 per cent, respectively. These percentages for women

TABLE 12.3 Employment in labor-market sectors: Kingston and St Andrew, Jamaica, 1977–89

Sector of Labor Market	Men			Women		
	1977	1983	1989	1977	1983	1989
Government and Services	23.7	23.1	14.0	28.2	24.8	17.7
Primary	10.2	7.8	10.3	11.8	12.0	13.6
Upper secondary	26.5	28.5	29.0	12.6	13.2	13.7
Lower secondary	17.4	18.2	26.0	12.2	11.0	18.2
Domestic	1.9	3.2	2.0	17.4	17.2	16.3
Informal manufacturing	5.6	5.3	6.1	0.5	0.7	0.3
Sellers	4.1	5.6	5.8	8.8	14.7	12.5
Small-scale services	2.5	2.1	1.8	5.3	4.0	6.1
Own-account farmers	5.6	5.8	3.7	1.4	1.8	0.7
Agricultural employers	2.6	0.4	1.3	1.9	0.5	0.9
Total (%)	100.1	100.1	100.1	100.0	99.9	100.0

Source: Gordon and Dixon (1992: 149).

TABLE 12.4 Economic activity in Kingston and St Andrew, Jamaica

	1977	1983	1989
Men (%)			
Employed	68.4	65.9	69.2
Unemployed	17.5	21.0	11.4
Economically active	82.9	83.5	78.1
Women (%)			
Employed	49.1	46.0	50.0
Unemployed	29.9	35.3	21.8
Economically active	70.1	71.1	64.0

Source: Gordon and Dixon (1992: 144).

are quite crucial for Kingston, given the prevalence of female household headship in downtown Kingston (more than 50 per cent in many areas) (Clarke, 1975). Although unemployment improved in the late 1980s, the proportion of the population that was economically active declined after a slight rise in 1983 – to 78 per cent for men and 64 per cent for women in 1989. This suggests the hopelessness of the task of gaining work and implies that Anderson's results conceal the involvement of men and women in informal illegal activities and/or their dependence on charity.

Low-income residents of Kingston are trapped in grinding informal employment with few opportunities for linkage into, let alone movement into, the formal sector. These circumstances characterize the life of most residents of 'down' town Kingston and reach their peak in the West Kingston ghetto – anathematized by the rest of the city not because it is black but because it is dangerous, mostly to itself. Here, drugs, guns and gratuitous violence commingle on a scale probably unparalleled in any other city in the Caribbean region – with the possible exception of some Colombian cities. Many gangs are affiliated to the two political parties that, in government, have been responsible for implementing the IMF-approved structural adjustment program which has set in aspic the Kingston economy and its accompanying socio-economic stratification.

In Oaxaca City, Mexico, in contrast to Kingston, the informal sector of employment has grown at the expense of the formal, precisely because of the curtailment of 'protected' formal sector employment required by national economic crises and the structural adjustment package. The figures given by Murphy and Stepick (1991) are 44 per cent of male household heads in the informal sector in 1977, rising to 66 per cent in 1987; for female household heads, the percentages were 63 per cent rising to 73 per cent. The net result has been to remove job security from men and even more dramatically from women. Workshops in Oaxaca are largely unlinked to the formal sector (largely because the latter is so small); rather than acting as counter-cyclical absorbers of labor in times of economic recession, they contract to the use of family labor (Holloway, 1998).

To survive the fiscal crises and the shift to neoliberal strategy Mexican firms have shed labor and gone into flexible accumulation regimes, which involve contract outworking, and this helps to account for some of the inflation of the informal sector. In this post-Fordist version of what is otherwise a traditional looking feature, the modern factory is modest in scale and its connections reach into the local small-scale urban economy, as in the case of the Nissan Company at Aguas Calientes. But these developments are likely to by-pass Mexico City, which is seen as too polluted and decayed, and to avoid Oaxaca City, which is thought as being too remote. Nevertheless, a great deal of casual labor in both these cities is kept alive by informal personal relationships which it has with members of the middle class. The latter, for example, may take on more household helpers than they really need, especially if those to be helped are the relatives of long-term employees.

With the growth of informality, the recent period has been marked by a rapid expansion of female participation in the urban labor force across Latin America (Table 12.5). Moreover, the development of nonmanual industries and export-orientated economies based on assembly-plant production has altered the nature of labor demand, in many cases increasing employment opportunities for young women. Nevertheless, women often remain segregated in the labor market – where they still concentrate in the domestic informal sector, a reflection of the traditional relations of patriarchy, which continue to influence domestic codes of practice. A major way in which the poor in Mexico have coped with the unfolding crisis of the past 15 years has been for women (and, clandestinely, children) to enter the labor force – usually the informal sector. Chant (1991a) has shown that in the medium-sized industrial town of Querétaro it is usual for nuclear households to be extended (usually by incorporating kin) to provide household support to permit the senior female to be absent, and for female headship plus extension to emerge as a viable alternative to male headship in circumstances where the diminished income resulting from male absence is offset by greater democracy over household expenditure from the common pot.

TABLE 12.5 Female labor force participation in selected Latin American countries, 1960–90

	Women as percentage of the total labor force							1960 distribution of the female labor force (%)			1980 distribution of the female labor force (%)			1980 distribution of the male labor force (%)		
	1960	1970	1975	1980	1985	1990	2000	Agric.	Manuf.	Serv.	Agric.	Manuf.	Serv.	Agric.	Manuf.	Serv.
Argentina	21.0	24.9	26.0	26.9	27.5	28.1	29.1	4.9	26.7	68.4	3.1	18.3	78.7	16.7	39.6	43.8
Bolivia	20.4	21.4	22.0	22.5	24.2	25.8	25.6	26.2	27.3	46.6	27.5	16.5	56.1	52.0	20.6	27.5
Brazil	17.5	21.7	24.4	26.9	27.2	27.4	28.8	27.8	24.3	48.0	15.3	19.0	65.7	37.0	29.4	33.6
Chile	21.7	22.4	24.8	27.2	28.0	28.5	28.9	3.6	20.3	76.1	2.3	16.4	81.3	21.8	28.4	49.9
Colombia	19.4	21.3	21.9	22.5	22.2	21.9	22.3	11.3	22.3	66.4	5.0	21.0	74.0	42.7	24.2	33.1
Costa Rica	15.8	18.1	19.7	21.3	21.6	21.8	22.6	7.5	17.0	75.5	4.0	20.0	76.0	38.0	24.0	38.0
Dominican Republic	10.0	10.9	11.7	12.4	13.7	15.0	17.9	13.2	16.2	70.7	7.8	7.6	84.6	51.1	16.6	32.4
Ecuador	16.3	16.3	17.7	19.3	19.3	19.3	19.7	21.5	30.8	47.7	12.8	18.0	69.2	44.7	20.3	35.0
El Salvador	16.8	20.4	22.6	24.9	25.1	25.1	25.3	7.3	25.0	67.7	5.0	18.2	76.8	55.8	19.8	24.4
Guatemala	12.3	13.1	13.5	13.8	15.1	16.4	19.5	11.0	22.9	66.1	9.4	20.0	70.6	64.5	16.0	19.0
Honduras	12.3	14.2	14.9	15.7	17.3	18.8	22.7	5.3	18.6	76.1	7.3	30.2	62.5	70.4	13.6	16.1
Mexico	15.3	17.8	22.6	27.0	27.0	27.1	27.7	32.8	14.0	53.3	19.3	27.9	52.8	42.9	29.4	27.7
Nicaragua	17.9	19.7	20.6	21.6	23.4	25.2	29.1	14.2	19.7	66.1	8.0	15.0	77.0	57.2	16.0	26.8
Panama	20.9	25.2	25.8	26.1	26.7	27.2	28.7	9.1	10.8	80.2	8.0	10.5	81.6	40.2	20.9	39.0
Paraguay	21.4	21.3	21.0	20.8	20.8	20.7	20.8	21.5	31.0	47.6	12.5	24.5	63.1	58.0	19.5	22.5
Peru	20.9	20.3	22.3	24.3	24.2	24.1	24.4	33.7	19.1	47.3	24.4	13.5	62.2	45.1	19.8	35.2
Uruguay	24.1	26.3	27.9	29.6	30.4	31.1	32.5	4.1	26.3	69.7	2.9	23.1	74.1	21.2	31.7	47.2
Venezuela	18.3	20.7	23.3	25.8	26.7	27.6	28.8	4.6	19.7	75.8	2.6	18.4	79.1	20.7	31.9	47.5
Latin America	19.2	21.7	na	26.1	na	26.6	27.5	24.4	21.0	54.7	14.9	19.9	65.2	38.6	27.8	33.6

Source: Korzeniewicz (1997: 239).
Note: Agric. = agriculture; Manuf. = manufacturing; Serv. = services.
na = not available.

Poverty, the household and housing

Poverty and household organization

The 1980s witnessed a widespread decline in real wages throughout Latin America and the Caribbean. ILO data suggest that the real minimum wage declined by 14 per cent between 1980 and 1987 (Roberts, 1995). Public sector employees were hit the hardest – many were driven into self-employment, while manufacturing incomes declined the least. Free-market policies after the 1980s in Latin American countries sought to encourage private sector investment at the expense of state intervention. The urban middle, and especially lower middle, classes were severely affected. While the professional classes did suffer a decline in real wages, their relatively high levels of formal training and education remained in constant demand given the persistent shortage of formal skills.

Widening income differentials between and within the informal and formal sectors highlight the social exclusion of lower-income groups from already limited public and social services. Surfaced and properly drained roads, piped water supply, sewerage, schools and health facilities are seldom located in areas of informal or low-income settlement, unless there has been government or NGO support for improvements. Insecurity of tenure is a key circumstance for many occupants of informal (often squatter) settlements, who live their daily lives under the shadow of uncertainty of possible eviction, which necessarily restricts their willingness or ability to upgrade their property.

Some commentators not only have emphasized the social exclusion of the urban poor but also have interpreted economic and political marginality as an unfortunate and unavoidable consequence of dependent development (Kay, 1989). A culture of poverty is deemed to be the inevitable result of rapid urban growth without corresponding industrialization, thus creating an excluded population of urban poor who will remain poverty stricken (Lewis, 1961, 1968). It is now widely appreciated that this pessimistic interpretation fails to acknowledge the ability of residents to change their own situation rather than to wait in vain for largesse from above; the poor may not be passive, yet they certainly are dominated (Perlman, 1976).

In recent years, household survival strategies have been a focus for attention following the shift in focus to grass-roots development and the emphasis on coping mechanisms among the urban poor. Networks of exchanges and obligations which frequently characterize the urban economy, especially within the informal sector, have been investigated by Lomnitz (1977), who emphasized [probably overemphasized according to Willis's work in Oaxaca City (1993)] the extent to which reciprocity underpinned survival strategies among the urban poor in Mexico City. Many low-income households make regular contributions to savings clubs or informal credit systems, whereby members take turns to receive the pooled amount.

Further innovative research at the level of the household has analysed changing household structures in response to neoliberal shifts in the macro-economy and has identified changing gender roles and relations in the new context. To enable households to survive the crisis, women have entered the labor force as informal sector workers, domestic servants and as employees in the newly founded light manufacturing industries. A major way in which the poor have accommodated to declining economic conditions has been to share housing with kin and to incorporate extended household members both as a safety net for them and to facilitate childcare when women work. This process of kin incorporation has been a major feature where female household headship occurs, particularly among the urban middle class (Willis, 1994).

Members of a household may adopt multiple employment strategies, with some in the formal, and others in the informal, sector to maximize income and gain access to the potential welfare benefits of formal employment. Alternatively, or in addition, workers may turn to their employers as a source of patronage, either to find work for other members of the household or to seek some form of nonfinancial assistance in times of extreme hardship (Ward, 1998). Given the growing importance of foreign investment and large-scale enterprise, such informal agreements between employer and employee are, however, less likely to operate than in the past, particularly if large firms are involved.

Housing tenure: squatting, ownership and renting

The deleterious impact of neoliberalism on labor markets, which were already split into formal and informal sectors, has been mirrored with regard to housing. As a generalization it can be said that the gap between rapid urban population growth and the inability of the urban economy to absorb labor in the formal sector after World War Two led to the expansion of makeshift, informal housing. The Latin American and Caribbean city in 1950 was characterized by wholesale renting; only the elite and upper-middle class owned the property in which they (and the rest of the urban population) lived. But these relatively compact settlements were totally unable to absorb rapid population growth within the formally owned and rented fabric of the city; the overspilling of population into marginal squatter settlements was an inevitable consequence of the speed and scale of population growth in the 1960s and 1970s, and a measure of the inability of the state to cope with this phenomenon. However, before examining the impact of neoliberalism on the housing sector, it is essential to understand its composition and problems dating back to the 1960s.

By the early 1960s, squatting was being stereotyped as involving flimsy, insanitary dwellings, constructed by marginal migrants who created rural environments within towns. Some informal settlements conform to this description; however, over the past 30 years attitudes towards squatting have changed considerably. Prior to the mid-1960s, squatting was viewed as an urban 'cancer', to be eradicated as swiftly as it appeared. Since then informal settlement has been interpreted more as a solution to the urban housing problem – especially for the poor (Mangin, 1968; Turner, 1968). A developmental trajectory over time was identified, based on the physical upgrading of the dwelling, from incipient to consolidating and consolidated, dependent on the granting of secure title to land. This interpretation, based on field evidence from Lima, was taken up by the World Bank and used as the model for aided upgrading and site-and-service schemes in the mid-to-late-1970s.

By the mid-1970s it had become apparent that other forms of illegality have to be taken into account in addition to squatting on vacant land. For example in Mexico City about half the informal settlements in 1975–80 were on illegal subdivisions, or *fraccionamientos illegales*. These subdivisions might have been deemed illegal for a number of reasons. The plot might not have been legally owned by the vendor, or might have been legally owned but sold without proper conveyancing or appropriate services (such as electricity, sewerage) and planning permission. Any of these deficiencies, singly or in combination, would have made the development illegal and would have necessitated government intervention to remedy what was wrong. In short, they would have deficiencies akin to, but probably less all-embracing than, those in squatter settlements (Ward, 1998).

Two additional aspects of the 1970s squatting debate require attention. First, the notion that squatter settlements and shantytowns are one and the same thing (they may or may not be); and second, the implication that governments are supportive of the consolidation process. While illegality of tenure is the hallmark of the squatter settlement, shantytowns (huts or mean dwellings) are defined by their fabric. Some squatters do occupy shantytowns, but shanties are more commonly located on small, rented plots. In Mexico City, rented shack yards make only a limited contribution to the housing stock, but in the Caribbean their residents vastly outnumber squatters, in the strict sense of the term (Clarke and Ward, 1980).

Eradication, rather than government support for titling and upgrading, of squatter camps was the norm in Kingston, until the Michael Manley government of the 1970s introduced site-and-service schemes (in East and West Kingston). Moreover, in Caracas, successive Venezuelan governments have refused to grant titles to illegal occupants of land, on the assumption that it would relocate all the residents in appropriate modern housing – a task now beyond its capacity, given the decline in world oil prices. Finally, in Mexico City successive governments had been increasingly hostile to the growth of new informal settlements after President Echeverría (1970–76), until Salinas (1988–94), through his *Solidaridad* program, targetted informal settlements and distributed land titles on a block-by-block basis in Mexico City in the early 1990s. *Solidaridad* represented an important attempt by Salinas to cushion some of the worst effects of the neoliberal shock on the Mexican poor by funding grass-roots projects defined by the beneficiaries themselves.

There is, in the literature, the idea that renting is inimical to shantytown upgrading. However, Eyre (1972, 1997) on Montego Bay and Hanson (1975) on Kingston in Jamaica, respectively, argued that shantytowns did upgrade over time. But the improvements were modest (from huts to carpentered sheds), and there was no identification of tenure change. Indeed, Eyre largely disregarded the issue of tenure (most of his original Montego Bay shantytowns were rent yards), while Hanson's argument about tenure change was based on his inclusion of different types of tenure such as rented land, government-purchased land and squatted land. Not only did these tenure types not change over time, but also they gave rise to only the most rudimentary improvements in fabric.

Even where consolidation of property occurred in former illegal settlements, however, the process did not necessarily end with the establishment of a property-owning proto-middle class, as Mangin and Turner had optimistically implied. Ward (1976) showed that the costs to residents of the upgrading process in Mexico City were so great that many had to leave the improving settlement, which was then invaded by middle-class owners and renters. Those who stayed had been able to share the plot (and cost of public utilities) with kin; or they had survived by letting (land or rooms). But the consolidation of informal settlements depended on the household having a surplus to invest, a factor which has been severely curtailed during the 1980s and 1990s, as low-income urban dwellers have faced increasing hardships of the kind outlined in the previous section. The result of this, and the withdrawal of the state from commitments to 'social' expenditure, means that the urban poor are condemned to static urban environments which are impoverished and, if anything, are experiencing densification (increased population and housing density).

Even prior to the neoliberal project, state housing in Latin America and the Caribbean had played a relatively minor role in shelter provision over the decades because of the vast scale of the requirement and the weak tax base in most countries. In Venezuela, for example, government housing provision currently is undermined because 50 per cent of the allotted funds are used to pay off its foreign debt, though Venezuela is one of the most urbanized countries in the Caribbean region, with more than 80 per cent of its 21 million population living in cities. This level of urbanization has been driven by Venezuela's economic growth, based almost entirely on the oil industry, which has faced severe problems in recent years. The fundamental dilemma for Venezuela is that the oil economy cannot support the undertakings entered into by previous governments. International oil prices have tumbled since the early 1980s, and, after a brief recovery in 1990 associated with the Iraq crisis, have fallen back again. Currently, only the most privileged members of the elite and upper middle class can afford newly constructed housing; over 80 per cent of households are left to their own devices (IADB, 1993b).

This Venezuelan situation has more general application; formal solutions to housing need can cater only to the middle classes. Government aid to mortgage housing schemes in Kingston in the 1960s and 1970s, the Nonoalco–Tlatelolco project in Mexico City in the 1950s, and middle-class filtering into the *superbloques* of Caracas since the 1960s all have conformed to, or created, this association of government housing with the more privileged and creditworthy sectors of urban society. However, as in the case of Caracas, financial markets under the pressure of rates of inflation and bank interest rates of 40–50 per cent per annum, cannot sustain a long-term borrowing system to assist with the purchase of an expensive item such as a house.

The consequence is that, given government's withdrawl from involvement, and rising household poverty, rental has to be a 'solution'. Whereas the urban population increases of the 1950s and 1960s could not be accommodated in the fabric of the comparatively compact city, and had to be housed through the expansion of informal housing (squatter settlements, rent yards, illegal subdivisions and the occupation of *ejidal* land), the city of the 1990s is a larger entity, with scope for subdivision and letting. Moreover, letting may offset the declining purchasing power of the poor (and the lower middle class) – so evident in Mexico City, Oaxaca City, Kingston and Caracas.

Gilbert and Varley (1991) have investigated the rental issue in the context of Guadalajara and Puebla in Mexico, and Gilbert has extrapolated this research to Caracas and Santiago (1993). They conclude that renting is a flexible and cheap solution for the urban poor who cannot now afford to upgrade property and who might otherwise be trapped in unserviced squatter settlements. The dwelling owner may use rental income to supplement his or her income (20 per cent of household heads in most cities are female, higher in the Caribbean). Petty landlordism is the norm; most landlords are resident in the property. The city in Mexico is reverting to the early-twentieth-century pattern of being a rented phenomenon; in the Caribbean, renting has remained characteristic throughout the twentieth century both in the form of tenements and rent yards.

To conclude this section, it is clear that economic stagnation has been accompanied by inflation in many parts of the region, by declining per capita income since the late 1970s and by severe problems in meeting debt repayments, incurred since the 1980s, despite neoliberal schemes of restructuring. Given failing domestic economies, the benefits of self-help housing are to be judged by survival rather than by petty capital accumulation, though the past upgrading of squatter settlements indicates that there was a time, especially in the 1970s and early 1980s, when improvements did occur without titles. This was simply because there was sufficient household surplus, in oil-propelled days, to fund improvements, notable in Port-of-Spain, Trinidad, and Caracas. Nowadays, swathes of urban territory in Latin American and Caribbean cities associated with the informal poor are composed of static informal–rental areas in which improvements have long been overtaken by physical decay (see Chapter 7). These are impenetrable by the police and 'polite society' and are often more 'cancerous' than the squatter settlements of the 1950s and early 1960s were supposed to have been!

SOCIAL INEQUALITY, CITIZENSHIP AND IDENTITY

Socio-economic inequality

Middle-income and upper-income groups (who were predominantly urban in location) generally benefited from the growth of Latin American economies from the 1960s through to the 1970s, and only the unskilled and informal sector workers suffered a decreased share of total income (Roberts, 1995). Since the 1970s, economic crises and the impact of structural adjustment have heightened economic hardship for a far wider range of the urban population. Increased reliance on capital-intensive technologies for industrial production have limited labor opportunities in the urban areas. State-interventionist approaches to development are no longer viable, yet neoliberal policies exacerbate hardships for the majority of citizens (Tardanico and Menjívar Larín 1997). Urban poverty increased during the 1980s and 1990s, and by the mid-1990s 36 per cent of urban households were classified as poor (ECLAC, 1996).

The impact of rural-to-urban migration and the concomitant shift of labor from agricultural to manufacturing or service-centered employment in the urban arena has made a significant mark on occupational structures in Latin American countries. Occupational mobility may therefore be closely associated with spatial fluidity, as populations adjust to shifts in macro-economic policy at the state level. Among the urban-born population, however, the transition to nonmanual employment in the majority of Latin American cities has elicited substantial changes in occupational structure.

Change in the nature of urban labor markets varies from country to country (Roberts, 1995). In Argentina, the growth of nonagricultural employment opportunities was evidenced most directly in self-employment and the expansion of nonmanual occupations between 1940 and 1980, but this increase was unusually focused on middle-income groups. Manual occupations during this period witnessed a general decline throughout the region, particularly those associated with transport and manufacturing. Since the 1960s, however, the construction and service industries have provided some increase in nonagricultural manual employment. Urban employment opportunities have increasingly involved nonmanual labor in both the formal and the informal spheres of activity.

Roberts posits that

> the degree of social mobility occurring in Latin American cities between 1940 and 1980 suggests a period in which the negative side of the dislocations produced by urbanisation and industrialisation were offset by a substantial improvement in people's life chances (1995: 149).

But, despite improvements in absolute levels of poverty 20 years on, a fifth of the population in Latin America still has insufficient income adequately to feed and house itself, and the absolute numbers of urban poor are still increasing (United Nations, 1994).

Citizenship

As neoliberalism decreases, or at least redefines, the role of the state and increases the role of the market there should be far-reaching effects on civil society and the state (Roxborough, 1997). A broad definition of citizenship focuses on the collective right and involvement of people (here urban dwellers) to make changes in the way resources are allocated and society is governed. Citizenship is thus a negotiated status, since active participation by citizens may alter their rights and obligations, while governments may concurrently seek to limit these changes and the nature of popular participation (Roberts, 1995: 184).

Traditionally, Latin American states have adhered to strong nationalist programs, directing parochial loyalties to centralized affiliation. Many liberal national constitutions sought to enhance the authority of private ownership, projecting equality of all before the law and creating the basis for a civil society independent of the state. In the era of neoliberal economic policies, these same notions form a potential basis for the cooptation of popular social movements by modern state and business interests. Urban social movements generally seek representation within, rather than removal from, the state system, and so may find solace, at least in theoretical terms, in the logic of neoliberal development policies.

Neoliberal transformations have partially displaced the locus of political concern from civil society to economic decision-making. They have produced some renewed development of social movements to defend previous privileges. But they have also stimulated a process of disarticulation in civil society: a reduction in the role of organized labor; a generalized sense of crisis and despair that may erode democracy, where it exists; and perplexity among political parties as old lines of social cleavage shift (Roxborough, 1997).

Among formal sector workers, trade unions in Latin American cities have been an important forum for the representation of citizenship. In the relatively more industrialized countries such as Argentina, Chile and Venezuela, union membership accounts for up to 40 per cent of the economically active population, though this figure drops to between 20 and 30 per cent for Colombia and Mexico and to less than 10 per cent for El Salvador and Guatemala (Roberts, 1995). While formal unions remain active, informal workers are essentially dispossessed of their right to representation at work. Workers without an established history of union membership, particularly women and children, have no precedent of labor representation to follow. In many instances, the development of foreign-controlled enterprises or export-orientated industries (as in Jamaica) involves the denial of union membership for the workforce that is involved.

The necessity of forging new alliances to enhance business and state interests under the impact of structural adjustment measures impinges on the development of citizenship, as the market economy is 'reinvigorated'. Arguably, a modernizing, economically developed state will be more able and likely to satisfy the basic requirements of its population. However, the means of achieving that development have created much economic hardship in the short and medium terms and potentially have strengthened popular resistance around issues of collective consumption, as Castells forewarned (1983).

Formation of identities

Urban contexts, coupled to the lack of social and economic provision for alienated low-income groups, are important catalysts for the creation of neighborhood and political identities. Local issues, and the focus on collective consumption, allow for the development of community leaders and organizations which seek to redress social inequalities and the marginalization of the urban poor (Castells, 1983; Naerssen and Schuurman, 1989). Urban social movements form the crucible for the formation of new social forces, but, despite initial optimism and support for radical agendas, low-income groups generally act within the confines of the existing socio-economic and political system rather than seeking fundamental change (Smith and Korzeniewicz, 1997).

The changing status of women in Latin American societies, the recognition of their activities both in the productive sphere and in the reproductive sphere, has heralded an increase in the participation of women in social and political movements. Gendered representation, however, may reinforce stereotypes, such as the automatic association of women with domestic issues, rather than breaking down traditional identities in order to forge new ones (Molyneux, 1984; Craske, 1994).

Increasing openness of markets and state reluctance to protect victims from the impact of structural adjustment have generated an impetus for ethnic identities to develop. Race and ethnicity frequently have the potential to project social cleavage and discord in many Latin American cities, as the result of the cityward migration of previously rural ethnic groups, usually Indians. Moreover, ethnic allegiance may form the basis for successful enclave economies or political groupings which actively improve the social status and economic well-being of minority groups, many of whom may have previously suffered or have been marginalized as a result of their stated or perceived group consciousness.

Measuring the size of ethnic groups is often a specious exercise, given the flexibility which is inherent in racial/ethnic boundaries. Someone who describes herself as *blanca*, may be perceived by another to be clearly *india*. Similarly, as Roberts notes, the ascription of ethnicity from a personal or group perspective is contingent on the social context, 'so that, for example, a relatively fair-skinned person dressed poorly by city standards may be labelled 'Indian' in a metropolis, a *mestizo* in a small provincial town and 'white' in the hamlet of his or her birth' (Roberts, 1995: 142).

Albó (1995) has studied the growing importance of Aymara ethnicity in relation to the development of the informal economy in La Paz, Bolivia. The sharing of ethnicity provides a crucial basis of trust for economic exchange in a sector which operates outside the formal regulated system. Similarly, ethnic allegiance may be reproduced in the city in the form of neighborhood clustering, particularly through the process of chain migration from rural areas. In Guatemala, ongoing migration from rural areas to Guatemala City has increased the total Indian population in the city to approximately 12 per cent. Rural migrants are often concentrated in forms of precarious employment, implying that the Indian populations in the cities are more likely to be without access to social welfare and be disadvantaged in terms of income, housing, education and health provision. Thus in Oaxaca City, indigenous language speakers (5 per cent of the population) shift to Spanish, as migrants adjust to urban circumstances that are likely to disadvantage avowed Zapotecs or Chatinos.

Finally, religion has long played an important role in fomenting social cohesion or hostility. De la Peña and de la Torre (1990) have compared the influence of parish associations, pentecostal groups and Christian base communities in Guadalajara, Mexico. The overall result, despite differences in organization and worship, was to bring together and strengthen neighborhood identities that transcended the spiritual sphere to activate and articulate social and economic demands. However, positive outcomes of this kind depend both on the quality of local leadership and on the permissiveness, or otherwise, of the political situation.

Conclusions

Structural adjustment has created winners and losers. Many large, polluted and crime-ridden cities are now being shunned by modern businesses in favor of smaller, more manageable urban environments. Women are often favored over men in modern light manufacturing (largely because they are thought to be more tolerant of sweatshop conditions). The state seems everywhere to be in retreat as an employer and regulator of labor, while nodes of affluence and decay coexist within its urban areas. State managers no longer have the levers, nor perhaps the will, to rectify spatial inequalities. Indeed, it may be that there will be variable responses from the state in Latin America and the Caribbean; some states will embrace neoliberalism in its pure form; others will intervene in social matters, thus producing a highly disparate and fragmented region. At a smaller scale, 'produced' urban space may be partitioned even more rigidly than in the past into policed residences for the 'haves' and insecure neighborhoods for the 'have-nots', who will be driven ever more into informal or illegal activities in an attempt to escape the grinding, constant pressure of urban poverty. Under these circumstances, grass-roots social movements may be stalled by nihilistic, intraclass violence, which undermines the apparent gains in citizenship rights that urbanization has facilitated over the past 50 years.

Further reading

Chant, S. 1991 *Women and survival in Mexican cities: perspectives on gender, labour markets and low-income households*. University of Manchester Press, Manchester. A key text which addresses the role of gender in Mexican cities, highlighting the importance of household survival strategies.

Potter, R. B., Conway, D. (eds) 1997 *Self-help housing, the poor, and the state in the Caribbean*. University of Tennessee Press, Knoxville, TN. A wide-ranging and informative collection of case studies from the Caribbean region.

Rakowski, C. A. (ed.) 1994 *Contrapunto: the informal sector debate in Latin America*. State University of New York Press, Albany, NY. Excellent overview of the informal sector and its significance for Latin American cities. Good coverage of theoretical and practical developments.

Tardanico, R., Menjívar Larín, R. (eds) 1997 *Global restructuring, employment, and social inequality in urban Latin America*. North–South Center Press, Miami, FL. A detailed collection of essays which address directly the influence of structural adjustment policies on urban employment in Latin America.

Ward, P. 1998 *Mexico City*. John Wiley, Chichester, Sussex. Revised second edition of the fluently written and highly informative account of Mexico City. An excellent analysis of the capital's contemporary development and the problems which confront the urban population in the 1990s.

BIBLIOGRAPHY

Acero, L. 1997 Conflicting demands of new technology and household work: women's work in Brazilian and Argentinian textiles. In Mitter, S., Rowbotham, S. (eds), *Women encounter technology: changing patterns of employment in the Third World*. Routledge, London, 70–92.

Agüero, F. 1998 Chile's lingering authoritarian legacy. *Current History* (February), 66–70.

Aguirre Beltrán, G. 1976 *Obra polémica*. Centro de Investigaciones Superiores, Instituto Nacional de Antropología e Historia, Mexico City.

Akin Aina, T. 1990 Understanding the role of community organisations in environmental and urban contexts. *Environment and Urbanization* **2**(1), 3–6.

Alba, F. 1989 The Mexican demographic situation. In Bean,F., Schmandt, J., Weintraub, S. (eds), *Mexican and Central American population and US immigration policy*. University of Texas Press, Austin, TX, 5–32.

Albó, X. 1995 La Paz/Cukiyawu: the two faces of a city. In Altamirano, T., Hirabayashi, L. (eds), *Migrants, regional identities, and Latin American cities*. Society for Latin American Anthropology.

Albó, X. 1996 Bolivia: making the leap from local mobilization to national politics. *NACLA Report on the Americas*, **29**(5), 15–20.

Almond, G.A., Bingham Powell, G. Jr 1966 *Comparative politics: system, process, and policy*. Little, Brown, Boston, MA.

Almond, G.A., Bingham Powell, G. Jr 1978 *Comparative politics: system, process, and policy*, 2nd edn. Little, Brown, Boston, MA.

Altimir, O. 1994 Income distribution and poverty through crisis and adjustment. *CEPAL Review* 52, 7–31.

Alvarez, S. 1990 *Engendering democracy: the women's movement in Brazil*. Princeton University Press, Princeton, NJ.

Alvarez, S., Dagnino, E., Escobar, A. 1998 Introduction: the cultural and the political in Latin American social movements. In Alvarez, S. *et al.* (eds), *Cultures of politics, politics of cultures*. Westview Press, Boulder, CO, 1–32.

Amis, P. 1995 Making sense of urban poverty. *Environment and Urbanisation* **7**(1), 145–7.

Anderson, P. 1987 Informal sector or secondary labour market? Towards a synthesis. *Social and Economic Studies* **36**(3), 149–76

Anderson, S., Cavanagh, J., Ranney, D., Schwalb, P. (eds) 1994 *Nafta's first year: lessons for the hemisphere*. The Institute for Policy Studies, Washington, DC.

Anglade, C., Fortín, C. (eds) 1985 *The state and capital accumulation in Latin America: volume I*. University of Pittsburgh Press, Pittsburgh, PA.

Annis, S. (ed.) 1992 *Natural resources and public policy in Central America*. Overseas Development Council, Washington, DC.

Apey, A. 1995 *Agricultural restructuring and co-ordinated policies for rural development in Chile*, unpublished PhD thesis, University of Birmingham, Birmingham.

Arevalo Torres, P. 1997 May hope be realised: Huaycan self-managing urban community in Lima. *Environment and Urbanization* **9**, 1.

Arguedas, A. 1975 *Pueblo Enfermo*. In Siles Guevara, J. (ed.), *Las cien obras capitales de la literatura Boliviana*. Editorial Los Amigos del Libro, La Paz.

Arizpe, L. 1982 Relay migration and the survival of the peasant household. In Safa, H. (ed.), *Towards a political economy of urbanization in developing countries*. Oxford University Press, Oxford, 19–46.

Arndt, H.W. 1987 *Economic development: the history of an idea,* University of Chicago Press, Chicago, IL.

Ascher, W., Healy, R. 1990 *Natural resource policymaking: a framework for developing countries.* Duke University Press, Durham, NC.

Asociación de Exportadores various years *Estadísticas de Exportaciónes HortoFrutícolas.* Santiago, Chile.

Assies, W. 1997 The extraction of non-timber forest products as a conservation strategy in Amazonia. *European Review of Latin American and Caribbean Studies* no. 62 (June), 33–53.

Asthana, S. 1994 Economic crisis, adjustment and the impact on health. In Phillips, D., Verhasselt, Y. (eds), *Health and development.* Routledge, London, 50–64.

Auty, R. 1993 *Sustaining development in mineral economies: the resource curse thesis.* Routledge, London.

Baden, S. 1993 *The impact of recession and structural adjustment on women's work in developing countries,* Bridge Report No.2. Institute of Development Studies, Sussex.

Baer, W. 1969 *The development of the Brazilian steel industry.* Vanderbilt University Press, Nashville, TN.

Bailey, A., Hane, J. 1995 Population in motion: Salvadorean refugees and circulation migration. *Bulletin of Latin American Research,* **14**(2), 171–200.

Bakacs, P. 1970 Public health problems in metropolitan areas. In Miles, S.R. (ed.), *Metropolitan problems: international perspectives.* Methuen, London.

Barham, B., Clark, M., Katz, E., Schurman, R. 1992 Non-traditional agricultural exports in Latin America. *Latin American Research Review* **27**(2), 43–82.

Barraclough S., Ghimire, K. 1995 *Forests and livelihoods: the social dynamics of deforestation in developing countries.* Macmillan, Basingstoke, Hants.

Barraclough, S. 1973 *Agrarian structure in Latin America.* D.C. Heath, Lexington, MA

Barrera, M. 1996 Las reformas económicas neoliberales y la representación de los sectores populares en Chile. *Estudios Sociales* 88.

Barrientos, A. 1996 Pension reform and pension coverage in Chile: Lessons for other countries. *Bulletin of Latin American Research,* **15**(3), 309–322.

Barrientos, S., Bee, A., Matear, A., Vogel, I. 1999 *Women and agribusiness: working miracles in the Chilean fruit export sector.* Macmillan, London.

Barrig, M. 1994 The difficult equilibrium between bread and roses: women's organisations and democracy in Peru. In Jaquette, J. (ed.), *The women's movement in Latin America.* Westview Press, Boulder, CO, 151–75.

Barrow, C. 1992 *Family land and development in St. Lucia.* Institute of Social and Economic Research, University of the West Indies, Cave Hill, Barbados.

Barry, T. 1987 *Roots of rebellion: land and hunger in Central America.* South End Press, Boston, MA.

Barton, J. 1997a *A political geography of Latin America.* Routledge, London.

Barton, J. 1997b ¿Revolución Azul? El impacto regional de la acuicultura del salmón en Chile. *Revista Eure,* **22**(68), 57–76.

Bartone, C.R. 1990 Water quality and urbanization in Latin America. *Water International* 15.

Batley, R. 1997 Social agency versus global determination in Latin American urban development. *Third World Planning Review* **19**(4), 333–46.

Baumol, W., Wolff, E. 1996 Catching up in the postwar period: Puerto Rico as the fifth 'Tiger'? *World Development* **24**(5), 869–85.

Bebbington, A. 1996 Debating 'indigenous' agricultural development: Indian organizations in the Central Andes of Ecuador. In Collinson, H. (ed.), *Green guerrillas: environmental conflicts and initiatives in Latin America and the Caribbean.* Latin American Bureau, London, 51–60.

Bebbington, A. et al. 1992 *Actores de una década ganada.* COMUNIDEC, Quito.

Bebbington, A., Thiele, G. 1993 *Non-governmental organizations and the state in Latin America: rethinking roles in sustainable agricultural development.* Routledge, London.

Becker, C., Morrison, A. 1997 Public policy and rural–urban migration. In Gugler, J. (ed.), *Cities in the developing world: issues, theory and policy.* Oxford University Press, Oxford, 74–87.

Bee, A., Vogel, I. 1997 Temporeras and household relations: seasonal employment in Chile's agro-export sector. *Bulletin of Latin American Research* **16**(1), 83–95.

Béjar, R. 1988 *El Mexicano, aspectos culturales y psicosociales.* UNAM, Mexico City.

Benería, L. 1991 Structural adjustment, the labour market and the household: the case of Mexico. In Standing, G., Tokman, V. (eds), *Towards social adjustment: labour market issues in structural adjustment*. International Labour Organization, Geneva, 161–83.

Benería, L., Roldan, M. 1987 *The crossroads of class and gender: industrial homework, subcontracting and household dynamics in Mexico City*. University of Chicago Press, Chicago, IL.

Berry, A. 1997 The income distribution threat in Latin America. *Latin American Research Review* **32**(2), 3–40.

Binswanger, H., Feder, G., Deininger, K. 1995 Power, distortions and reform in agricultural land relations. In Behrman, J., Srinivasan, T.N. (eds), *Handbook of development economics*, Vol. 3. North-Holland, Amsterdam, 2661–761.

Birkbeck, C. 1979 Garbage, industry and the 'vultures' of Cali, Columbia. In Bromley, R., Gerry, C. (eds), *Casual work and poverty in Third World cities*. John Wiley, Chichester, Sussex, 161–83.

Bitar, S. 1988 Neo-liberalism versus neo-structuralism in Latin America. *CEPAL Review* no. 34, 45–62.

Blokland, K. 1992 *Participación campesina en el desarrollo económico. La Unión Nacional de Agricultores y Ganaderos de Nicaragua durante la revolución Sandinista*. Paulo Freire Stichting, Doetinchem.

Blomström, M., Hettne, B. 1984 *Development theory in transition. The dependency debate and beyond: Third World resposes*. Zed Books, London.

Blum, W. 1986 *The CIA: a forgotten history*. Zed Books, London.

Booth, J., Walker, T. 1993 *Understanding Central America*, 2nd edn. Westview Press, Boulder, CO.

Bradshaw, S. 1995a Women's access to employment and the formation of women-headed households in rural and urban Honduras. *Bulletin of Latin American Research* **14**(2), 143–5.

Bradshaw, S. 1995b Female-headed households in Honduras: perspectives on rural–urban differences. *Third World Planning Review* **17**(2), 117–31.

Bresser Pereira, L.C. 1996 *Economic crisis and state reform in Brazil: toward a new interpretation of Latin America*. Lynne Rienner, Boulder, CO.

Brockett, C.D. 1988 *Land, power, and poverty: agrarian transformations and political conflict in rural Central America*. Unwin Hyman, Boston, MA.

Bromley, R. 1978 The informal sector: why is it worth discussing? *World Development* **6**(9–10), 1033–9.

Bromley, R. 1997 Working in the streets of Cali, Colombia: survival strategy, necessity or unavoidable evil? In Gugler, J. (ed.), *Cities in the developing world: issues, theory and policy*. Oxford University Press, Oxford, 124–38.

Browder, J. 1989 *Fragile lands of Latin America: strategies for sustainable development*. Westview Press, Boulder, CO.

Browner, C. 1989 Women, household and health in Latin America. *Social Science and Medicine* **18**(5), 461–73.

Brundtland, H. 1987 *Our common future*. Oxford University Press, Oxford.

Brunner, J. 1988 *El espejo trizado*. FLACSO, Santiago.

Brunner, J.J. 1994 *Cartografías de la modernidad*. Dolmen Ediciones, Santiago.

Bulmer-Thomas, V. (ed.) 1996a *The new economic model in Latin America and its impact on income distribution and poverty*. Macmillan, London.

Bulmer-Thomas, V. 1994 *The economic history of Latin America since independence*. Cambridge University Press, Cambridge.

Bulmer-Thomas, V. 1996b Conclusions. In Bulmer-Thomas, V. (ed.), *The new economic model in Latin America and its impact on income distribution and poverty*. Macmillan, London.

Bulmer-Thomas, V. 1998 The Central American Common Market: from closed to open regionalism. *World Development* **26**(2), 313–22.

Bunge, C.O. 1926 *Nuestra América*. Espasa Calpe, Madrid.

Bunker, S. 1985. *Underdeveloping the Amazon: extraction, unequal exchange and the failure of the modern state*. University of Chicago Press, Chicago, IL.

Burgwal, A. 1995 *Struggle of the poor*. CEDLA, Amsterdam.

Buvinic, M. 1995 *Investing in women*. International Center for Research on Women, Washington, DC.

Cairncross, S., Hardoy, J., Satterthwaite, D. 1990 The urban context. In Hardoy, J., Cairncross, S., Satterthwaite, D. (eds), *The poor die young: housing and health in Third World cities*. Earthscan, London.

Calderón, F., Piscitelli, A., Reyna, J.L. 1992 Social movements: actors, theories, expectations. In Escobar, A., Alvarez, S. (eds), *The making of social movements in Latin America*. Westview Press, Boulder, CO, 19–36.

Cammack, P. 1997 *Cardoso's political project in Brazil: the limits of social democracy*. In Panitch, L. (ed.), *Ruthless criticism of all that exists: socialist register 1997*. Merlin Press, London, 223–43.

Cardoso, E., Helwege, A. 1992 *Latin America's economy: diversity, trends, and conflicts*. MIT Press, Cambridge, MA.

Cardoso, F.H. 1973 Associated-dependent development: theoretical and practical implications. In Stepan, A. (ed.), *Authoritarian Brazil: origins, policies, and future*. Yale University Press, New Haven, 142–70.

Cardoso, F.H. 1979 On the characteristics of authoritarian regimes in Latin America. In Collier, D. (ed.), *The new authoritarianism Latin America*. Princeton University Press, Princeton, NJ.

Cardoso, F.H., Faletto E. 1969 *Dependencia y desarrollo en América Latina: ensayo y interpretación sociológica*. Siglo Veintiuno Editores, Mexico City.

Cardoso, F.H., Faletto E. 1979 *Dependency and development in Latin America*, University of California Press, Los Angeles and Berkeley, CA; translation of 1969 reference with new introduction and post-scriptum.

Carley M., Christie, I. 1993 *Managing sustainable development*. University of Minnesota Press, Minneapolis, MN.

Carrière, J. 1991 The crisis in Costa Rica: an ecological perspective. In Redclift, M., Goodman, D. (eds), *Environment and development in Latin America: the politics of sustainability*. Manchester University Press, Manchester.

Carter, M., Mesbah. D. 1993 Can land market reform mitigate the exclusionary aspects of rapid agro-export growth? *World Development* **21**, 7, 1085–1100.

Carter, M.R., Barham, B.L. 1996 Level playing fields and *laissez faire*: postliberal development strategy in inegalitarian agrarian economies. *World Development* **24**, 7, 1133–40.

Carter, M.R., Barham, B.L., Mesbah, D. 1996 Agricultural export booms and the rural poor in Chile, Guatemala and Paraguay. *Latin American Research Review* **31**(1), 33–65.

Casper, G., Taylor, M.M. 1996 *Negotiating Democracy: Transitions from Authoritarian Rule*. Pittsburgh (PA): University of Pittsburgh Press.

Castañeda, J.G. 1994 *Utopia unarmed: the Latin American left after the Cold War*. Vintage Books, New York.

Castells, M. 1983 *The city and the grassroots*. University of California Press, Berkeley, CA.

Castells, M. 1997 *Power of identity*, vol. II of Network society. Basil Blackwell, Oxford.

Castells, M., Laserna, R. 1995 The new dependency: technological change and socioeconomic restructuring in Latin America. In Kincaid, A.D., Portes, A. (eds), *Comparative national development: society and economy in the new global order*. University of North Carolina Press, Chapel Hill, NC, 57–83.

Centeno, M.A., Silva, P. (eds) 1998 *The politics of expertise in Latin America*. MacMillan, London.

CEPAL 1994 El sector informal urbano desde la perspectiva de género: el caso de México. Paper presented at the workshop 'El sector informal urbano desde la perspectiva de género: el caso de México', Mexico City, 28–29 November.

CEPAL 1995 *Annuario estadístico de América Latina y el Caribe*. Comisión Económica para América Latina y el Caribe, Santiago, Chile.

CEPAL/FAO 1986 *El crecimiento productivo y la heterogeneidad agraria*. División Agrícola Conjunta CEPAL/FAO, Santiago.

Chalmers, D.A., Vilas, C.M., Hite, C. (eds) 1997 *The new politics of inequality in Latin America: rethinking participation and representation*. Oxford University Press, Oxford.

Chant, S. 1991a *Women and survival in Mexican cities: perspectives on gender, labour markets and low-income households*. Manchester University Press, Manchester.

Chant, S. 1991b Gender, households and seasonal migration in Guanacaste, Costa Rica. *European Review of Latin American and Caribbean Studies* no. 50, 51–85.

Chant, S. 1992 Migration at the margins: gender, poverty and population movement on the Costa Rican periphery. In Chant, S. (ed.), *Gender and migration in developing countries*. Belhaven Press, London, 49–72.

Chant, S. 1996 Women's roles in recession and economic restructuring in Mexico and the Philippines. *Geoforum* **27**(3), 297–327.

Chant, S. 1997a *Women-headed households: diversity and dynamics in the developing world.* Macmillan, Basingstoke, Hants.

Chant, S. 1997b Men, households and poverty in Costa Rica: a pilot study. Final report to the Economic and Social Research Council, UK (award no. R000222205).

Chant, S. 1998 Households, gender and rural–urban migration: reflections on linkages and considerations for policy. *Environment and Urbanisation* **10**(1), 5–21.

Chant, S., Radcliffe, S. 1992 Migration and development: the importance of gender. In Chant, S. (ed.), *Gender and migration in developing countries.* Belhaven Press, London, 1–29.

Chauvin, L. 1998 *Smoking economy.* LatinAmerica Press, Lima, Peru.

Chevannes, B., Ricketts, H. 1997 Return migration and small business development and Jamaica. In Pessar, P.R. (ed.), *Caribbean circuits: new directions in the study of migration.* Center for Migration Studies, New York.

Chickering, A.L., Salahdine, M. 1991 Introduction. In Chickering, A.L., Salahdine, M. (eds), *The silent revolution: the informal sector in five Asian and near Eastern countries.* International Center for Economic Growth, San Francisco, CA, 1–14.

Chilcote, R.H. 1981 *Theories of comparative politics: the search for a paradigm.* Westview Press, Boulder, CO.

Chilcote, R.H. 1982 *Dependency and marxism: toward a resolution of the debate.* Westview Press, Boulder, CO.

Chiriboga, M. 1992: Modernización democrática e incluyente. *Revista Latinoamericana de Sociología Rural,* 1, 27–37.

CISAS 1997 Adjusting health care: the case of Nicaragua. Oxfam, Oxford, 86–93.

Clapp, R.A. 1995 Creating competitive advantage; forest policy as industrial policy in Chile. *Economic Geography* **71**(3), 273–96.

Clapp, R.A. 1998 Waiting for the forest law: resource-led development and environmental politics in Chile. *Latin American Research Review* **33**(2), 3–36.

Clark, M.A. 1997 Transnational alliances and development policy in Latin America: nontraditional export promotion in Costa Rica. *Latin American Research Review* **32**(2), 71–97.

Clarke, C. 1975 *Kingston, Jamaica: urban development and social change, 1692–1962.* University of California Press, Berkeley, CA, and London.

Clarke, C., Ward, P. 1980 Stasis in makeshift housing: prespectives from Mexico and the Caribbean. *Comparative Urban Research* **8**(1), 117–27.

Cockburn, A., Hecht, S.B. 1988 *The fate of the forest: developers, destroyers, and defenders of the Amazon.* Verso, London.

Colburn, F. 1998 The INCAE–Harvard project for Central America. *LASA Forum* **28**(4), 15–16.

Collier, D. (ed.) 1979 *The new authoritarianism in Latin America.* Princeton University Press, Princeton, NJ.

Collier, D., Levitsky, S. 1997 Democracy with adjectives: conceptual innovation in comparative research. *World Politics* **49**(3), 430–51.

Collier, G.A. 1994 *Basta! Land and the Zapatista rebellion in Chiapas.* Institute for Food and Development Policy, Oakland, CA.

Comisión Nacional del Medio Ambiente 1992 *Chile: Informe nacional a la conferencia de las Naciones Unidas sobre Medio Ambiente y Desarrollo.* Ministerio de Bienes Nacionales, Santiago, Chile.

Commoner, B. 1990 *Making peace with the planet.* Pantheon, New York.

Congdon, T. 1988 *The debt threat.* Basil Blackwell, Oxford.

Conger Lind, A. 1992 Power, gender and development: popular women's organizations and the politics of needs in Ecuador. In Escobar, A., Alvarez, S. (eds), *The making of social movements in Latin America.* Westview Press, Boulder,CO, 134–49.

Conroy, M., Murray, D., Rosset, P. 1996 *A cautionary tale: failed U.S. development policy in Central America.* Lynne Rienner, Boulder, CO.

Conway, D. 1998 Misguided directions, mismanaged models, or missed paths? In Klak, T. (ed.), *Globalization and neoliberalism: the Caribbean context.* Rowman and Littlefield, Lanham, MD, 29–50.

Corbridge, S. 1993 *Debt and development*. Basil Blackwell, Oxford.

Cordera Campos, R., González Tiburcio, E. 1991 Crisis and transition in the Mexican economy. In González de la Rocha, M., Escobar, A. (eds), *Social responses to Mexico's economic cisis of the 1980s*. Contemporary Perspectives Series No.1. Center for US Mexican Studies, San Diego, CA, 19–56.

Cordero-Guzman, H. 1993 Lessons from Operation Bootstrap. *NACLA Report on the Americas* **27**(3), 7–10.

Cornelius, A. 1991 'Los Migrantes de la crisis': the changing profile of Mexican migration to the United States. In González de la Rocha, M., Escobar, A. (eds), *Social responses to Mexico's economic crisis of the 1980s*, Contemporary Perspectives Series No.1. Center for US Mexican Studies, San Diego, CA, 155–93.

Cornia, G. 1987 An overview of the alternative approach. In Cornia, G., Jolly, R., Stewart, F. (eds), *Adjustment with a human face: protecting the vulnerable and promoting growth*. Clarenden Press, Oxford.

Corrêa, S., Reichmann, R. 1994 *Population and reproductive rights: feminist perspectives from the South*. Zed Press, London.

Cousiño, C., Valenzuela, E. 1994 *Politización y monetarización en América Latina*. Pontificia Universidad Católica de Chile, Santiago.

CPT 1998 Assembly factories in Haiti: a wage of sacrifice. Report by the Christian Peacemaker Teams, 23 February, available at www.prairienet.org/cpt/haiti.html.

Craske, N. 1993 Women's political participation in colonias populares in Guadalajara, Mexico. In Radcliffe, S., Westwood, S. (eds), *Viva: women and popular protest in Latin America*. Routledge, London, 112–35.

Craske, N. 1994 Women and regime politics in Guadalajara's low-income neighbourhoods. *Bulletin of Latin American Research* **13**(1), 61–78.

Cristi, R. 1992 Estado nacional y pensamiento conservador en la obra madura de Mario Góngora. In Cristi, R., Ruiz, C. (eds), *El pensamiento conservador en Chile*. Editorial Universitaria, Santiago.

Crozier, M., Huntington, S.P., Watanuki, J. 1975 *The crisis of democracy: report on the governability of democracies to the trilateral commission*. New York University Press, New York.

Cubitt, T. 1995 *Latin American society*, 2nd edn. Longman, Harlow, Essex.

Cuddington, J.T. 1992 Long-run trends in 26 primary commodity prices: a disaggregated look at the Prebisch–Singer hypothesis. *Journal of Development Economics* **39**(2), 207–27.

Cuenya, B., Armus, D., DiLoreto, M., Penalva, S. 1990 Land invasions and grassroots organisation: the Qualmes settlements in Greater Buenos Aires, Argentina. *Environment and Urbanization* **2**(1), 61–74.

Curto de Casas, S. 1994 Health care in Latin America. In Phillips, D., Verhasselt, Y. (eds), *Health and development*. Routledge, London, 234–48.

Dahl, R. 1971 *Polyarchy: participation and opposition*. Yale University Press, New Haven, CT.

Daly, H.E., Townsend, K.N. (eds) 1993 *Valuing the earth: economics, ecology, ethics*. MIT Press, Cambridge, MA.

Dawson, E. 1992 District planning with community participation in Peru: the work of the Institute of Local Democracy (IPADEL). *Environment and Urbanization* **4**(2), 90–100.

de Barbieri, T., de Oliveira, O. 1989 Reproducción de la fuerza de trabajo en América Latina: àlgunas hipótesis. In Schteingart, M. (ed.), *Las ciudades Latinoamericanas en la crisis*. Editorial Trillas, Mexico, DF, 19–29.

de Blij, H., Muller, P. 1998 Puerto Rico's clouded future. In de Blij, H., Muller, P. *Geography: realms, regions, and concepts*, 8th edn, updated and revised. John Wiley, New York.

de Janvry, A. 1981 *The agrarian question and reformism in Latin America*. Johns Hopkins University Press, Baltimore, MD.

de Janvry, A. 1994 Social and economic reforms: the challenge of equitable growth in Latin American agriculture. In Muchnik E., Niño de Zepeda, A. (eds), *Apertura económica, modernización y sostenibilidad de la agricultura*. ALACEA, Santiago.

de Janvry, A., Gordillo, G., Sadoulet, E. 1997 *Mexico's second agrarian reform: household and community responses, 1990–1994*. Center for US–Mexican Studies, University of California, San Diego, La Jolla, CA.

de Janvry, A., Marsh, R., Runsten, D., Sadoulet, E., Zabin, C. 1989a *Rural development in Latin America: an evaluation and a proposal*. IICA, San José de Costa Rica.

de Janvry, A., Radwan, S., Sadoulet, E., Thorbecke, E. (eds) 1995 *State, market and civil organizations. New theories, new practices and their implications for rural development.* Macmillan, Basingstoke, Hants. and London.

de Janvry, A., Sadoulet, E., Young, L.W. 1989b Land and labour in Latin American agriculture from the 1950s to the 1980s. *The Journal of Peasant Studies* **16**, 396–424.

de la Peña, G., de la Torre, R. 1990 Religión y política en los barrios populares de Guadalajara. *Estudios Sociológicos* **24**, 571–602.

de Mattos, C.A. 1996 Avances de la globalización y nueva dinámica metropolitana: Santiago de Chile, 1975–1995. *EURE, Revista Latinoamericana de Estudios Urbano Regionales* **22**(65), 39–64.

de Oliveira, O. 1991 Migration of women, family organization and labour markets in Mexico. In Jelin, E. (ed.), *Family, household and gender relations in Latin America.* Kegan Paul, London/UNESCO, Paris, 101–18.

de Soto, H. 1989 *The other path: the invisible revolution in the Third World.* Harper Row, New York.

de Souza, R.-M. 1998 Bridging isolated proximities: transnationalism and the Caribbean migrant. Paper presented at the Association of American Geographers annual meeting, March 25–29, Boston, MA.

Deere, C., Melendez, E. 1992 When export growth isn't enough: US trade policy and Caribbean Basin economic recovery. *Caribbean Affairs* **5**, 61–70.

Deere, C.D. 1985 Rural women and state policy: the Latin American agrarian reform experience. *World Development* **13**(9), 1037–53.

Deere, C.D. 1987 The Latin American agrarian reform experience. In Deere, C.D., León, M. (eds), *Rural women and state policy: feminist perspectives on Latin American agricultural development.* Westview Press, Boulder, CO.

Deere, C.D., et al., 1990 *In the shadows of the sun: Caribbean development alternatives and U.S. policy.* Westview Press, Boulder, CO.

Demas, W. 1997 *West Indian development and the deepening and widening of the Caribbean community.* Ian Randle, Kingston.

Demery, L., Ferroni, M., Grootaert, C., Wong-Valle, J. 1993 *Understanding the social effects of policy reform.* World Bank, Washington, DC.

Detwyler, R., Melvin, M. 1972 *Urbanisation and environment: the physical geography of the city.* Duxbury Press, Belmont, MA.

Devlin, R. 1989 *Debt and crisis in Latin America: the supply side of the history.* Princeton University Press, Princeton, NJ.

DeWalt, B.R., Rees, M.W., Murphy, A.D. 1994 *The end of agrarian reform in Mexico: past lessons and future prospects.* Center for US–Mexican Studies, University of California, San Diego, La Jolla, CA.

Diakosavvas, D., Scandizzo, P.L. 1991 Trends in the terms of trade of primary commodities, 1900–1982: the controversy and its origins. *Economic Development and Cultural Change* **39**(2), 231–64.

Diamond, L., Linz, J., Lipset, S.M. (eds) 1989 *Democracy in developing countries: Latin America.* Lynne Rienner and Adamantine Press, Boulder, CO.

Díaz Polanco, H. 1997 *Indigeneous people in Latin America: the quest for self-determination.* Westview Press, Boulder, CO.

Diaz, V.J. 1992 Landslides in the squatter settlements of Caracas: towards a better understanding of causative factors. *Environment and Urbanization* **4**(2), 80–9.

Diaz-Briquets, S. 1989 The Central American demographic situation: trends and implications. In Bean, F., Schmandt, J., Weintraub, S. (eds), *Mexican and Central American population and US immigration policy.* University of Texas Press, Austin, TX, 33–64.

Dicken, P. 1992 *Global shift: the internationalization of economic activity*, 2nd edn. The Guilford Press, New York.

Dicken, P., Peck, J., Tickell, A. 1997 Unpacking the global. In Lee, R., Wills. J. (eds), *Geographies of economies.* Edward Arnold, London, 158–66.

Dickenson, J.P. 1978 *Brazil.* Dawson, London.

Dierckxsens, W. 1992 Impacto del ajuste estructural sobre la mujer trabajadora en Costa Rica. In Acuña-Ortega, M. (ed.), *Cuadernos de política económica.* Universidad Nacional de Costa Rica, Heredia, 2–59.

Dietz, J.L. (ed.) 1995 *Latin America's economic development: confronting crisis.* Lynne Rienner, Boulder, CO.

Dinges, J. 1988 *Our man in Panama.* Random House, New York.

Dore, E. 1995: Latin America and the social ecology of capitalism. In Halebsky, S., Harris, R.L. (eds), *Capital, power, and inequality in Latin America.* Westview Press, Boulder, CO, 253–78.

Dore, E., Molyneux, M. (eds) 1999 *The hidden history of gender and the state in Latin America.* Duke University Press, Durham, NC and London.

Dorner, P. 1992 *Latin American land reforms in theory and practice: a retrospective analysis.* University of Madison Press, Madison, WI.

Dos Santos, T. 1970 The sructure of dependency. *American Economic Review* **60** (2), 231–6.

Duncan, K., Rutledge, I. (eds) 1977 *Land and labour in Latin America: essays on ther development of agrarian capitalism in the nineteenth and twentieth centuries.* Cambridge University Press, Cambridge.

Dunkerley, J. 1994 *The pacification of Central America: political change in the Isthmus, 1987–1993.* Verso, London.

Dupuy, A. 1997 *Haiti in the new world order: the limits of the democratic revolution.* Westview, Press, Boulder, CO.

Durand, J., Massey, D. 1992 Mexican migration to the United States: a critical view. *Latin American Research Review* **27**(2), 3–42.

Durand, J., Parrado, E.A., Massey, D.S. 1996 Migradollars and development: a reconsideration of the Mexican case. *International Migration Review* **30**(2), 423–44.

Easton, D. 1965 *A framework for political analysis.* Prentice Hall, Englewood Cliffs, NJ.

Eckstein, S. (ed.) 1989 *Power and popular protest: Latin American social movements.* University of California Press, Berkeley,CA.

ECLA 1968 *Economic survey of Latin America 1966*, Economic Commission for Latin America. United Nations, New York.

ECLAC 1990 *Changing production patterns with social equity.* Economic Commission for Latin America and the Caribbean, Santiago.

ECLAC 1991 *Sustainable development: changing production patterns, social equity and the environment*, Economic Commission for Latin America and the Caribbean. United Nations, Santiago.

ECLAC 1992a *Social equity and changing production patterns: an integrated approach.* Economic Commission for Latin America and the Caribbean, Santiago.

ECLAC 1992b *Major changes and crisis. The impact on women in Latin America and the Caribbean.* Economic Commission for Latin America and the Caribbean, Santiago.

ECLAC 1993 *Statistical yearbook for Latin America and the Caribbean 1992.* Economic Commission for Latin America and the Caribbean, Santiago.

ECLAC 1994a *Open regionalism in Latin America and the Caribbean.* Economic Commission for Latin America and the Caribbean, Santiago.

ECLAC 1994b *Social Panorama of Latin America.* Economic Commission for Latin America and the Caribbean, Santiago.

ECLAC 1995 *Latin America and the Caribbean: policies to improve linkages with the global economy.* Economic Commission for Latin America and the Caribbean, Santiago.

ECLAC 1996 *Statistical yearbook for Latin America and the Caribbean, 1996*, Economic Commission for Latin America and the Caribbean. United Nations, Santiago.

Edwards, A. 1987 *La fronda aristocrática en Chile.* Editorial Universitaria, Santiago.

Edwards, S. 1995 *Crisis and reform in Latin America: from despair to hope.* Oxford University Press, Oxford and New York.

Ekins, P. 1992 *A new world order: grassroots movements for global change.* Routledge, New York.

Emmanuel, A. 1972 *Unequal exchange: a study of the imperialism of trade.* New Left Books, London.

Enríquez, L.J. 1991 *Harvesting change: labor and agrarian reform in Nicaragua, 1979–1990.* University of North Carolina Press, Chapel Hill.

Enríquez, L.J. 1997 *Agrarian reform and class consciousness in Nicaragua.* University Press of Florida, Gainesville, FL.

Escobar Latapí, A., González de la Rocha, M. 1995 Crisis, restructuring and urban poverty in Mexico. *Environment and Urbanization* **7**(1), 57–76.

Escobar, A. 1992 Culture, economics, and politics in Latin American social movement theory and research. In Escobar, A., Alvarez, S. (eds), *The making of social movements in Latin America*. Westview Press, Boulder, CO, 62–85.

Escobar, A., Alvarez, S. (eds) 1992 *The making of social movements in Latin America*. Westview Press, Boulder,CO.

Espinal, R. 1991 Review of Diamond, Linz, and Lipset 1989. *Journal of Latin American Studies* 23(2), 431–3.

Espinal, R. 1992 Development, neoliberalism and electoral politics in Latin America. *Development and Change* 23(4), 27–48.

Evans, P.B. 1979. *Dependent development: the alliance of multinational, state, and local capital in Brazil*. Princeton University Press, Princeton, NJ.

Evans, P.B., Rueschemeyer, D., Skocpol, T. (eds) 1985 *Bringing the state back in*. Cambridge University Press, New York.

Evans, R. 1993 Banking on the black economy. *The Economist* September, 40–2.

Evans, T., Castro, C., Jones, J. 1995 *Structural adjustment and the public sector in Central America and the Caribbean*. Coordinadora Regional de Investigaciones Económicas y Sociales, Managua.

Evers, T. 1985 Identity: the hidden side of new social movements in Latin America. In Slater, D. (ed.), *New social movements and the state in Latin America*. CEDLA, Amsterdam, 43–72.

Eyre, L.A. 1972 The shanty towns of Montego Bay. *Geographical Review* 62(3), 394–413.

Eyre, L.A. 1997 In Potter, R.B., Conway, D. (eds), *Self-help housing, the poor, and the state in the Caribbean*. University of Tennessee Press, Knoxville,TN, 75–101.

Eyzaguirre, J. 1947 *Hispanoamérica del Dolor*. Instituto de Estudios Políticos, Madrid.

Ezcurra, E., Mazari-Hiriart, M. 1996 Are megacities viable?A cautionary tale fromMexico City. *Environment* 38(1), 6–16.

Fajnzylber, F. 1990 Sobre la impostergable transformación productiva de América Latina. *Pensamiento Iberoamericano* no. 16, 85–129.

Falabella, G. 1991 Organizarse para sobrevivir en Santa María. Democracia social en un sindicato de temporeros y temporeras. Paper presented at the 47th International Congress of Americanists, New Orleans, 7–11 July.

FAO 1996 *Trade Yearbook*, United Nations Food and Agriculture Organisation, Rome.

Feder, E. 1971 *The rape of the peasantry: Latin America's landholding system*. Doubleday, Garden City, NY.

Félix, D. 1992 Privatizing and rolling back the Latin American state. *CEPAL Review*, 46, 31–46.

Ferguson, B. 1996 The environmental impacts and public costs of unguided informal settlement: the case of Montego Bay. *Environment and Urbanization* 8(2), 171–93.

Ferguson, B., Maurer, C. 1996 Urban management for environmental quality in South America. *Third World Planning Review* 18(2), 117–54.

Fernández-Kelly, M.P. 1983 Mexican border industrialization, female labor force participation and migration. In Nash, J., Fernández-Kelly, M.P. (eds), *Women, men and the international division of labor*. State University of New York Press, Albany, NY, 205–23.

Ffrench-Davis, R. 1988 An outline of a neo-structuralist approach. *CEPAL Review* no. 34, 37–44.

Figueroa, A. 1977 Agrarian reforms in Latin America: a framework and an instrument of rural development. *World Development* 5(1/2), 155–68.

Figueroa, A. 1993 Agricultural development in Latin America. In Sunkel, O. (ed.), *Development from within: towards a neostructuralist approach for Latin America*. Lynne Rienner, Boulder, CO, and London, 287–314.

Fisher, J. 1993a *The road from Rio: sustainable development and the nongovernmental movement in the Third World*. Praeger Publishers, Westport, CT.

Fisher, J. 1993b *Out of the shadows: women, resistance and politics in South America*. Latin American Bureau, London.

Flores Galindo, A. 1994 *Buscando un Inca*. Editorial Horizonte, Lima.

Foweraker, J. 1995 *Theorising social movements*, Critical Studies on Latin America. Pluto Press, London

Foxley, A. 1996 Preface. In Bulmer-Thomas, V. (ed.), *The new economic model in Latin America and its impact on income distribution and poverty*. Macmillan, London, 1–6.

Frank, A.G. 1966 The development of underdevelopment. *Monthly Review* **18**(4), 17–31.

Frank, A.G. 1967 *Capitalism and underdevelopment in Latin America*. Monthly Review Press, New York.

Frank, A.G. 1991 Latin American development theories revisited: a participant review essay. *European Journal of Development Research* **3**(2), 146–59; a longer version was published in 1991 in *The Scandinavian Journal of Development Research* **10**(3), 133–50.

Frank, A.G. 1992, Latin American development theories revisited: a participant review essay. *Latin American Perspectives* **19**(3), 125–39.

Freudenheim, M. 1992 Tax credits of $8.5 Billion received by 22 drug makers. *The New York Times* 15 May, C3.

Freyre, G. 1946 *The master and the slaves: a study in the development of Brazilian civilization*. Alfred A. Knopf, New York.

Frías, P., Ruiz-Tagle, J. 1995 Free market economics and belated democratization: the case of Chile. In Thomas, H. (ed.), *Globalization and Third World trade unions*. Zed Press, London, 130–48.

Frieden, J. 1991 *Debt, development, and democracy: modern political economy and Latin America*. Princeton University Press, Princeton, NJ.

Friedland, W.H. 1994 The global fresh fruit and vegetable: an industrial organization analysis. In McMichael, P. (ed.), *The global restructuring of agro-food systems*. Cornell University Press, Ithaca, NY.

Friedmann, J., Rangan, H. (eds) 1993 *In defense of livelihood: comparative studies on environmental action*. Kumarian Press, West Hartford, CT.

Fuentes, C. 1990 *Valiente mundo nuevo: épica, utopía y mito en la Novela Hispanoamericana*. Narrativa Mondadori, Madrid.

Fukuyama, F. 1992 *The end of history and the last man*. Hamish Hamilton, London.

Furedy, F. 1997 *Population and development*. Polity Press, Cambridge.

Furley, P. 1996 Environmental issues and the impact of development. In Preston, D. (ed.), *Latin American development – geographical perspectives*, 2nd edn. Longman, Harlow, Essex.

Furtado, C. 1970 *Economic development of Latin America*. Cambridge University Press, Cambridge.

Galeano, E. 1991 The Blue Tiger and the Promised Land. *Report on the Americas* **5**, 13–17.

García Canclini, N. 1989 *Culturas híbridas: estrategias para entrar y salir de la modernidad*. Grijalbo, Mexico.

Garcia, B., de Oliveira, O. 1997 Motherhood and extradomestic work in urban Mexico. *Bulletin of Latin American Research* **16**(3), 367–84.

Garcia, N.E. 1993 *Ajuste, reformas y mercado laboral: Costa Rica, Chile, Mexico*. PREALC, Santiago.

Gayle, D. 1998 Trade policies and the hemispheric integration process. In Klak, T. (ed.), *Globalization and neoliberalism: the Caribbean context*. Rowman and Littlefield, Lanham, MD, 65–85.

Geertz, C. 1963 *Peddlars and princes: social development and economic change in two Indonesian towns*. University of Chicago Press, Chicago, IL.

Geras, N. 1987 Post-marxism? *New Left Review* no. 163, 40–77.

Gereffi, G. 1994 Rethinking development theory: insights from East Asia and Latin America. In Kincaid, A.D., Portes, A. (eds) *Comparative national development: society and economy in the new global order*. University of North Carolina Press, Chapel Hill, NC, 26–56.

Gereffi, G. 1996 Mexico's 'old' and 'new' maquiladora industries: contrasting approaches to North American integration. In Otero, G. (ed.), *Neo-liberalism revisited: economic restructuring and Mexico's political future*. Westview Press, Boulder, CO, 85–106.

Gereffi, G., Korzeniewicz, M. (eds), 1994 *Commodity chains and global capitalism*. Praeger, Westport, CT.

Gereffi, G., Wyman, D.L. (eds) 1990 *Manufacturing miracles: paths of industrialization in Latin America and East Asia*. Princeton University Press, Princeton, NJ.

Germani, G. 1965 *Política y sociedad en una época de transición*. Editorial Paidos, Buenos Aires.

Gerschenkron, A. 1962 *Economic backwardness in historical perspective*. Belknap Press of the University of Harvard University Press, Cambridge, MA.

Gerth, H.H., Mills, C.W. (eds) 1958 *From Max Weber: essays in sociology*. Oxford University Press, New York.

Ghai, D. (ed.) 1994 Development and the environment: sustaining people and nature. *Development and Change* 25, 1, Special Issue.

Ghai, D., Kay, C., Peek, P. 1988 *Labour and development in rural Cuba*. Macmillan, Basingstoke, Hants, and London.

Ghai, D., Vivian, J.M. (eds) 1992 *Grassroots environmental action: people's participation in grassroots development*. Routledge, London.

Gibb, R., Michalak, W. (eds) 1994 *Continental trading blocs: the growth of regionalism in the world economy*. John Wiley, Chichester.

Giddens, A. 1990 *The consequences of modernity*. Stanford University Press, Stanford CA.

Gil Fortoul, J. 1983 La Raza. In Teran, O. (ed.), *América Latina: positivismo y nación*. Editorial Katún, Mexico.

Gilbert, A. (ed.) 1996 *The mega-city in Latin America*. United Nations University Press, Tokyo.

Gilbert, A. 1990 *Latin America*. Routledge, London.

Gilbert, A. 1993 *In search of a home: rental and shared housing in Latin America*. University College London Press, London.

Gilbert, A. 1994 *The Latin American city*. Latin American Bureau, London.

Gilbert, A. 1995a Debt, poverty and the Latin American city. *Geography* 80(4), 323–33.

Gilbert, A. 1995b Globalization, employment and poverty: the case of Bogotá, Colombia. Seminar, Geography and Planning Research Series, London School of Economics, London, 30 November.

Gilbert, A. 1997 Mining, manufacturing and services. In Blouet, B.W., Blouet, O.M. (eds), *Latin America and the Caribbean: a systematic and regional survey*. John Wiley, New York.

Gilbert, A. 1998 *The Latin American city*, revised edn. Latin America Bureau, London.

Gilbert, A., Gugler, J. 1992 *Cities, poverty and development: urbanization in the Third World*. Oxford University Press, Oxford.

Gilbert, A., Varley, A. 1991 *The landlord as tenant*. UCL Press, London.

Gilbert, A., Ward, P. 1984 Community participation in upgrading settlements: the community response. *World Development* 12(9), 913–22.

Gills, B., Rocamora, J. 1992 Low intensity democracy. *Third World Quarterly* 13(3), 501–24.

Gledhill, J. 1988 Agrarian social movements and forms of consciousness. *Bulletin of Latin American Research* 7(2), 257–76.

Gledhill, J. 1995 *Neoliberalism, transnationalisation and rural poverty: a case study of Michoacán*. Westview Press, Boulder, CO.

Gledhill, J. 1997 The challenge of globalization: reconstruction of identities, transnational forms of life and the social sciences. English translation of keynote address to the XIX Coloquio de Antropología e Historia Regionales 'Fronteras Fragmentadas: Género, Familia e Identidades en la Migración Mexicana al Norte', El Colegio de Michoacán, Zamora, 22 October.

Gligo, N. 1993 Environment and natural resources in Latin American development. In Sunkel, O. (ed.), *Development from within: towards a neostructuralist approach for Latin America*. Lynne Rienner, Boulder, CO, and London, 185–222.

Goldin, I., Winters, L.A. 1995 *The economics of sustainable development*. Cambridge University Press, Cambridge.

Goldsworthy, D. 1988 Thinking politically about development. *Development and Change* 19(3), 505–30.

Gomes, E. 1994 Choice or authorized crime? an epidemic of caesareans and sterilizations in Brazil. In Panos (ed.), *Private decisions, public debate: women, reproduction and population*. Panos, London, 69–80.

Góngora, M. 1981 *Ensayo histórico sobre la noción de Estado en Chile en los siglos XIX y XX*. Ediciones La Ciudad, Santiago.

González de la Rocha, M. 1988 Economic crisis, domestic reorganization and women's work in Guadalajara, Mexico. *Bulletin of Latin American Research* 7(2), 207–23.

González de la Rocha, M. 1991 Family well-being, food consumption and survival strategies during Mexico's economic crisis. In González de la Rocha, M., Escobar, A. (eds), *Social responses to Mexico's economic crisis of the 1980s*, Contemporary Perspectives Series No.1, Center for US Mexican Studies, San Diego, CA, 115–27.

González de la Rocha, M. 1994 *The resources of poverty: women and survival in a Mexican City*. Basil Blackwell, Oxford.

González de la Rocha, M. 1997 The erosion of the survival model: urban responses to persistent poverty. Paper prepared for the UNRISD.UNDP/CDS Workshop 'Gender, Poverty and Well-being: Indicators and Strategies', Trivandrum, Kerala, 24–27 November.

González de la Rocha, M., Escobar, A. Martinez Castellanos, M. 1990 Estrategias versis conflictos: reflexiones para el estudio del grupo doméstico en epoca del crisis. In de la Peña, G., Durán, J.M., Escobar, A., García de Alba, J. (eds), *Crisis, conflicto y Sobrevivencia: estudios sobre la sociedad urbana en México*. Universidad de Guadalajara/CIESAS, Guadalajara, 351–67.

Gonzalez Prada, M. 1964 *Horas de lucha*. Fondo de Cultura Económica, Lima.

Goodman, D., Redclift, M. 1981 From peasant to proletarian: capitalist development and agrarian transformations. Basil Blackwell, Oxford.

Gordon, D., Dixon, C. 1992 La urbanización en Kingston, Jamaica: años de crecimiento y años de crisis. In Portes, A., Lungo, M. (eds), *Urbanización en el Caribe*. Flasco, San José.

Gore, C. 1996 Methodological nationalism and the misunderstanding of East Asian industrialisation. *The European Journal of Development Research* **8**(1), 77–122.

Gourevitch, P.A. 1986 *Politics in hard times: comparative responses to international crises*. Cornell University Press, Ithaca, NY.

Grabowski, R., Shields, M. 1996 *Development economics*. Basil Blackwell, Oxford.

Grant, R.D. 1993 *Against the grain: agricultural trade policies of the US, the European Community and Japan at the GATT*. Political Geography **12**(3), 247–62.

Green, D. 1991 *Faces of Latin America*. Latin America Bureau, London.

Green, D. 1995 *Silent revolution: the rise of market economics in Latin America*. Cassell in asociation with Latin America Bureau, London.

Green, D. 1996 Latin America: neoliberal failure and the search for alternatives. *Third World Quarterly* **17**(1), 109–22.

Griffith, I. 1997 *Drugs and security in the Caribbean*. Pennsylvania State University Press, State College, PA.

Grilli, E., Yang, H. 1988 Primary commodity prices, manufactured goods prices and the terms of trade of developing countries: what the long-run shows. *World Bank Economic Review* **2**, 1–47.

Grugel, J. 1995. *Politics and development in the Caribbean Basin: Central America and the Caribbean in the new world order*. Indiana University Press, Bloomington, IN.

Grzybowski, C. 1990 Rural workers and democratisation in Brazil. In Fox, J. (ed.), *The challenge of rural democratisation*. Frank Cass, London, 15–43.

Guoymer, H., Mahe, L.P., Munk, K.J., Roe, T.L. 1993 Agriculture in the Uruguay Round. *Journal of Agricultural Economics* **27**(1), 230–46.

Gutmann, M. 1996 *The meanings of macho: being a man in Mexico City*. University of California Press, Berkeley, CA.

Gutmann, M. 1997 The ethnographic (g)ambit: women and the negotiation of masculinity in Mexico City. *American Ethnologist* **24**(4), 833–55.

Gwynne, R.N. 1976 *Economic development and structural change: the Chilean case, 1970–73*, OP-2, Department of Geography, University of Birmingham, Birmingham.

Gwynne, R.N. 1978 Government planning and the location of the motor vehicle industry in Chile. *Tijdschrift voor Economische en Sociale Geografie* **69**, 130–40.

Gwynne, R.N. 1985 *Industrialisation and urbanisation in Latin America*. Johns Hopkins University Press, Baltimore, MD.

Gwynne, R.N. 1990 *New horizons? Third World industrialization in an international framework*. Longman, Harlow, Essex.

Gwynne, R.N. 1993a Non-traditional export growth and economic development: the Chilean forestry sector since 1974. *Bulletin of Latin American Research* **12**(2), 147–169.

Gwynne, R.N. 1993b Outward orientation and marginal environments: the question of sustainable development in the Norte Chico, Chile. *Mountain Research and Development* **13**(3), 281–293.

Gwynne, R.N. 1995 Regional integration in Latin America: the revival of a concept?. In Gibb, R., Michalak, W. (eds), *Continental trading blocs: the growth of regionalism in the world economy*. John Wiley, Chichester, Sussex.

Gwynne, R.N. 1996a Direct foreign investment and non-traditional export growth in Chile: the case of the forestry sector. *Bulletin of Latin American Research* **15**(3), 341–357.

Gwynne, R.N. 1996b Industrialization and urbanization. In Preston, D. (ed.), *Latin American development*. Longman, Harlow, Essex.

Gwynne, R.N., Kay, C. 1997 Agrarian change and the democratic transition in Chile. *Bulletin of Latin American Research* **16**, 1, 3–10.

Gwynne, R.N., Meneses, C. 1994 Climate change and sustainable development in the Norte Chico, Chile: land, water and the commercialisation of agriculture. OP-34, School of Geography, University of Birmingham, Birmingham.

Gwynne, R.N., Ortiz, J. 1997 Export growth and development in poor rural regions: a meso scale analysis of the Upper Limarí. *Bulletin of Latin American Research* **16**(1), 25–41.

Haas, P.M. 1990 *Saving the Mediterranean: the politics of international environmental cooperation*. Columbia University Press, New York.

Haas, P.M., Keohane, R.O., Levy, M.A. (eds) 1993 *Institutions for the earth: sources of international environmental protection*. MIT Press, Cambridge:, MA.

Habermas, J. 1989 *The structural transformation of the public sphere*. Polity Press, Cambridge.

Haggard, S. 1986 The newly industrializing countries in the international system. *World Politics* **38**(4), 343–70.

Haggard, S. 1990 *Pathways from the periphery: the politics of growth in the newly industrializing countries*. Cornell University Press, Ithaca, NY.

Haggard, S., Kaufman, R.R. 1995 *The political economy of democratic transitions*. Princeton University Press, Princeton.

Hall, A. 1996 Did Chico Mendes die in vain? Brazilian rubber tappers in the 1990s. In Collinson, H. (ed.), *Green guerrillas: environmental conflicts and initiatives in Latin America and the Caribbean*. Latin American Bureau, London, 93–102.

Hall, A. 1997 Peopling the environment: a new agenda for research, policy, and action in Brazilian Amazonia. *European Review of Latin American and Caribbean Studies* 62 (June), 9–31.

Hall, S., Jacques, M. (eds) 1983 *The politics of Thatcherism*. Lawrence & Wishart, London.

Handelman, H. 1996 *The challenge of third world development*. Prentice Hall, Englewood Cliffs, NJ.

Hanson, G.T. 1975 *Shantytown stage development: the case of Kingston, Jamaica*. PhD thesis, Louisiana State University, Baton Rouge, LA.

Hardon, A. 1997a A review of national family planning policies. In Hardon, A., Hayes, E. (eds), *Reproductive rights in practice: a feminist report on the quality of care*. Zed, London, 15–21.

Hardon, A. 1997b Setting the stage: health, fertility and unmet need in eight countries. In Hardon, A., Hayes, E. (eds), *Reproductive rights in practice: a feminist report on the quality of care*. Zed Press, London, 22–9.

Hardoy, J., Mitlin, D., Satterthwaite, D. 1992 *Environmental problems in Third World cities*. Earthscan, London.

Hardoy, J., Satterthwaite, D. 1989 *Squatter citizen: life in the urban Third World*. Earthscan, London.

Harrison, D. 1988 *The sociology of modernization and development*. Unwin Hyman, London.

Hart, K. 1973 Informal income opportunities and urban employment in Ghana. *Journal of Modern African Studies* **1**, 61–81.

Hartmann, B. 1987 *Reproductive rights and wrongs: the global politics of population control and reproductive choice*. Harper and Row, New York.

Hartmann, B. 1997 Population control in the new world order. In Hill, E. (ed.), *Development for health*, Development in Practice Reader. Oxfam, Oxford, 80–5.

Hartshorn, G. 1989 Sustained yield management of natural forests: the Palcazú production forest. In Browder, J. (ed.) *Fragile lands of Latin America: strategies for sustainable development*. Westview Press, Boulder, CO, 130–38.

Harvey, N. 1994 Rebellion in Chiapas: rural reforms, campesino radicalism, and the limits to Salinismo. In *The transformation of rural Mexico*, no. 5, Center for US–Mexican Studies, University of California at San Diego, La Jolla, CA, 1–49.

Hecht, S.B. 1985 Environment, development and politics: capital accumulation and the livestock sector in Eastern Amazonia. *World Development* **13**(6), 663–84.

Held, D. 1996. *Models of democracy*, 2nd edn. Stanford University Press, Stanford, CA.

Herbert, B. 1996 Banana Bully. *The New York Times* 13 May, A15.

Higley, J., Gunther, R. (eds) 1992 *Elites and democratic consolidation in Latin America and southern Europe*. Cambridge University Press, Cambridge.

Hirschman, A.O. 1979 The turn to authoritarianism in Latin America and the search for its economic determinants. In Collier, D. (ed.), *The new authoritarianism in Latin America*. Princeton University Press, Princeton, NJ.

Hirst, P., Thompson, G. 1996 *Globalization in question: the international economy and the possibilities of governance*. Polity Press, Cambridge.

Hitiris, T. 1989 *European Community economics*. Harvester Wheatsheaf, Hemel Hempstead, Herts.

Hobsbawm, E. 1994 *Age of extremes. The short twentieth century 1914–1991*. Michael Joseph, London.

Hobsbawm, E. 1996 Identity politics and the left. *New Left Review* no. 217, 38–47.

Hojman, D.E. 1995 Too much of a good thing? Macro and microeconomics of the Chilean peso appreciation. In Hojman D.E. (ed.), *Neoliberalism with a human face? The politics and economics of the Chilean model*, Monograph Series no. 20, Institute of Latin American Studies, University of Liverpool, Liverpool.

Holloway, A. 1998 *An investigation into the nature of 'micronegocios' in Oaxaca City Mexico, with particular reference to urban workshops*. Unpublished MPhil thesis in Latin American Studies, Oxford University, Oxford.

Hout, W. 1993 *Capitalism and the Third World: development, dependency and the World system*. Edward Elgar, Aldershot, Hants.

Huber, E., Safford, F. 1995 *Agrarian structure and political power: landlord and peasant in the making of Latin America*. Pittsburgh University Press, Pittsburgh, PA.

Humphrey, J. 1997 Gender divisions in Brazilian industry. In Gugler, J. (ed.), *Cities in the developing world: issues, theory and policy*. Oxford University Press, Oxford, 171–83.

Huntington, S. 1968 *Political order in changing societies*. Yale University Press, New Haven, CT.

Huntington, S.P. 1991. *The third wave of democratization in the late twentieth century*. University of Oklahoma Press, Norman, OK.

Hurrell, A. 1991 The politics of Amazonian deforestation. *Journal of Latin American Studies* **23**(1), 197–215.

Hurrell, A., Kingsbury, B. (eds) 1992 *The international politics of the environment: actors, interests, and institutions*. Clarendon Press, Oxford.

IADB 1991 *Our own agenda*, Inter-American Development Bank, Washington, DC.

IADB 1993a *Economic and social progress in Latin America: 1993 report*. The Johns Hopkins University Press, Baltimore, MD, for the Inter-American Development Bank (IADB), Washington, DC.

IADB 1993b *Hacia una política social efectiva en Venezuela*. Inter-American Development Bank, Washington, DC.

IADB 1996 *Economic and social progress in Latin America: 1996 report*, Johns Hopkins University Press, Baltimore, MD for the Inter-American Development Bank, Washington, DC.

IADB 1997 *Latin America after a decade of reforms: economic and social progress, 1997 report*, Inter-American Development Bank. Johns Hopkins University Press, Baltimore, MD.

Ibáñez, G. 1990 *América Latina y el Caribe: pobreza rural persistente*. Serie Documentos de Programas no.17. IICA, San José.

IEA 1996 *Energy balances and statistics of the non-OECD countries*. International Energy Agency, Paris.

Imaz de, J.L. 1984 *Sobre la identidad Iberoamericana*. Editorial Sudamericana. Buenos Aires.

Izazola Conde, H., Marquette, C. 1995 Migration in response to the urban environment: outmigration by middle class women and their families from Mexico City after 1985. *Geographia Polonica*, **64**, 224–56.

Jackson, S., Russett, B., Snidal, D., Sylvan, D. 1979 An assessment of empirical research on dependencia. *Latin American Research Review* **14**(3), 7–27.

Jansen, K., Roquas, E. 1998 Modernizing insecurity: the land titling project in Honduras, *Development and Change* **29**(1), 81–106.

Jaquette, J. (ed.) 1994 *The women's movement in Latin America: participation and democracy*, 2nd edn. Westview Press, Boulder, CO.

Jarvis, L.S. 1992 The unravelling of the agrarian reform. In Kay, C., Silva, P. (eds), *Development and social change in the Chilean countryside: from the pre-land reform period to the democratic transition*. CEDLA, Amsterdam, 189–213.

Jeffrey, P. 1998 Central America: a growing role in drug trade, **30**(1 and 2). Latinamerica Press, Lima, Peru.

Jelin, E. (ed.) 1990 *Women and social change in Latin America.* UNRISD/Zed Press, London.

Jelin, E., Hershberg, E. (eds) 1996 *Constructing democracy: human rights, citizenship, and society in Latin America.* Westview Press, Boulder, CO.

Jenkins, R. 1997 Structural adjustment and Bolivian industry. *European Journal of Development Research*, **9**(2), 107–28.

Jenkins, R.O. 1977 *Dependent industrialization in Latin America.* Praeger, New York.

Jenkins, R.O. 1991 The political economy of industrialization: a comparison of Latin American and East Asian newly industrialising countries. *Development and Change* **22**, 197–231.

Johnston, R.J., Taylor, P.J., Watts, M.J. (eds) 1995 *Geographies of global change: remapping the world in the late twentieth century.* Basil Blackwell, Oxford.

Jordán, F., Miranda, C. de, Reuben, W., Sepúlveda, S. 1989 La economía campesina en la reactivación y el desarrollo agropecuario. In Jordán, F. (ed.), *La economía campesina: crisis, reactivación y desarrollo.* IICA, San José.

Kaimovitz, D. 1996 Social pressure for environmental reform in Latin America. In Collinson, H. (ed.), *Green guerrillas: environmental conflicts and initiatives in Latin America and the Caribbean.* Latin American Bureau, London, 20–32.

Kant, T.L. 1997 *The paradox of plenty: oil booms and petro-states.* University of California Press, Los Angeles, CA.

Kaplinsky, R. 1995 A reply to Willmore. *World Development* **23**(3), 537–40.

Kaufman, R.R. 1979 Industrial change and authoritarian rule in Latin America: a concrete review of the bureaucratic-authoritarian model. In Collier, D. (ed.), *The new authoritarianism in Latin America.* Princeton University Press, Princeton, NJ.

Kay, C. 1974 Comparative development of the European manorial system and the Latin American hacienda system. *The Journal of Peasant Studies* **2**(1), 69–98.

Kay, C. 1977 Review of 'Agrarian Reform and Agrarian Reformism' edited by D. Lehmann. *The Journal of Peasant Studies* **4**(2), 241–4.

Kay, C. 1978 Agrarian reform and the class struggle in Chile. *Latin American Perspectives* **5**(3), 117–140.

Kay, C. 1981 Political economy, class alliances and agrarian change in Chile. *The Journal of Peasant Studies* **8**(4), 485–513.

Kay, C. 1982 Achievements and contradictions of the Peruvian agrarian reform. *Journal of Development Studies* **18**(2), 141–70.

Kay, C. 1983 The agrarian reform in Peru: an assessment. In Ghose, A.K. (ed.), *Agrarian reform in contemporary developing countries.* Croom Helm, London, 185–239.

Kay, C. 1988a The landlord road and the subordinate peasant road to capitalism in Latin America. *Etudes Rurales*, 77, 5–20.

Kay, C. 1988b Cuban economic reforms and collectivisation. *Third World Quarterly* **10**(3), 1239–66.

Kay, C. 1989 *Latin American theories of development and underdevelopment.* Routledge, London and New York.

Kay, C. 1993 For a renewal of development studies: Latin American theories and neoliberalism in the era of structural adjustment. *Third World Quarterly* **14**(4), 691–702.

Kay, C. 1995 Rural development and agrarian issues in contemporary Latin America. In Weeks, J. (ed.), *Structural adjustment and the agricultural sector in Latin America and the Caribbean.* Macmillan, London, 9–44.

Kay, C. 1997 Globalisation, peasant agriculture and reconversion. *Bulletin of Latin American Research*, **16**(1), 11–24.

Kearney, M. 1995 The local and the global: the anthropology of globalization and transnationalism. *Annual Review of Anthropology* **24**, 547–65.

Kearney, M., Varese, S. 1995 Latin America's indigenous peoples: changing identities and forms of resistance. In Halebsky, S., Harris, R.L. (eds), *Capital, power, and inequality in Latin America.* Westview Press, Boulder, CO, 207–31.

Keck, M. 1995 Social equity and environmental politics in Brazil: lessons from the rubber tappers of Acre. *Comparative Politics* **27**(4), 409–24.

Kiely, R. 1995 *Sociology and development: the impasse and beyond.* University College London Press, London.

Killick, T., Malik, M. 1992 Country experience with IMF programmes in the 1980s. *The World Economy* **15**, 599–632.

Klak, T. (ed.) 1998 *Globalization and neoliberalism: the Caribbean context*. Rowman and Littlefield, Lanham, MD.

Klak, T. 1996 Distributional impacts of the 'free zone' component of structural adjustment: the Jamaican experience. *Growth and Change* **27**(Summer), 352–87.

Kling, M. 1968 Toward a theory of power and political instability in Latin America. In Petras, J., Zeitein, M. (eds), *Latin America: reform or revolution?* Fawcett, Greenwich, CT, 76–93.

Koechlin, T. 1995 The globalization of investment. *Contemporary Economic Policy* **13**(January), 92–100.

Koonings, K., Kruijt, D., Wils, F. 1995 The very long march of history. In Thomas, H. (ed.), *Globalization and Third World trade unions*. Zed Press, London, 99–129.

Korzeniewicz, R.P. 1997 The deepening differentiation of states, enterprises and households in Latin America. In Smith, W.C., Korzeniewicz, R.P.(eds), *Politics, social change, and economic restructuring in Latin America*. North–South Center Press, Miami, FL, 215–50.

Kowarick L. 1979 Capitalism and urban marginality in Brazil. In Bromley, R., Gerry, C. (eds), *Casual work and poverty in the Third World*. John Wiley, Chichester, 69–85.

Kuznets, S. 1976 *Modern economic growth: rate, structure, and spread*. Yale University Press, New Haven, CT.

Laclau, E. 1985 New social movements and the plurality of the social. In Slater, D. (ed.), *New social movements and the state in Latin America*. CEDLA, Amsterdam, 27–42.

Laclau, E., Mouffe, C. 1985 *Hegemony and socialist strategy: towards a radical democratic politics*. Verso, London.

Langer, A., Lozano, R., Bobadilla, J.L. 1991 Effects of Mexico's economic crisis on the health of women and children. In González de la Rocha, M., Escobar, A. (eds), *Social responses to Mexico's economic crisis of the 1980s*, Contemporary Perspectives Series No. 1. Center for US Mexican Studies, San Diego, CA, 195–219.

Lara Flores, S.M. (ed.) 1995 *Jornaleras, temporeras y Bóias-Frias: el rostro femenino del mercado de trabajo rural en América Latina*. Nueva Sociedad, Caracas.

Larraín, J. 1989 *Theories of development: capitalism, colonialism and dependency*. Polity Press, Cambridge.

Larraín, J. 1996 *Modernidad: razón e identidad en América Latina*. Editorial Andrés Bello, Santiago.

Larranaga, O., Sanhueza, G. 1994 *Descomposición de la pobreza en Chile*, WP-I-79, ILADES, Santiago.

Latinamerica Press 1997 Help from afar. 18 September, 7.

Le Heron, R. 1993 *Globalised agriculture*. Pergamon, Oxford.

Lehmann, D. 1982 After Lenin and Chayanov: new paths of agrarian capitalism. *Journal of Development Economics* **11**, 2, 133–61.

Lehmann, D. 1990 *Democracy and development in Latin America*. Polity Press, Cambridge.

Lehmann, D. 1996 *The struggle for the spirit*. Polity Press, Cambridge.

Leitmann, J., Bartone, C., Bernstein, J. 1992 Environmental management and urban development: issues and options for Third World cities. *Environment and Urbanization* **4**(2), 131–40.

LeoGrande, W. 1997 Enemies evermore: US policy towards Cuba after Helms–Burton. *Journal of Latin American Studies* **29**, 211–21.

Leonard, H.J. (ed.) 1989 *Environment and the poor: development strategies for a common agenda*. Transaction Books, New Brunswick, NJ.

Leonard, H.J. 1987 *Natural resources and economic development in Central America*. Transaction Books, New Brunswick, NJ.

Levitas, R. (ed.) 1986 *The ideology of the New Right*. Polity Press, Cambridge.

Lewis, O. 1961 *The children of Sanchez*. Random House, New York.

Lewis, O. 1968 *La vida: a Puerto Rican family in the culture of poverty – San Juan and New York*. Random House, New York.

Lewis. W.A. 1954 Economic development with unlimited supplies of labour. *The Manchester School of Economic and Social Studies* **22**(2), 139–91.

Leys, C. 1996 *The rise and fall of development theory*. James Curry, London.

Lindblom, C. 1977 *Politics and markets*. Basic Books, New York.

Linz, J.J. 1978 *The breakdown of democratic regimes: crisis, breakdown, and reequilibration.* The John Hopkins University Press, Baltimore, MD.

Linz, J.J., Valenzuela, A. (eds) 1994 *The failure of presidential democracy.* The Johns Hopkins University Press, Baltimore,MD.

Lipschutz, R., Conca, K. (eds) 1993 *The state and social power in global environmental politics.* Columbia University Press, New York.

Lira, O. 1985 *Hispanidad y mestizaje.* Editorial Covadonga, Santiago.

Little, W. 1997 Democratization in Latin America, 1980–95. In Potter, D. *et al.* (eds), *Democratization.* Polity Press, Cambridge, 174–94.

Little, W., Posada-Carbó, E. (eds) 1996 *Political corruption in Europe and Latin America.* Macmillan, Macmillan.

Llambí, L. 1988 The small modern farmers: neither peasants nor fully-fledged capitalists? *The Journal of Peasant Studies* **15**, 3, 350–72.

Lomnitz, L. 1977 *Networks and marginality: life in a Mexican shanty town.* Academic Press, New York.

López Cordovez, L. 1982 Trends and recent changes in the Latin American food and agricultural situation. *CEPAL Review,* 16, 7–41.

López, M., Izazola, H. 1995 *El perfil censal de los hogares y las familias en México.* Instituto Nacional de Estadística, Geografía e Informática, Aguascalientes.

López, M., Izazola, H., Gómez de León, J. 1993 Characteristics of female migrants according to the 1990 census of Mexico. In United Nations (ed.), *Internal migration of women in developing countries.* United Nations, New York, 133–53.

Love, J.L. 1994 Economic ideas and ideologies in Latin America since 1930. In Bethell, L. (ed.), *Latin America since 1930. Economy, society and politics. The Cambridge History of Latin America, Vol. VI, Part I.* Cambridge University Press, Cambridge, 393–460.

Loveman, B. 1994 *The constitution of tyranny: regimes of exception in Spanish America.* University of Pittsburgh Press, Pittsburgh, PA.

Lumbreras, L.G. 1991 Misguided development. *Report on the Americas* **5**, 18–22.

Lungo, M. 1997 Costa Rica: dilemmas of urbanization in the 1990s. In Portes, A., Dore-Cabral, C., Landoff, P. (eds), *The urban Caribbean: transition to a new global economy.* John Hopkins University Press, Baltimore, MD, 57–86.

Lustig, N. 1991 From structuralism to neostructuralism: the search for a heterodox paradigm. In Meller, P. (ed.), *The Latin American development debate: neostructuralism, neomonetarism, and adjustment processes.* Westview Press, Boulder, CO, 27–42.

Mahon, J.E. 1996 *Mobile capital and Latin American development.* Pennsylvania State University Press, University Park, PA.

Mainwaring, S., O'Donnell, G., Valenzuela, J.S. (eds) 1992 *Issues in democratic consolidation: the new South American democracies in comparative perspective.* University of Notre Dame Press, South Bend, IN.

Malloy, J., Seligson, M. (eds) 1987 *Authoritarians and democrats: regime transition in Latin America.* University of Pittsburgh Press, Pittsburgh, PA.

Mandel, E. 1978 *Late capitalism.* Verso, London.

Mandle, J.R. 1996 *Persistent underdevelopment: change and economic modernization in the West Indies.* Gordon and Breac, Amsterdam.

Mangin, W. 1967 Latin American squatter settlements: a problem and a solution. *Latin American Research and Review* **2**(3), 65–98.

Marcel, M., Solimano, A. 1994 The distribution of income and economic adjustment. In Bosworth, B.P. *et al.* (eds), *The Chilean economy: policy lessons and challenges.* The Brookings Institution, Washington, DC, 217–55.

Mariátegui, J.C. 1976 *Siete ensayos de interpretación de la realidad Peruana.* Editorial Crítica, Barcelona.

Marín, C. 1997 *Modernity and mass communication: the Latin American case.* Unpublished PhD thesis, University of Birmingham, Birmingham.

Marshall, J. 1998 The political viability of free market experimentation in Cuba: evidence from Los Mercados Agropecuarios. *World Development* **26**(2), 277–88.

Martínez Estrada, E. 1968 *Meditaciones sarmientinas.* Editorial Universitaria, Santiago.

Martínez Estrada, E. 1946 *Radiografía de la Pampa.* Editorial Losada, Buenos Aires.

Martínez, J., Díaz, A. 1996 *Chile: the great transformation*. Brookings Institution, Washington, DC.

Massey, D., Arango, J., Hugo, G., Kouaouci, A., Pellegrino, A., Taylor, J.E. 1993 Theories of international migration: a review and appraisal. *Population and Development Review* **19**(3), 431–66.

Mazumdar, D. 1976 The urban informal sector. *World Development* **4**(8), 655–79.

McBain, H. 1990 Government financing of economic growth and development in Jamaica: problems and prospects. *Social and Economic Studies* **39**, 179–212.

McClenaghan, S. 1997 Women, work and empowerment: romanticizing the reality. In Dore, E. (ed.), *Gender politics in Latin America: debates in theory and practice*. Monthly Review Press, New York, 19–35.

McIlwaine, C. 1997 Vulnerable or poor? A study of ethnic and gender disadvantage among Afro-Caribbeans in Limón, Costa Rica. *European Journal of Development Research* **9**(2), 35–61.

McMichael, P. 1993 The restructuring of the world food system. *Political Geography* **12**(3), 200–14.

McMichael, P. 1996 Globalization: myths and realities. *Rural Sociology* **61**(1), 25–55.

Mejías, J.A., Vos, R. 1997 *Poverty in Latin America and the Caribbean. An inventory, 1980–95*. Working Paper Series I-4. Inter-American Development Bank, Washington, DC.

Melhuus, M., Stolen, K.A. (eds) 1996 *Machos, mistresses and madonnas: contesting the power of Latin American gender imagery*. Verso, London.

Melucci, A. 1989 *Nomads of the present: social movements and individual needs in contemporary society*. Temple University Press, Philadelphia, PA.

Melucci, A. 1995 The new social movements revisited: reflections on a sociological understanding. In Maheu, L. (ed.), *Social movements and social classes*. Sage, London, 107–19.

Mendez-Rivero, D. 1995 Decline of an oil economy: Venezuela and the legacy of incorporation. In Thomas, H. (ed.), *Globalization and Third World trade unions*. Zed Press, London, 149–65.

Migdal, J.S., Kohli, A., Shue, V. (eds) 1994 *State power and social forces: domination and transformation in the Third World*. Cambridge University Press, Cambridge.

Miller, M. 1991 *Debt and the environment: converging crises*. United Nations, New York.

Millet, R.L., Gold-Biss, M. (eds) 1996 *Beyond praetorianism: the Latin American military in transition*. North–South Center, Miami, FL.

Minority Rights Group (ed.) 1995 *No longer invisible: Afro-Latin Americans today*. Minority Rights Publications, London.

Miraftab, F. 1994 (Re)production at home: reconceptualising home and family. *Journal of Family Issues*, **15**(3), 467–89.

Mitlin, D. 1992 Sustainable cities. *Environment and Urbanization* **4**(2), 3–8.

Mitlin, D. 1996 City inequality. *Environment and Urbanization* **8**(2), 3–7.

Mitter, S. 1997 Information technology and working women's demands. In Mitter, S., Rowbotham, S. (eds), *Women encounter technology: changing patterns of employment in the Third World*. Routledge, London, 19–43.

Moghadam, V. 1995 Gender aspects of employment and unemployment in global perspective. In Simai, M., Moghdadam, V., Kuddo, A. (eds), *Global employment: an international investigation into the future of work*. Zed Press, London, 111–39.

Molyneux, M. 1984 Mobilisation without emancipation? Women's interests, state and revolution in Nicaragua. *Critical Social Policy* **10**(4), 59–75.

Molyneux, M. 1996 *State, gender and institutional change in Cuba's 'Special Period': the Federación de Mujeres Cubanas*. Research Paper 43, Institute of Latin American Studies, University of London, London.

Montbiot, G. 1993. Brazil: landownership and the flight to Amazonia. In Colchester, M., Lohmann, L. (eds), *The struggle for land and the fate of the forests*. Zed, London, 139–63.

Monteón, M. 1995 Gender and economic crises in Latin America: reflections on the Great depression and the debt crisis. In Blumberg, R.L., Rakowski, C., Tinker, I., Monteón, M. (eds), *Engendering wealth and well-being: empowerment for global change*. Westview Press, Boulder, CO, 39–62.

Moore, B. Jr 1966 *The social origins of dictatorship and democracy: lord and peasant in the making of the modern world*. Beacon Press, Boston, MA.

Morandé, P. 1984 *Cultura y modernización en América Latina*. Universidad Católica de Chile, Santiago.

Morris, A. 1981 *Latin America: economic development and regional differentiation*. Hutchinson, London.

Morris, A. 1995 *South America: a changing continent*. Hodder and Stoughton, London.

Moser, C. 1978 Informal sector or petty commodity production? Dualism or dependence in urban development. *World Development* **6**, (9–10), 135–78.

Moser, C. 1992 Adjustment from below: low-income women, time and the triple role in Guayaquil, Ecuador. In Afshar, H., Dennis, C. (eds), *Women and adjustment policies in the Third World*. Macmillan, Basingstoke, Hants, 87–116.

Moser, C. 1997 *Household responses to poverty and vulnerability volume 1: confronting crisis in Cisne Dos, Guayaquil, Ecuador*. Urban Management and Poverty Reduction Series No. 21. World Bank, Washington, DC.

Moser, C., Holland, J. 1997 *Urban poverty and violence in Jamaica*. The World Bank, Washington, DC.

Moser, C.O.N. 1994 The informal sector debate, part 1: 1970–1983. In Rakowski, C.A. (ed.), *Contrapunto: the informal sector debate in Latin America*. State University of New York Press, Albany, NY, 11–29.

Moßrucker, H. 1997 Amerindian migration in Peru and Mexico. In Gugler, J. (ed.), *Cities in the developing world: issues, theory and policy*. Oxford University Press, Oxford, 74–87.

Moulian, T. 1997 *Chile actual: anatomía de un mito*. LOM-ARCIS, Santiago.

Mouzelis, N. 1986 *Politics in the semi-periphery*. Macmillan, London.

Mullings, B. 1995 Telecommunications restructuring and the development of export information processing services in Jamaica. In Dunn, H. (ed.), *Globalization, communications and Caribbean identity*. Ian Randle, Kingston, 174–91.

Mullings, B. 1998 Jamaica's information processing services: neoliberal niche or structural limitation? In Klak, T. (ed.), *Globalization and neoliberalism: the Caribbean context*. Rowman and Littlefield, Lanham, MD, 135–54.

Murmis, M. 1994 Incluidos y excluidos en la reestructuración del agro latinoamericano, *Debate Agrario*, 18, 101–33.

Murphy, A., Stepick, A. 1991 *Social inequality in Oaxaca: a history of resistance and change*. Temple University Press, Philadelphia, PA.

Murray, D., Hoppin, P. 1990 *Pesticides and nontraditional agriculture: a coming crisis for US development policy in Latin America*. Institute of Latin American Studies, Austin, TX.

Murray, W.E. 1996 *Neoliberalism, restructuring and non-traditional fruit exports in Chile: implications of export-orientation for small-scale farmers*. Unpublished PhD thesis, University of Birmingham, Birmingham.

Murray, W.E. 1997 Competitive global fruit export markets: marketing intermediaries and impacts on small-scale growers in Chile. *Bulletin of Latin American Research* **16**(1), 43–55.

Murray, W.E. 1998 The globalisation of fruit, neoliberalism and the question of sustainability: lessons from Chile. *European Journal of Development Research* **10**(1), 201–27.

Myers, N. 1992 *The primary source: tropical forests and our future*. Norton, New York.

Naerssen, T.V., Schuurman, F.J. (eds) 1989 *Urban social movements in the Third World*. London, Routledge.

Nash, J. 1995 Latin American women in the world capitalist crisis. In Bose, C., Acosta-Belén, E. (eds), *Women in the Latin American development process*. Temple University Press, Philadelphia, PA, 151–66.

Neto, F. 1990 Development planning and mineral mega-projects: some global considerations. In Goodman, D., Hall, A. (eds), *The future of Amazonia: destruction or sustainable development?* St. Martin's Press, New York: .

Nettl, J.P. 1968 The state as a conceptual variable. *World Politics* **20**(4), 559–92.

Nicholls, S. 1998 Measuring trade creation and trade diversion in the Central American Common Market: a Hicksian alternative. *World Development* **26**(2), 323–35.

Nickson, A. 1995 *Local government in Latin America*. Lynne Rienner, Boulder, CO.

Nielson, D.L., Stern, M.A. 1997 Endowing the environment: multilateral development banks and environmental lending in Latin America. In MacDonald, G.J., Nielson, D.L., Stern, M.A. (eds), *Latin American environmental policy in international perspective*. Westview Press, Boulder, CO, 130–55.

Nisbet, R. 1969 *The quest for community*. Oxford University Press, Oxford.

NLC 1997 Setting the record straight/the real Disney in Burma, Haiti, Indonesia, China. 17 January. National Labor Committee.

O'Brien, P., Cammack, P. 1985 *Generals in retreat: the crisis of military rule in Latin America*. Manchester University Press, Manchester.

O'Brien, R. 1992 *Global financial integration: the end of geography*. Council on Foreign Relations, New York.

O'Donnell, G. 1973 *Modernization and bureaucratic-authoritarianism: studies in South American Politics*. Institute of International Studies, Univesity of California, Berkeley, CA.

O'Donnell, G. 1979 Tensions in the bureaucratic-authoritarian state and the question of democracy. In Collier, D. (ed.), *The new authoritarianism in Latin America*. Princeton University Press, Princeton, NJ.

O'Donnell, G. 1992 *Delegative democracy*. University of Notre Dame Press, South Bend, IN.

O'Donnell, G., Schmitter, P., Whitehead, L. (eds) 1986 *Transitions from authoritarian rule*. Johns Hopkins University Press, Baltimore, MD.

Ocampo, J.A. 1993 Terms of trade and center–periphery relations. In Sunkel, O. (ed.), Development from within: toward a neostructuralist approach for Latin America. Lynne Rienner, Boulder, CO, 333–57.

ODEPA 1996 *Mercados fruticolas*. Oficina de Estudios y Políticas Agrarias, Santiago, Chile.

Oliveira, O.D., Roberts, B. 1994 Urban growth and urban social structure in Latin America, 1930–1990. In Bethell, L. (ed.), *The Cambridge History of Latin America*, vol. VI. Cambridge University Press, Cambridge, 253–324.

Olwig, K. 1993 *Global culture, island identity: continuity and change in the Afro-Caribbean community of Nevis*. Harwood, Philadelphia.

Organski, A.F.K. 1965 *The stages of political development*. Alfred A. Knopf, New York.

Ortega, E. 1992 Evolution of the rural dimension in Latin America and the Caribbean. *CEPAL Review*, 47, 115–36.

Otero, G. (ed.) 1996a *Neo-liberalism revisited: economic restructuring and Mexico's political future*. Westview Press, Boulder, CO.

Otero, G. 1996b Neoliberal reform and politics in Mexico: an overview. In Otero, G. (ed.), *Neo-liberalism revisited: economic restructuring and Mexico's political future*. Westview Press, Boulder, CO, 1–26.

Ozório de Almeida, A.L., Campari, J.S. 1995 *Sustainable settlement in the Brazilian Amazon*. Oxford University Press, New York.

Pacheco, M. 1992 Recycling in Bogotá: developing a culture for urban sustainability. *Environment and Urbanization* 4(2), 74–9.

Packenham, R.A. 1992 *The dependency movement: scholarship and politics in development studies*. Harvard University Press, Cambridge, MA.

Paige, J.M. 1996 Land reform and agrarian revolution in El Salvador: comment on Seligson and Diskin. *Latin American Research Review* 31(2), 127–39.

Painter M., Durham, W.H. 1995 *The social causes of environmental destruction in Latin America*. The University of Michigan Press, Ann Arbor.

Pantojas-García, E. 1990 *Development strategies as ideology: Puerto Rico's export-led industrialization experience*. Lynne Rienner, Boulder, CO.

Paolisso, M., Gammage, S. 1996 *Women's responses to environmental degradation: poverty and demographic constraints. Case studies from Latin America*. International Center for Research on Women, Washington, DC.

Paré, L., Bray, D., Burstein, J., Vázquez, S.M. (eds) 1997 *Semillas para el cambio en el campo: medio ambiente, mercados, y organización campesina*. UNAM, Mexico City.

Parker, C. 1993 *Otra lógica en América Latina: religión popular y modernización capitalista*. Fondo de Cultura Económica, Santiago.

Parras, M., Morales, M.J. 1997 Reproductive rights on paper: four Bolivian cities. In Hardon, A., Hayes, E. (eds), *Reproductive rights in practice: a feminist report on the quality of care*. Zed Press, London, 77–94.

Parsons, T. 1951 *The social system*. The Free Press, Glencoe, IL.

Pastor, R. 1989 Migration and development in the Caribbean Basin: implications and recommendations for policy. Working papers of the Commission for the Study of International Migration and Cooperative Economic Development, Washington, DC.

Patterson, O. 1994 Ecumenical America: global culture and the American cosmos. *World Policy Journal* 11(2), 103–17.

Pattullo, P. 1996 *Last resorts: the cost of tourism in the Caribbean*. Monthly Review Press, New York.

Paz, O. 1959 *El laberinto de la soledad*. Fondo de Cultura Económica, Mexico City.

Paz, O. 1979 *El ogro filantrópico*. Joaquín Hortiz, Mexico City.

Pearce, D.W., Turner, R.K. 1990 *Economics of natural resources and the environment*. The Johns Hopkins University Press, Baltimore, MD.

Peattie, L. 1990 Participation: a case study of how invaders organize, negotiate and interact with government in Lima, Peru. *Environment and Urbanization* **2**(1), 19–30.

Peet, R. 1991 *Global capitalism: theories of societal development*. Routledge, London.

Pelupessy, W. 1995 *Agrarian transformation and economic adjustment in El Salvador, 1960–1990*. Unpublished PhD dissertation, Katholieke Universiteit Brabant, Brabant, The Netherlands.

Peña Saint Martin, F. 1996 *Discriminación laboral femenina en la industria del vestido de Mérida, Yucatán*. Serie Antropología Social. Instituto Nacional de Antropología e Historia, Mexico City.

Perlman, J.E. 1976 *The myth of marginality: urban poverty and politics in Rio de Janeiro*. University of California Press, Berkeley, CA.

Petras, J. 1997 The peasantry strikes back. Latin America: the resurgence of the left. *New Left Review* no. 223, 17–47.

Petras, J., Morley, M. 1992 *Latin America in the time of cholera: electoral politics, market economy, and permanent crisis*. Routledge, New York.

Picari, N. 1996 Ecuador: taking on the neoliberal agenda. *NACLA Report on the Americas* **29**(5), 23–32.

Pile, S., Keith, M. (eds) 1997 *Geographies of resistance*. Routledge, London.

Pitanguy, J., de Mello e Souze, C. 1997 Codes of honour: reproductive life histories of domestic workers in Rio de Janeiro. In Harcourt, W. (ed.), *Power, reproduction and gender: the intergenerational transfer of knowledge*. Zed Press, London, 72–97.

Poggi, G. 1978 *The development of the modern state*. Yale University Press, New Haven, CT.

Poggi, G. 1990 *The state: it's nature, development and prospects*. Stanford Univesity Press, Stanford, CA.

Portes, A. 1985 Latin American class structures. *Latin American Research Review* **20**(3), 7–39.

Portes, A. 1989 Latin American urbanization during the years of the crisis. *Latin American Research Review* **24**(3), 7–44.

Portes, A., Guarnizo, L.E. 1991 Tropical capitalists. In Diaz-Briquets, S., Weintraub, S. (eds), *Migration, remittances, and small business development: Mexico and Caribbean Basin countries*. Westview Press, Boulder, CO, 101–31.

Portes, A., Schauffer, R. 1993 Competing perspectives on the Latin American informal sector. *Population and Development Review* **19**(3), 33–60.

Poulantzas, N. 1973 *Political power and social classes*. New Left Books, London.

Prado, J. 1983 Estado social del Perú. In Teran, O. (ed.), *América Latina: positivismo y nación*. Editorial Katún, Mexico City.

PREALC 1993 Latin America: economic growth that generates more jobs of inferior quality. Newsletter No. 32.

Prebisch, R. 1950 *The economic development of Latin America and its principal problems*. Economic Commission for Latin America, Santiago.

Prebisch, R. 1962 The economic development of Latin America and its principal problems. *Economic Bulletin of Latin America* **7**(1), 1–22.

Preston, D. (ed.) 1996 *Latin American development: geographical perspectives*. Longman, Harlow, Essex.

Preston, P.W. 1996 *Development theory: an introduction*. Basil Blackwell, Oxford.

Przeworski, A., Wallerstein, M. 1988 Structural dependence of the state on capital. *American Political Science Review* **82** (March), 11–29.

Quijano, A. 1988 *Modernidad, identidad y utopía en América Latina*. Ediciones Sociedad Política, Lima.

Quijano, A. 1991 Recovering utopia. *Report on the Americas* **5**, 34–88.

Qureshi, Z. 1996 Globalization: new opportunities, tough challenges. *Finance and Development* (March), 30–3; [this publication summarizes the main points from World Bank. 1995a].

Radcliffe, S. 1991a The role of gender in peasant migration: conceptual issues from the Peruvian Andes. *Review of Radical Political Economy* **23**(3–4), 148–73.

Radcliffe, S. 1991b Ethnicity, patriarchy, and incorporation into the nation: female migrants as domestic servants in Peru. *Environment and Planning D: Society and Space* **8**, 379–93.

Radcliffe, S. 1992 Mountains, maidens and migration: gender and mobility in Peru. In Chant, S. (ed.), *Gender and migration in developing countries*. Belhaven Press, London, 30–48.

Radcliffe, S. 1993a 'People have to rise up like the great women fighters': the state and peasant women in Peru. In Radcliffe, S., Westwood, S. (eds), *Viva: women and popular protest in Latin America*. Routledge, London, 197–218.

Radcliffe, S.A. 1993b Women's place/*El lugar de las mujeres*: Latin America and the politics of gender identity. In Keith, M., Pile, S. (eds), *Place and the politics of identity*. Routledge, London, 102–17.

Radcliffe, S.A., Westwood, S. 1996 *Remaking the nation: place, identity and politics in Latin America*. Routledge, London.

Rakowski, C.A. (ed.) 1994 *Contrapunto: the informal sector debate in Latin America*. State University of New York Press, Albany, NY.

Ramos, J., Sunkel, O. 1993 Toward a neostructuralist synthesis. In Sunkel, O. (ed.), *Development from within: toward a neostructuralist approach for Latin America*. Lynne Rienner, Boulder, CO, 5–19.

Randall, L. 1996 *Reforming Mexico's Agrarian Reform*. M.E. Sharpe, Armonk, NY.

Redclift, M. 1987 *Sustainable development: exploring the contradictions*. Methuen, London.

Redclift, M. 1992 Sustainable development, popular participation, empowerment and local resource management. In Ghai, D., Vivian, J. (eds), *Grassroots environmental action: people's participation in sustainable development*. Routledge, New York.

Reich, R. 1991 The work of nations: preparing ourselves for 21st century capitalism. Vintage Books, New York.

Reuters 1998 Cuban official calls tourism heart of economy. *News Wire* 28 February.

Rey de Marulanda, N. 1996 Prospects for Latin America in the new world economic order. In Karlsson, W., Malaki, A. (eds), *Growth, trade and integration in Latin America*. Stockholm University, Institute of Latin American Studies, Stockholm, 17–33.

Ribeiro, D. 1992 *Las Américas y la civilización*. Biblioteca Ayacucho, Caracas.

Ribeiro, G.L. 1998 Cybercultural politics: political activism at a distance in a transnational world. In Alvarez, S. et al. (eds), *Cultures of politics, politics of cultures*. Westview Press, Boulder, CO, 325–52.

Richardson, B. 1992. *The Caribbean in the wider world, 14921992. A regional geography*. Cambridge University Press, New York.

Rieff, D. 1993 Multiculturalism's silent partner. *Harper's Magazine* (August), 62–70.

Roberts, B. 1991 The changing nature of informal employment: the case of Mexico. In Standing, G., Tokman, V. (eds), *Towards social adjustment: labour market issues in structural adjustment*. International Labour Organization, Geneva, 115–40.

Roberts, B. 1994 Informal economy and family strategies. *International Journal of Urban and Regional Research*, **18**(1), 6–23.

Roberts, B.R. 1995 *The making of citizens: cities of peasants revisited*. Arnold, London.

Roberts, K. 1985 Household labour mobility in a modern urban economy. In Standing, G. (ed.), *Labour circulation and the labour process*. Croom Helm, London, 358–81.

Roberts, K. 1997 Beyond romanticism: social movements and the study of political change in Latin America. *Latin American Research Review* **32**(3), 137–51.

Roberts, S.M. 1994 Fictitious capital, fictitious spaces? The geography of off-shore financial flows. In Corbridge, S., Martin, R., Thrift, N. (eds), *Money, power and space*. Basil Blackwell, Oxford, 88–120.

Rodgers, G. 1989 Introduction: trends in urban poverty and labour market access. In G. Rodgers (ed.), *Urban poverty and the labour market*. International Labour Office, Geneva, 1–33.

Rodó, J.E. 1976 *Ariel*. Anaya, Salamanca.

Rojas Mix, M. 1991 Reinventing identity. *Report on the Americas* **5**, 29–33.

Rosales, O. 1988 An assessment of the structuralist paradigm for Latin American development and the prospects for its development. *CEPAL Review* no. 34, 19–36.

Rosen, F. 1997 Back on the agenda: ten years after the debt crisis. *NACLA Report on the Americas* **31**(3), 21–4.

Rosenberg, R.L. 1994 Trade and the environment: economic development versus sustainable development. *Journal of Interamerican Studies and World Affairs* **36**, 3.

Ross, J. 1997 'Zapata's children defending the land and human rights in the countryside. *NACLA Report on the Americas* **30**(4), 30–35.

Rostow, W.W. 1960 *The stages of economic growth: a non-communist manifesto*. Cambridge University Press, Cambridge.

Rowe, W., Schelling, V. 1991 *Memory and modernity: popular culture in Latin America*. Verso, London.

Roxborough, I. 1997 Citizenship and social movements under neoliberalism. In Smith, W.C., Korzeniewicz, R.P. (eds), *Politics, social change and economic restructuring in Latin America*. North–South Center Press at the University of Miami, Miami, FL.

Rueda-Junquera, F. 1998 Regional integration and agricultural trade in Central American. *World Development* **26**(2), 315–62.

Rueschemeyer, D., Stephens, E.H., Stephens, J.D. 1992 *Capitalist development and democracy*. University of Chicago Press, Chicago, IL.

Safa, H. 1995a *The myth of the male breadwinner: women and industrialisation in the Caribbean*. Westview Press, Boulder, CO.

Safa, H. 1995b Economic restructuring and gender subordination. *Latin American Perspectives*, **22**(2), 32–50.

Sánchez-Ayéndez, M. 1993 Women as primary support providers for the elderly: the case of Puerto Rico. In Gómez Gómez, E. (ed.), *Gender, women and health in the Americas*. Pan American Health Organization, Washington, DC, 263–8.

Santana, R. 1995 *Ciudadanos en la etnicidad*. Abya-Yala, Quito.

Santos, M. 1979 *The shared space: the two circuits of the urban economy in underdeveloped countries*. Methuen, London.

Saporta Sternbach, N., Navarro-Aranguren, M., Chuchryk, P., Alvarez, S. 1992 Feminisms in Latin America: from Bogotá to San Bernardo. In Escobar, A., Alvarez, S. (eds), *The making of social movements in Latin America*. Westview Press, Boulder, CO, 207–39.

Satterthwaite, D. 1989 Guide to the literature: environmental problems of Third World cities. *Environment and Urbanization* **1**(1), 76–83.

Satterthwaite, D. 1995 The underestimation of urban poverty and of its health consequences. *Third World Planning Review* **17**(4), iii–xii.

Satterthwaite, D. et al. 1996 *The environment for children*. Earthscan, London.

Sayavedra, G. 1997 Fulfilling providers' preferences: four Mexican states. In Hardon, A., Hayes, E. (eds), *Reproductive rights in practice: a feminist report on the quality of care*. Zed Press, London, 95–111.

Schamis, H.E. 1991 Reconceptualizing Latin American authoritarianism in the 1970s: from bureaucratic-authoritarianism to neoconservatism. *Comparative Politics* **23**(2), 201–20.

Schejtman, A. 1994 *Economía política de los sistemas alimentarios en América Latina*. FAO and División Agrícola Conjunta FAO/CEPAL, Santiago.

Schild, V. 1998 New subjects of rights? women's movements and the construction of citizenship in the 'New Democracies'. In Alvarez, S. et al. (eds) *Cultures of politics, politics of cultures*. Westview Press, Boulder, CO, 93–117.

Schirmer, J. 1993 The seeking of truth and the gendering of consciousness: the Comadres of El Salvador and the CONAVIGUA widows of Guatemala. In Radcliffe, S., Westwood, S. (eds), *Viva!* Routledge, London, 30–64.

Schmink, M., Wood, C.H. 1992 *Contested frontiers in Amazonia*. Columbia University Press, New York.

Schneider, B.R. 1998 The material bases of technocracy: investor confidence and neoliberalism in Latin America. In Centeno, M.A., Silva, P. (eds) *The politics of expertise in Latin America*. Macmillan, London, 77–95.

Schoepfle, G., Pérez-López, J. 1992 Export-oriented assembly operations in the Caribbean. In Tirado de Alonso, I. (ed.), *Trade issues in the Caribbean*. Gordon and Breach, Philadelphia, PA.

Schrieberg, D. 1997 Dateline Latin America: the growing fury. *Foreign Policy* **106**, 161–75.

Schteingart, M. 1989 The environmental problems associated with urban development in Mexico City. *Environment and Urbanization* **1**(1), 40–50.

Schumacher, E.F. 1973 *Small is beautiful: a study of economics as if people mattered*. Blond and Briggs, London.

Schuurman, F. 1993 Modernity, post-modernity and the new social movements. In Schuurman, F. (ed.), *Beyond the impasse: new directions in development theory*. Zed Press, London, 187–206.

Schwartzman, S. 1991 Deforestation and popular resistance in Acre: from local social movement to global network. *The Centennial Review* **35**(2), 397–422.

Scott, A. 1990 *Ideology and the new social movements*. Unwin Hyman, London.

Scott, A.M. 1994 *Divisions and solidarities: gender, class and employment in Latin America*. Routledge, London.

Scott, C. 1996 The distributive impact of the new economic model in Chile. In Bulmer-Thomas, V. (ed.), *The new economic model in Latin America and its impact on income distribution and poverty*. Macmillan, London, 147–84.

Scott, C.D. 1985 Transnational corporations, comparative advantage and food security in Latin America. In Abel, C., Lewis, C. (eds), *Latin America, economic imperialism and the state*. Athlone Press, London.

Scott, J. 1986 *Weapons of the weak*. Yale University Press, New Haven, CT.

Segre, R., Coyula, M., Scarpaci, J. 1997 *Havana: two faces of the Antillean metropolis*. John Wiley, New York.

Selby, H., Murphy, A., Lorenzen, S. 1990 *The Mexican urban household: organizing for self-defence*. University of Texas Press, Austin.

Seligson, M.A. 1995 Thirty years of transformation in the agrarian structure of El Salvador, 1961–1991. *Latin American Research Review* **30**(3), 43–74.

Sen, K. 1994 *Ageing: debates on demographic transition and social policy*. Zed Press, London.

Serra, J. 1979 Three mistaken theses regarding the connection between industrialization and authoritarian regimes. In Collier, D. (ed.), *The new authoritarianism in Latin America*. Princeton University Press, Princeton, NJ.

Serrill, M.S. 1996 Of land and death. *Time* **147**(19), 6 May, 35.

Shaiken, H. 1994 Advanced manufacturing and Mexico: a new international division of labour. *Latin American Research Review* **29**(2), 39–72.

Shallat, L. 1994 Rites and rights: Catholicism and contraception in Chile. In Panos (ed.), *Private decisions, public debate: women, reproduction and population*. Panos, London, 149–62.

Sheahan, J. 1987 *Patterns of development in Latin America: poverty, repression, and economic strategy*. Princeton University Press, Princeton, NJ.

Sheahan, J. 1997 Effects of liberalization programs on poverty and inequality: Chile, Mexico and Peru. *Latin American Research Review* **32**(3), 7–37.

Shearer, E.B., Lastarria-Cornhiel, S., Mesbah, D. 1990 The reform of rural land markets in Latin America and the Caribbean: research, theory, and policy implications. LTC Paper No. 141. Land Tenure Center, University of Wisconsin, Madison, WI.

Sheffner, J. 1995 Moving the wrong direction in social movement theory. *Theory and Society* **24**, 595–612.

Silva, E. 1994 Thinklng politically about sustainable development in the forests of Latin America. *Development and Change* **25**(4), 697–721.

Silva, E. 1996–97 Democracy, market economics, and environmental policy in Chile. *Journal of Interamerican Studies and World Affairs* **38**(4), 1–33.

Silva, E. 1997a Sustainable development and the plight of the forest in Chile. *North–South Issues* **6**(2), 1–8.

Silva, E. 1997b The Politics of sustainable development: native forest policy in Chile, Venezuela, Costa Rica, and Mexico. *Journal of Latin American Studies* **29**(2), 457–95.

Silva E. (Forthcoming) Forest policy and sustainable development in Costa Rica. *North–South Issues*.

Silva, P. 1995 Modernization, consumerism and politics in Chile. In Hojman, D.E. (ed.), *Neo-liberalism with a human face? The politics and economics of the Chilean model*, University of Liverpool Monograph Series No. 20, University of Liverpool, Liverpool, 118–32.

Simmons, A., Guengant, J.P. 1992 Caribbean exodus and the world system. In Kritz, M., Lim, L., Zlotnik, H. (eds), *International migration systems: a global guide*. Clarendon Press, Oxford.

Singer, H.W. 1991 Terms of trade: new wine and new bottles? *Development Policy Review* **9**(4), 339–52.

Skeldon, R. 1990 *Population mobility in developing countries*. Belhaven Press, London.

Skidmore, T. 1995 Dependency by any other name? *Brown Journal of World Affairs*. **2**(2), 227–9.

Sklair, L. (ed.) 1994 *Capitalism and Development*. Routledge, London.

Sklair, L. 1989 *Assembling for development: the maquila industry in Mexico and the United States.* Unwin Hyman, Boston.

Skocpol, T. 1979 *States and social revolutions: a comparative analysis of France, Russia, and China.* Cambridge University Press, Cambridge.

Slater, D. 1989 *Territory and state power in Latin America.* Macmillan, London.

Slater, D. 1990 Development theory at the crossroads. *European Review of Latin American and Caribbean Studies* no. 48, 116–26.

Slater, D. 1998 Rethinking the spatialities of social movements: questions of (b)orders, culture and politics in global times. In Alvarez, S. et al. (eds), *Cultures of politics, politics of cultures.* Westview Press, Boulder, CO, 380–404.

Smith, W.C., Acuña, C.H. Gamarra, E. (eds) 1994 *Latin American political economy in the age of neoliberal reform: theoretical and comparative perspectives for the 1990s.* Transaction Press, New Brunswick, NJ.

Smith, W.C., Korzeniewicz, R.P. (eds) 1997 *Politics, social change, and economic restructuring in Latin America.* North–South Center Press, Miami, FL.

Smyth, I. 1994 *Population policies: official responses to feminist critiques.* DP-14, School of Economics, Centre for the Study of Global Governance, London.

So, A.Y. 1990 *Social change and development: modernization, dependency, and world-system theories.* Sage, Newbury Park, CA.

Sørensen, N.N. 1985 Roots, routes and transnational attractions: Dominican migration, gender and cultural change. In Peek, P., Standing, G. (eds), *State policies and migration: studies in Latin America and the Caribbean.* Croom Helm, London, 173–205.

Spybey, T. 1992 *Social change, development and dependency: modernity, colonialism and the development of the West.* Polity Press, Cambridge.

Stallings, B. 1992 International influence on economic policy: debt, stabilization, and structural reform. In Haggard, S.S., Kaufman, R.R. (eds), *The politics of economic adjustment.* University of Princeton Press, Princeton, NJ, 41–88.

Standing, G. 1989 Global feminization through flexible labour. *World Development* **17**(7), 1077–95.

Stanfield, D.J. 1992: Titulación de tierra: alternativa a la reforma agraria en un contexto de ajuste estructural. In Noé Pino, H., Thorpe, A. (eds), *Honduras: el ajuste estructural y la reforma agraria.* CEDOH–POSCAE, Tegucigalpa, 181–206.

Starn, O. 1992 'I dreamed of foxes and hawks': reflections on peasant protest, new social movements and the *Rondas Campesinas* of Northern Peru. In Escobar, A., Alvarez, S. (eds), *The making of social movements in Latin America.* Westview Press, Boulder, CO, 89–111.

Stephen, L. 1993 Challenging gender inequality. Grassroots organizing among women rural workers in Brazil and Chile. *Critique of Anthropology* **13**(1), 33–55.

Stewart, F. 1995 *Adjustment and poverty: options and choices.* Routledge, London.

Stewart, S. 1996 The price of a perfect flower: environmental destruction and health hazards in the Columbian flower industry. In Collinson, H.(ed.), *Green guerrillas: environmental conflicts and initiatives in Latin America and the Caribbean.* Latin American Bureau, London, 132–39.

Stolcke, V. 1991 Conquered women. *Report on the Americas* **5**, 23–28.

Storper, M. 1991 *Industrialization, economic development and the regional question in the Third World.* Pion, London.

Stren, R.E., White, R., Whitney, J. 1992 *Sustainable cities: urbanization and the environment in international perspective.* Westview Press, Boulder, CO.

Sunkel, O. (ed.) 1993 *Development from within: towards a neostructuralist approach for Latin America.* Lynne Rienner, Boulder,CO, and London.

Sunkel, O. 1973 Transnational capitalism and national disintegration in Latin America. *Social and Economic Studies* **22**, 132–76.

Sunkel, O. 1993a From inward-looking development to development from within. In Sunkel, O. (ed.) *Development from within: toward a neostructuralist approach for Latin America.* Lynne Rienner, Boulder, CO, 23–59.

Sunkel, O. 1994 Un enfoque neoestructuralista de la reforma económica, la crisis social y la viabilidad democrática en América Latina. Paper presented at the XVIII International Congress of the Latin American Studies Association (LASA), Atlanta, 10–12 March.

Sunkel, O., Zuleta, G. 1990 Neo-structuralism versus neo-liberalism in the 1990s. *CEPAL Review* no. 42, 35–41.

Sunshine, C.A. 1988 *The Caribbean: survival, struggle and sovereignty*. Ecumenical Program on Central America and the Caribbean, Washington, DC.

Susman, P. 1998 Cuban socialism in crisis: a neoliberal solution?. In Klak, T. (ed.), *Globalization and neoliberalism: the Caribbean context*. Rowman and Littlefield, Lanham, MD, 179–208.

Szirmai, A. 1997 *Economic and social development*. Prentice Hall, Hemel Hempstead, Herts.

Tardanico, R., Menjívar Larín, R. (eds) 1997 *Global restructuring, employment, and social inequality in urban Latin America*. North–South Center Press, Miami, FL.

Teubal, M. 1992 Food security and regimes of accumulation: with reference to the case of Argentina. ISS Rural Development Research Seminars, 29 April, Institute of Social Studies, The Hague.

The Economist 1997a Banana row: the Eastern Caribbean, 31 May, 36.

The Economist 1997b Expelled from Eden, 20 December, 45–48.

The Economist 1998a Importing Asian energy, 14 February, 88.

The Economist 1998b The road from Santiago, 11 April, 49–51.

The World Factbook 1997 Available on-line at: http://odci.gov/cia/publications/factbook/index.html.

Therborn, G. 1995 *European modernity and beyond*. Sage, London.

Thiesenhusen, W.C. (ed.) 1989 *Searching for agrarian reform in Latin America*. Unwin Hyman, Boston, MA.

Thiesenhusen, W.C. 1995 *Broken promises. Agrarian reform and the Latin American campesino*. Westview Press, Boulder, CO.

Thomas, C.Y. 1988 *The poor and powerless: economic policy and change in the Caribbean*. Latin American Bureau, London.

Thomas, J.J. 1995 *Surviving in the city: the urban informal sector in Latin America*. Pluto Press, London.

Thomas, J.J. 1996 The new economic model and labour markets in Latin America. In Bulmer-Thomas, V. (ed.), *The new economic model in Latin America and its impact on income distribution and poverty*. Macmillan, Basingstoke, Hants.

Thomas, J.J. 1997 The urban informal sector and social policy: some Latin American contributions to the debate. Paper presented at Workshop for the Social Policy Study Group, Institute of Latin American Studies, University of London, London, 28 November.

Thrupp, L.A. 1995 *Bittersweet harvests for global supermarkets: challenges in Latin America's agricultural export boom*. World Resources Institute, Washington, DC.

Thrupp, L.A. 1996 New harvests, old problems: the challenge facing Latin America's agro-export boom. In Collinson, H. (ed.), *Green guerrillas: environmental conflicts and initiatives in Latin America and the Caribbean*. Latin American Bureau, London, 122–31.

Tilly, C. (ed.) 1975 *The formation of national states in Western Europe*. Princeton University Press, Princeton, NJ.

Tinsman, H.E. 1996 *Unequal uplift: the sexual politics of gender, work, and community in the Chilean agrarian reform, 1950–1973*. Unpublished PhD dissertation, Yale University, New Haven, CT.

Tironi, E., Lagos, R. 1991 The social actors and structural adjustment. *CEPAL Review* no. 44, 35–50.

Tokman, V. 1978 Informal-formal sector relationship: an exploration into their nature. *CEPAL Review*, No. 5.

Tokman, V. 1989 Policies for a heterogeneous informal sector in Latin America. *World Development* 17(7), 1067–76.

Tokman, V. 1991 The informal sector in Latin America: from underground to legality. In Standing, G., Tokman, V. (eds), *Towards social adjustment: labour market issues in structural adjustment*. International Labour Organization, Geneva, 141–57.

Touraine, A. 1981 *The voice and the eye: an analysis of social movements*. Cambridge University Press, Cambridge.

Turner, J. 1968 Architecture that works. In Bell, G., Tyrwhitt, J. (eds), *Human identity in the urban environment*. Penguin, Harmondsworth, Middlesex, 352–65.

UN 1995 *The world's women 1995: trends and statistics*. United Nations, New York.

UN 1997 *Demographic yearbook 1995*. United Nations, New York.

UN various years. *Statistical yearbook for Latin America and the Caribbean*. Washington, DC.

UNCHS 1996 *An urbanizing world: global report on human settlements 1996*, United Nations Centre for Human Settlements. Oxford University Press, New York.
UNDP 1995 *Human development report 1995*. United Nations Development Programme. Oxford University Press, New York.
UNDP 1996 *Human development report*. United Nations. Oxford University Press, New York.
UNDP 1997 *Human development report 1997*. United Nations Development Programme. Oxford University Press, New York: .
UNIRDAP 1995 New challenges: Cairo international conference on population and development, United Nations International Committee on Integrated Rural Development for Asia and the Pacific. *Poverty Alleviation Initiatives*, **4**(2), 8–13.
United Nations 1994 *Population, environment and development*. United Nations, New York.
Uribe-Echevarria, F. 1996 Reestructuración económica y desigualdades interregionales: el caso de Chile, *EURE, Revista Latinoamericana de Estudios Urbano Regionales* **22**(65), 11–38.
Utting, P. 1992 The political economy of food pricing and marketing reforms in Nicaragua, 1984–87. *The European Journal of Development Research* **4**(2), 107–31.
Utting, P. 1993 *Trees, people, and power*. Earthscan, London.
Valcárcel, L.E. 1925 *Del ayllu al imperio*. Editorial Garcilaso, Lima.
Valcárcel, L.E. 1972 *Tempestad en los Andes*. Editorial Universo, Lima.
Valdés, A., Muchnik, E., Hurtado, H. 1990 Trade, exchange rate, and agricultural pricing policies in Chile. In *The political economy of agricultural pricing policy*. 2 vols. World Bank Comparative Studies. World Bank, Washington, DC.
Valdés, A., Siamwalla, A. 1988 Foreign trade regimes, exchange rate policy, and the structure of incentives. In Mellor, J.W., Ahmed, R. (eds), *Agricultural price policy for developing countries*. Johns Hopkins University Press, Baltimore, NJ, 103–23.
Valdés, J.G. 1995 *Pinochet's economists: the Chicago School in Chile*. Cambridge University Press, New York and Cambridge.
Valdés, T., Gomariz, E. 1995 *Latin American women: comparative figures*. FLACSO, Madrid and Santiago.
Valenzuela, J.S. 1997 Hacia la formación de instituciones democráticas: prácticas electorales en Chile durante el siglo XIX. *Estudios Públicos* 66.
van der Borgh, C. 1995 A comparison of four development models in Latin America. *The European Journal of Development Research* **7**(2), 276–96.
Van Scott, D.L. 1995 *Indigenous peoples and democracy in Latin America*. St. Martin's Press, New York.
Vargas, G. 1991 The women's movement in Peru: streams, spaces and knots. *European Review of Latin American and Caribbean Studies* no. 50, 7–50.
Vasconcelos, J. 1927 *La raza cósmica*. S.A., Barcelona.
Véliz, C. 1994 *The new world of the gothic fox: culture and economy in English and Spanish America*. University of California Press, Berkeley.
Vellinga, M. (ed.) 1993 *Social democracy in Latin America: prospects for change*. Westview Press, Boulder, CO.
Veltmeyer, H. 1997 New social movements in Latin America: the dynamics of class and identity. *The Journal of Peasant Studies* **25**, 1, 139–69.
Veltmeyer, H., Petras, J., Vieux, S. 1997 *Neoliberalism and class conflict in Latin America*. Macmillan, Basingstoke, Hants, and London.
Vergara, P. 1994 Market economy, social welfare and democratic consolidation in Chile. In Smith, W., Acuña, C., Gamarra, E. (eds), *Democracy, markets and structural reform in Latin America: Argentina, Bolivia, Brazil and Chile and Mexico*. North–South Center, University of Miami, Miami, FL, 237–61.
Vogelgesang, F. 1996 Property rights and the rural land market in Latin America. *CEPAL Review*, 58, 95–113.
Wade, R. 1990 *Governing the market: economic theory and the role of government in East Asian industrialization*. Princeton University Press, Princton, NJ.
Wade. R. 1996 Japan, the World Bank and the art of paradigm maintenance: the East Asian miracle in political perspective. *New Left Review* no.217, 3–36.
Wagner, P. 1994 *A sociology of modernity, liberty and discipline*. Routledge, London.
Walker, T. (ed.) 1997 *Nicaragua without illusions: regime transition and structural adjustment in the 1990s* Scholarly Resources, Wilmington, DE.

Ward, K., Pyle, J. 1995 Gender, industrialization, transnational corporations and development: an overview of trends and patterns. In Bose, C., Acosta-Belén, E. (eds), *Women in the Latin American development process*. Temple University Press, Philadelphia, PA, 37–64.

Ward, P. 1976 The squatter settlement as slum or housing solution: evidence from Mexico City. *Land Economics* **52**, 330–46.

Ward, P. 1998 *Mexico City*. John Wiley, Chichester, Sussex.

Warren, K. 1998 Indigenous movements as a challenge to the unified social movement paradigm in Guatemala. In Alvarez, S. *et al.* (eds), *Cultures of politics, politics of cultures*. Westview Press, Boulder, CO, 165–95.

Watson, H.A. 1996 Globalization, new regionalization, restructuring and NAFTA. *Caribbean Studies* **29**(1), 5–48.

Watts, M. 1996 Mapping identities: place, space and community in an African city. In Yeager, P. (ed.), *The geography of identity*. University of Michigan Press, Ann Arbor, MI, 59–97.

Weeks, J. 1995 Macroeconomic adjustment and Latin American agriculture since 1980. In Weeks, J. (ed.), *Structural adjustment and the agricultural sector in Latin America and the Caribbean*. Macmillan, London, 61–91.

Welch, B. 1994 Banana dependency: albatross or liferaft for the Windwards. *Social and Economic Studies* **43**(1), 123–49.

West, R., Augelli, J. 1989 *Middle America: its lands and people*. Prentice Hall, Englewood Cliffs, NJ.

Whatmore, S. 1995 From farming to agribusiness: the global agro-food system. In Johnson R.J., Taylor, P.J., Watts, M. (eds), *Geographies of global change*. Basil Blackwell, Oxford, 36–49.

White, R.R. 1993 *North, south, and the environmental crisis*. University of Toronto Press, Toronto.

Whitehead, L. 1985 Whatever happened to the Southern Cone model? In Hojman, D.E. (ed.), *Chile after 1973: elements for the analysis of military rule*. Centre for Latin American Studies, Liverpool University, Liverpool, 9–30.

Wiarda, H.J. 1990 *The democratic revolution in Latin America: history, politics, and U.S. policy*. Holmes and Meier, New York and London.

Wilentz, A. 1989 *In the rainy season: Haiti since Duvalier*. Touchstone, New York.

Wiley, J. 1996 The European Union's single market and Latin America's banana exporting countries. *Conference of Latin Americanist Geographers (CLAG) Yearbook 1996*. Volume 22. CLAG, Muncie, IN.

Wiley, J. 1998 Dominica's economic diversification: microstates in a neoliberal era? In Klak, T. (ed.), *Globalization and neoliberalism: the Caribbean context*. Roman and Littlefield, Lanham, MD, 155–178.

Williamson, J. (ed.) 1990 *Latin American adjustment: how much has happened?* Institute for International Economics, Washington, DC.

Williamson, J. (ed.) 1993 *The political economy of policy reform*. Institute for International Economics, Washington, DC.

Willis, K. 1993 Women's work and social network use in Oaxaca City, Mexico. *Bulletin of Latin American Research* **12**(1), 65–82.

Willis, K. 1994 *Women's work and social network use in Oaxaca city, Mexico: an analysis of class differences*. Unpublished DPhil Thesis, Oxford University, Oxford.

Willmore, L. 1994 *Export processing in the Caribbean: lessons from four case studies*. Report LC/CAR/G.407, United Nations Economic Commission for Latin America and the Caribbean, Santiago.

World Bank 1980 *World development report, 1980*. Oxford University Press, New York.

World Bank 1986 *World development report, 1986*. Oxford University Press, New York.

World Bank 1987 *World development report, 1987*. Oxford University Press, New York.

World Bank 1988 *World development report, 1988*. Oxford University Press, New York.

World Bank 1992 *World development report, 1992: development and the environment*. Oxford University Press, New York.

World Bank 1993 *The East Asian miracle: public policy and economic growth*. Oxford University Press, New York and Oxford.

World Bank 1995a *Global economic prospects and the developing countries. Annual report*. The World Bank, Washington DC.

World Bank 1995b *World development report 1995*. Oxford University Press, New York.

World Bank 1996 *World development report 1996.* Oxford University Press, New York.
World Bank 1997 *World development report, 1997.* Oxford University Press, New York.
World Resources Institute 1994 *World resources 1995: a guide to the global environment.* Oxford University Press, New York.
Wratten, E. 1995 Conceptualising urban poverty. *Environment and Urbanisation* 7(1), 11–36.
Yoffie, D.B. 1983 *Power and protection: strategies of newly industrializing countries.* Columbia University Press, Ithaca, NY.
Young, O.R. 1994. *International governance: protecting the environment in a stateless society.* Cornell University Press, Ithaca, NY.
Yúdice, G. 1998 The globalization of culture and the new civil society. In Alvarez, S. *et al.* (eds), *Cultures of politics, politics of cultures.* Westview Press, Boulder,CO, 353–79.
Zamosc, L. 1994 Agrarian protest and the Indian movement in the Ecuadorean highlands. *Latin American Research Review* **29**, 3, 37–68.
Zea, L. 1990 *Descubrimiento e identidad Latinoamericana.* UNAM, Mexico.
Zimbalist, A., Brundenius, K. 1989 *The Cuban economy.* Johns Hopkins University Press, Baltimore, MD.

INDEX

abortion 236
actors 292
   external 44
   international 162–4, 284
   state 164–6
Africa 6, 50, 184–5, 187
agrarian reform 6, 26, 28, 147, 272–85
   see also reformed sector
agrarian structure 50, 272, 274
   see also land tenure
agribusiness 147–8, 158, 286, 288, 302
agriculture 18, 20, 247–9, 272–304
agricultural
   exports 3, 22, 135–7, 245, 283–4, 303
   growth 273, 286–7
   production 280–1
   labor see workers: agricultural
agro-industry see agribusiness
Allende 3, 43, 275, 277–8, 282–83
Alliance for Progress 273, 275
alliances 37, 38, 42
Amazon 158, 160, 163, 287, 299–300
   indigenous organizations 299
Andean Community 11–12, 69, 93–4, 96
Argentina 10, 12–6, 18–19, 37, 41, 43, 46, 49, 52–5, 69, 74, 79, 81–2, 84, 129, 135–9, 194, 200, 206, 213, 220, 215, 228, 232, 249, 251, 253, 255, 259, 273, 310–12, 321–23
armed forces 42, 54–5, 63, 199
   see also military
Asia 5, 36, 50, 65, 185, 187, 198, 191, 195–6, 233
Association of Caribbean States (ACS) 100, 121
Australia 184–5, 187
authoritarian 25, 65
   breakdown 47
   capitalist states 35, 36
   enclaves 63
   governance 24
   government 16–18, 25
authoritarianism 32–3, 39, 41, 43, 47, 49, 191, 197–9, 209–10, 215
Aylwin, Patricio 17, 55, 148, 254

banana exports 102, 104–8
basic needs 166–70, 176–8

Belize 134–6, 248
black Latin Americans 213, 218, 219
Bolivia 25, 46, 60–1, 69, 74, 81–2, 134–6, 141, 154, 190, 216, 229, 232, 235–6, 238, 245, 249, 252, 265, 273, 275–7, 292, 298, 300
   New Economic Policy 252, 265
bourgeoisie 39, 46–7, 275–6
Brady Plan 79
Brazil 6–8, 10, 12–13, 15, 16, 18–20, 27, 37–8, 41, 43, 45–6, 49, 53, 55, 61, 69, 71–2, 74, 79, 81–2, 84, 129, 132, 134–8, 154, 159–60, 163, 165, 168, 174–5, 191, 195, 198, 205–6, 208, 212, 220, 218, 232, 237–8, 254–5, 262–3, 273, 277, 287–8, 289, 292, 298, 300, 310–12
breakdown of democracy 44
Britain 129, 191
British colonies 47
British Empire 185
Buenos Aires 220
bureaucratic authoritarianism 41–3
business community 58–60

campesinos
   definition 285
   see also peasants *and* agricultural *under* workers
Canada 10
capital 6, 7
   expropriation 141
   flight 75–6
   foreign 75–7, 85, 191, 309, 312, 322
   mobility 250
   national 191
   structural power 46
   see also financial
capitalists 34–5, 46–7, 294
   see also class *and* farmers
Caracas 144–5, 172, 243, 319–21
Carajas 159, 285
Cardoso, Fernando Henrique 6, 13, 15, 19, 26, 37, 58
Caribbean Basin Initiative (CBI) 99, 105–6
Caribbean 45, 98–126, 129, 311, 319–20
Caricom (the Caribbean Community and Common Market) 93–4, 121–4
Cavallo, Domingo 58

Central America 10, 12, 43, 45, 62, 82, 98–126, 129, 213, 215, 246, 250
Central American Common Market (CACM) 11, 92–4, 121–4
centre 5, 6, 27, 192
    see also periphery and structuralism for centre-periphery
Chamorro, Violeta 252
Chiapas 143, 212, 216, 219, 298–9, 304
Chicago Boys 146
Chile 7, 13–15, 17–21, 23–7, 37, 41, 43, 45–6, 49, 52, 54–5, 57–60, 62–3, 69, 70–1, 73–4, 77–82, 84–7, 129, 134–9, 141–2, 146–50, 159–60, 171, 178, 187, 193, 198, 200, 213, 215, 220, 228, 232, 235, 245, 251, 253–5, 259, 273, 275–11, 282–3, 286, 288–90, 295, 310, 312, 322
cholera 233
citizenship 51, 57, 61, 63–4, 322
civil society 27, 44, 50, 56, 59, 145–6, 161, 199, 212–13, 275, 282, 296, 302
class 3, 199, 209, 218–19, 250, 311–15
    alliances 34, 39, 42, 45–6, 312
    bourgeois 199
    capitalist 13
    class-based groups 34–7, 45
    conflict 34, 37, 42–3, 46, 276, 303
    lower 42, 45, 49, 238
    middle 13, 24, 34–7, 40, 44, 46–7, 189, 238–9, 246, 311
    structure 38–9
    tensions 35–8
    upper 38, 42–5, 46–7, 49
    working 13, 17, 37, 189, 239
    see also bourgeoisie, industrialists, landlords, peasants
clientelism 59, 189, 193, 196–7, 201, 207, 209
coalition-building 49
collective farms 278–11
collective memories 64
Collor de Mello, Fernando 56, 61
Colombia 37, 43, 69, 74, 78–9, 81–2, 84, 134–6, 200, 203, 213, 229, 232, 238, 249, 253, 255, 259, 263, 273, 276–7, 289–90, 298, 300, 310, 322
colonialism 129, 311
colonization 158, 160
commercial elites 34, 36
comparative advantage 12, 130, 312
competitive advantage 12
competitiveness 6, 10, 14, 20, 26, 197, 295
community 205, 207, 210–11, 214–15, 277
comunero
    definition of 277, 281–82
    see also communities *under* indigenous
condoms 237
    see also contraception; family planning
conservation 157
consolidation of democracy 32, 44–5, 47–8

consumerism 22, 25, 57–8
consumers 57
consumption
    conspicuous 196
    patterns of 57
    societies 57–8
contestation 35–6, 41
Contra War 232, 275, 281
contraception 228, 235–8
    see also Family Planning
contract farming see agribusiness
cooperatives
    agricultural 278–9, 282, 297
copper
    Chile Stabilization Fund 139
Costa Rica 20, 52, 60–1, 81, 135–6, 141, 154, 161, 226, 229, 230–2, 245, 247–50, 252–3, 255–7, 264–5, 277, 310
criminality 62
Cuba 3, 68, 104–5, 117–20, 129, 229, 236, 256, 263–4, 273, 275–9, 282
    Council for Mutual Economic Assistance (CMEA) 117
    'The Special Period in the Time of Peace' 117–19
Cubatao 168
cultural capital 214
cultural change 8, 39
cultural transplant 185, 187
culture 3, 182–203, 205, 207, 210, 212–13, 215

data processing 114, 116
debt
    crisis 3, 16–18, 24, 75–7, 154
    debt-equity schemes 79–80
    foreign 98, 108–9, 112, 117
decentralization 218
democracy 16, 25, 32–50, 65, 192, 197–201
    defined 35
    forms 35
    limited 46–7
    protected 48
    technocratic 51
democratic
    capitalist states 35, 37
    consolidation 2, 32, 44, 50, 65
    governance 16–17, 22
    government 18, 187
    regime 25
    restoration 51–7, 59
    transition 2, 16, 24–5, 51, 54–6, 62, 64–5
democratization 25, 40, 43–6, 65, 200, 294
    see also transition to democracy
depeasantization see proletarianization
dependency 4, 6, 12, 22, 27–8, 37–8, 42–3, 49, 52, 192, 195, 198–9, 311
    trade and economic 98, 100–4, 105–7, 110, 112–14, 123–5, 131,

dependency theory 3–10, 29, 37–9, 131, 304
depoliticization 54–7, 200
Depression 70–1, 130–1
deregulation 25, 259–60
development 5, 33, 36–8, 40
    grassroots 160–2, 164–6, 317, 319
    political (see political development)
    socioeconomic 43–5
    theory 129–32
dictatorship 10, 195, 201, 213
disease 170–1, 232–3
division of labor 20
domestic service 263
doctrine of national security 42
Dominican Republic 12, 46, 232, 249, 276, 292
drug-trafficking 61–2, 119, 124–5, 249
Dutch Disease 139–40
dysentery 232

East Asia 6, 9, 13, 26, 29, 38, 86–7, 93, 185, 187
    development model 72
Economic Commission for Latin America (ECLA) 4, 60, 130–1, 192, 199
Economic Commission for Latin America and the Caribbean (ECLAC) 28
economic
    crisis 3, 23, 32, 34, 37, 47–8
    development 32, 34, 39–45, 48, 68–84, 104–17
    integration 12, 26–7, 92–6, 121–4
    growth 2, 4, 5, 10, 17, 20–3, 25–6, 34, 36–7, 42–3, 48, 84–5
    modernization 32, 40, 42
    sectors 14, 34–5, 39, 134–7
economic stabilization plans 26
    Austral 16, 53
    Cruzado 16, 53
    Inti 53
Ecuador 21, 46, 52, 61, 63, 69, 74, 79, 81–2, 84, 133, 134–8, 154, 221–2, 232, 241, 248, 273, 277, 289, 291, 298–10
    indigenous organizations 215–16, 218–19, 221
education 24, 187–8, 195, 197, 209, 233, 237–8, 260–1, 267
ejido 278, 282
    definition of 276
El Salvador 55–6, 82, 134–6, 212, 232, 248, 250, 273, 276–9, 281, 292, 298, 322
electoral systems 44
elite settlement 55
elites 25, 36, 44–5, 47–9, 65, 185, 194–5
employment 14, 22, 88–91, 195, 226–69, 275
    female 247, 261–5, 288–90, 313–16
    male 313–16
    rural 272, 290–2
    urban 251–65, 308–16, 321
    see also deregulation; formal sector; informal sector; gender; underemployment

enclave economy 129, 140–1
Enlightenment 182–3, 186–9, 191–2, 198, 203–4, 210, 213
entrepreneurs 22, 52–3
environment 20, 194, 233–4, 246, 272
environmental
    degradation 142, 178, 289, 302
    movements 298–11
    problems 170–3
    regulations 176–8
estancias 274, 278
    definition of 274
estate *see estancia, hacienda*, latifundia, plantation
ethnic/ethnicity 51, 272, 299, 323
Europe 9, 10, 36, 47, 50, 182, 185–8, 191, 194, 198, 201, 205, 210, 215, 217
    Eastern 4, 9, 15, 26, 40, 185, 187
    Western 184–5, 187
    Southern 65
European Union (EU) 102, 106–8, 110, 121–2
exchange rates 73, 87
exclusion 24, 193, 294–5
exports
    agriculture *see* agricultural exports
    energy 134–5
    forestry 137
    growth of 73, 86–7, 134–7, 262
    manufacturing 20, 110–11, 114–15, 309
    metal and mineral 135
    natural resource 86–7, 134–7, 138–143
    nontraditional (NTE) 4, 19, 20, 110, 114–16, 122, 124, 126
    nontraditional agricultural (NTAE) 20, 146–50, 288–9, 293–6
    -orientated *see* outward-orientated
    primary product 85, 131–8
    windfalls 139
extractive reserves 163

family 250
    see also household
family planning 230, 233–5
    see also contraception; population
farm workers *see* workers *under* agricultural
farmers
    capitalist 272, 275, 283–4, 287, 290–2, 293, 302
    large 242 *see also* latifundistas
    small 242 *see also* minifundistas *and* peasant economy
fertility 228–30, 237, 239–40
    see also family planning
financial 6, 14, 22–6
    experts 60
    see also international financial institutions
food 245, 274, 277, 291, 292
    security 142, 296
foreign capital *see* capital
    see also transnational corporations

foreign sector 39, 42–3
forestry 89, 137, 161
　Plan Piloto Forestal 161
　plantations 159
formal sector 23, 88, 252–4, 260, 263, 307–13
Foxley, Alejandro 59
free-market reforms 43, 45
　see also neoliberal reforms
Free Trade Area of the Americas (FTAA) 93, 95
Frei, Eduardo (Jr.) 7, 148
Frei, Eduardo (Sr.) 147, 275, 278
Friedman, Milton 146
Fujimori, Alberto 13, 18–19, 56, 59
fusion effect 209

García, Alan 6
GATT 140
gender 210, 212, 219, 221, 226–69, 261–5, 304
　divisions of labor 20
　identities 323
　inequalities 263
　relations 272, 282, 289
　roles 317
　see also employment; migration; marriage; women
geopolitics 217
global 9, 10, 15, 19, 21, 25–6, 28, 32, 37
globalization 2, 3, 7–13, 16, 21, 25–6, 28, 32, 37, 61, 68, 109–15, 128, 150, 153, 165, 185, 193, 251
governance 3
Guatemala 20, 55–6, 61–3, 81, 135–6, 141, 227, 277, 292, 295, 302, 322–3
Guatemala City 256, 323
Guayaquil (Ecuador) 262

hacienda 199, 272–4, 278–9, 286, 291
　definition of 272
　see also latifundia
Haiti 63, 101–4, 109, 112–14, 120–1, 123, 292
health 24, 191, 195, 232, 289–90
　see also disease; mortality
health care 232
Hispanism 190, 249
Honduras 62, 81, 134–6, 229, 247–8, 276, 292
household
　and environment 170–1
　female-headed 240–1, 248, 264, 266
　size 241
　networks 315
　structure 265
　survival strategies 260, 263–6, 317
　see also fertility; marriage; migration
housing 317–21
Human Development Index (HDI) 101, 103, 107
human rights 54–5, 61, 63, 193, 195, 200, 205, 213, 234

identity 182–203, 205–7, 215–17, 242, 304, 323–4
　baroque 207
　colonial 188
　crisis of 201
　cultural 183, 185, 190–2, 194, 196, 201, 203, 207, 215, 217
　developmentalist 196, 201, 207
　Hispanic 183, 201, 215
　indian 183–4, 299, 323
　Indo-Iberian 182, 186–7, 198
　Latin American 183–4, 196, 204, 207
　modern 183
　true 184, 203
ideological change 2, 54, 56–8, 60
ideological traditionalism 209
imaginative geographies 217
immigration 188, 192, 210, 198
immunization 232
imperialism 192
import substitution industrialization (ISI) 4, 6, 17, 28, 42–3, 46, 53, 100, 122, 131, 189, 192, 309, 311–12
income distribution 24, 51, 57, 61, 89–92, 272, 281–82
incomes 230, 248
　see also wages
India 187
indian 198
Indianism 183, 189–90, 193–4
indigenous
　communities 194, 275, 281, 283, 292, 302
　movements 215–16, 221, 218, 298–303
　people 61, 64, 273, 299–301, 304
industrialists 276
industrialization 2, 42–3, 53, 71–2, 189–91, 306–11
　see also import substitution industrialization
inequality 22, 24–5, 88–9, 321–22
inflation 16–17, 23, 74–5, 84–5
informal sector 22–3, 254–61, 263, 266–7, 306–13, 317–18, 323
　see also deregulation; employment; neoliberalism
Inquisition 191, 210
institutions 34, 40, 45, 47–8
　political 48–50
　state 47
Inter-American Development Bank (IADB or IDB) 24, 60, 81–2
interests 213–14
　class 209, 217
　gendered 221
　strategic/practical (Molyneux) 221
international 4, 6, 7, 9, 10, 13, 20, 26–8
　banking 8
　corporations 191
　economy 38–9, 50
　factors 45–6, 48
　markets 13, 14, 20, 22

international financial institutions (IFIs) 6, 7, 52–3, 59, 61
International Labour Organization (ILO) 307, 317
International Monetary Fund (IMF) 9, 15, 52, 59, 76–9, 251–2, 260, 309
International Planned Parenthood Federation (IPPF) 237
inward-orientated 12, 17–19, 20–1, 70–5
  see also import substitution industrialization

Jamaica 101, 105, 109, 114, 116, 119–20, 311–14, 322
Japan 9, 10, 184–7, 189, 191, 198, 313

Kingston (Jamaica) 311, 313–15, 319–20

labor 34–7, 39–40, 44, 46–7
  agricultural *see* workers
  markets 20–3, 88–9, 237, 250–65, 308–11, 313–14, 321
  seasonal 148–9
  tenants 274 *see* tenants
  rural 274, 289 *see* agricultural *under* workers
land reform *see* agrarian reform
land
  conflicts 285
  invasion or occupation 277, 284–5, 298, 301
  markets 21, 284, 302
  policy 283, 297, 304
  tenure 274–5, 284, 303
  titling 284
landless peasants *see* agricultural *under* workers
landlords 36, 42, 46–7, 50, 192, 272–7, 280, 282–3, 286–7, 293, 302
La Paz (Bolivia) 323
latifundia 129, 272–5, 284
  see also hacienda
latifundista 272, 274
  definition of 274
legitimacy 18–19, 52, 57–61, 63–4
  crisis 44
liberalism 186–7, 191, 209
liberalization 2, 7, 14, 20, 22, 25, 52–3, 83–4, 226, 286, 295
life expectancy 231–2
Lima 167, 169, 171–2, 174–5, 211, 220, 244, 246, 307, 318
local 19, 20, 25
  communities 172–4
  governments 174–7
Lomé Conventions 106

malnutrition 232–3
Malthusian 234
Managua (Nicaragua) 256
manufacturing 9, 10, 19, 20, 111, 252, 262–3
maquiladora 263, 309
marginal groups 51, 64

marginalization 259, 272, 283, 293, 298
markets
  international 191
  see also labor
marriage 240
Marxism 42, 192–3, 195, 242
mass media 191
Mayas 297
  political organization 207, 218
Menem, Carlos 13, 19, 49, 58–9
Mercosur 11–12, 61, 93–6, 140
mestizaje 189–90, 199, 203
Meso-America 249
Mexico 6, 7, 9, 10, 14–15, 20–1, 38, 43, 53, 57, 69, 71–2, 74, 78–9, 81–2, 84, 132, 134–6, 141, 161, 167, 169, 173–4, 206, 207, 209, 211, 220, 226, 229, 230–2, 236, 238, 240–1, 246–50, 253, 255, 257, 258, 259, 262, 264, 273, 275–8, 280, 282–3, 291–2, 298, 304, 309–12, 315, 320, 323
Mexico City 167, 171, 175, 239, 246, 265, 309, 315, 318–20, 322
migrant associations 211
migration 111, 120, 209, 226–69, 307, 266–7
  gender-selective 246–7
  international 111, 120, 241, 243, 249–50
  rural-urban 230, 241, 243–5, 247–50, 266, 292–3, 299, 302, 323
  temporary 241, 248–9
  theoretical approaches to 242
  see also remittances
military 10, 46, 198
  see also armed forces
military government or dictatorship 32, 37, 41, 43–4, 46, 48, 193, 203, 206, 288
minifundia 272, 274–5, 278, 287
minifundista 274, 281–82, 291
  definition of 274
mining 156, 252
modernity 51, 58, 182–203, 205–7, 209, 212, 215, 217
  African 185, 187
  Enlightened 183, 184
  European 183–5, 189, 194, 201, 215
  Latin American 184–5, 187, 191, 196–201, 207, 212–13, 215, 217
  oligarchic 186–191, 194
  organised 189
  political cultures of 210, 216, 219, 221–2
  pseudo 183
  trajectory to 185, 187, 189, 198, 205, 215
modernization 2, 3, 26, 32, 40, 62, 64, 187, 191, 195, 272, 286, 293, 302
  access to 51, 61, 64
  as a project 52–4
  discourse of 52
  fragmented 51
modernization theory 39–41, 43–5, 47–9, 191
'motherist' politics *see* women's organizations

Mothers of the Plaza de Mayo, Argentina 206, 215, 220
mortality 231
  infant 230, 233
  maternal 236

nation 213, 218
National Population Council (Bolivia) 236
nationalism 209, 213
natural resources 156–66
  non-renewable 156–8
  renewable 156–8
neoclassical 5, 129, 272
neoliberal 12, 16, 64, 132, 265, 293, 297, 301–303
  economic policy 77–84, 210, 216–17, 219, 222
  ideology 52, 59, 64, 83–4
  land policy 283–4
  model 14, 18–19, 22, 24–6, 52, 75, 84–92, 112–17, 312, 315, 317–22
  order 52–4
  reforms 13, 15–16, 21–2, 24–7, 83–4
neoliberalism 3, 4, 13–15, 17–29, 53, 194, 197, 202, 209, 226, 232, 285, 295, 302
neomarxist 4, 131
neostructuralism 4, 27–9
newly industrializing countries (NICs) 5–7, 13, 28–9
Nicaragua 3, 56, 61, 104–5, 109, 112–13, 121–2, 134–6, 189, 221, 227, 233, 249, 251, 256, 273, 275–8, 281–83, 292
non-governmental organizations (NGOs) 24–5, 28, 53–4, 161–6, 169, 212, 219, 222, 237, 292–8, 304
nontraditional agricultural exports *see* NTAE *under* exports
nontraditional exports *see* NTE *under* exports
North America 9, 10, 36, 182–7, 189, 191–2, 194, 196, 198–9, 212, 235, 249, 298
  see also United States
North American Free Trade Area (NAFTA) 9–12, 20, 92–6, 98, 121, 123–4, 140, 298

offshore finance 116–17
oil 143–6
open regionalism 96
orthodox adjustment programs 59
outward-orientated 15, 20–2, 85–7, 189, 192, 272, 285, 303, 312
overcrowding 171

Pacific Asia 137
Panama 134–8, 249, 252–3, 276–7
Paraguay 12, 63, 69, 74, 81–2, 84, 134–6, 222, 249, 253, 275, 292, 295, 298
parceleros 281, 295–6
participation 17, 25, 35–6, 41, 282
party system 37, 48
peasant economy 272, 290–92

peasant movements/organizations 21, 207, 211, 218, 212, 275, 283–5, 288, 294, 298–303
peasants or peasantry 34, 36–7, 39, 50, 272–304
Pérez, Carlos Andrés 61
periphery 6, 7, 21, 27, 110, 192
  resource 128, 132–7
  semi- 132
  see also centre *and* structuralism for centre-periphery
Peru 6, 13–15, 18, 37, 53, 55–6, 59, 63, 69, 74, 79, 81–2, 84, 134–9, 141, 143, 169, 171, 188, 198, 200, 212, 218, 233, 235, 244, 247, 259, 272, 275–8, 281–83, 292, 312
  Highlands 207, 212–13, 218,
pensions 24
petty commodity production 259, 308, 313
Pinochet, Augusto 13, 49, 55, 63, 146–7, 198, 312
place 216–18
plantations 249, 272, 276–7, 281
pluralism 40
political
  apathy 62
  corruption 54, 61–4
  culture 44–5, 50
  demands 40
  development 39–40, 42–3
  economy 2, 3, 33, 37–9, 42, 45–9, 65, 153–66, 177–9
  elites 55, 60, 64
  institutions 45, 48, 50
  leaders 54–6
  legitimation 52, 57
  participation 25, 190
  parties 35, 37, 40, 42, 53–6, 58–9, 204–5, 208, 213, 220
  stability 34–6, 41, 46, 48
politics of needs 208
pollution 160, 172–3
polyarchy 33, 45
popular religiosity 184, 194, 203, 205
population 24, 69, 188, 226–69
  agricultural 290
  and development 234–5
  growth 227, 229, 244
  policies 234–8, 266
  rural 272, 290
  urban 306
  see also family planning; fertility; migration; mortality
populism 42–3, 56, 59, 64, 190, 193
Port-of-Spain (Trinidad) 321
Portugal 186, 189, 201
positivism 183, 190
postmodernism 193, 201
poverty 24–5, 89–92, 234, 245
  rural 281–82, 292, 302
  urban 167–77, 260, 262, 266, 317–21
preservation 157

primary products 132–7
  price volatility 139
private sector 35, 162, 168–9
privatization 14, 22, 52–3, 59, 88, 174, 195–6, 283, 293
proletariat *see* rural *and* workers
proletarianization 286, 290–92, 302, 304
prostitution 248
protectionism 70–1, 140
public sector 166–7, 208, 211–12, 215–16, 219–21, 252
  public-private action 168
Puerto Rico 105–6, 125, 236, 262–4, 311
  Operation Bootstrap 105–6, 311

Quito (Ecuador) 218, 221

race 187–9, 192, 194, 203, 209–10, 212–13, 218–19
  cosmic 189, 194
  inferior 187, 192
racism 187, 191–2, 199, 210, 212
  masked 210
ranching 156, 158, 274
reformed sector 278–11
  see also agrarian reform
regime loyalties 46
regional 20–1
regional trading blocs 81, 92–6, 100, 121–4, 140
religious culture 44, 323–4
remittances 249–50
renting 318–21
rent-seeking 139
rents
  mineral 139
repression 35–6, 43
resource curse thesis 128, 139, 143–5
Roman Catholicism 235, 237, 266
  factor in social mobilization 211–12, 220
  comunidades eclesiales de base 212
  Liberation theology 211
  see also Vatican
rubber tappers 299–300
rural 3, 24, 242, 287
  workers *see* agricultural *under* workers

San José (Costa Rica) 245–6
San Salvador (El Salvador) 256
sanitation 171–4
Santiago (Chile) 21, 173, 245, 320
São Paolo (Brazil) 168, 218, 262
secondary markets 79–80
services 89, 262–3, 307, 312
sexuality 203, 219, 221
social
  change 34, 36, 44
  classes 34, 47
  conflict 143, 275
  consciousness 190
  debt 24, 26, 53
  equity 24, 25
  groups 33–4, 38, 40–1, 45, 47
  justice 41, 48, 299, 301
  mobility 321
  order 35–7, 39, 42, 46
  participation 48
  security 14, 191,
social movements 26, 203, 213, 222, 322
  autonomy 206–8
  'life-worlds' 205, 210
  networks 205, 211–14, 218
  new social movements theory (NSM) 213–14, 217
  regional 218
  repertoires 205–6, 209, 216
  resource mobilization theory (RMT) 213–14, 217
  transnational 212, 218, 221
  urban 174, 208, 212, 217, 221, 322–3
  see also indigenous movement
social welfare 24
socialism 34, 104–5, 109, 113, 117–20, 125, 192–5, 278, 301
South East Asia 5, 184–5
Southern Cone 16, 33, 41, 50, 55, 228
South Korea 5, 6, 9, 313
Soviet Union 9, 15, 185, 187
Spain 47, 182–3, 186, 189, 196, 201
squatter settlements 285, 318–20
stabilization plan *see* economic plans
stages theory 39–41
state farms 277–80
state managers 34–6, 46–7
state 207, 222, 292, 260, 311–16, 322
  authoritarian 32, 50, 206, 208, 210, 212, 220
  autonomy 33, 38, 45–7
  citizenship 203, 205, 209, 215–16, 219
  defined 33
  democratic 32, 50
  developmentalist 208
  form 35, 37–8, 45, 47–8
  functions 34–7, 39, 42–3, 49
  in transition 203, 213, 215
  judicial process 210, 211, 207
  municipalities 208
  populist 208, 217
  women's voting rights 220, 222
sterilization 230, 235, 237
structural adjustment programmes (SAPs) 6, 22–4, 53–3, 64, 132, 238, 244, 285, 292, 302, 304, 309, 312–13
structuralism 3–10, 12, 28, 75, 130–1, 242
Structuralists 8, 22, 27, 29, 272
support coalition 44, 48
sustainable development 138–43, 149–50, 153–66, 170, 173–9, 234, 280, 284, 293, 296, 304
systems theory 39

technocratic government 25, 27
technocrats 17–18, 59–60, 64
technocratization 58, 60, 64
Tegucigalpa (Honduras) 256
tenants 271–3, 277, 285–6, 299
trade 73–4, 110–12
   liberalization 78, 81–2, 85–7, 132, 309, 312
   terms of 8, 130, 138–9
   theory 85, 130
trade-offs 157–8, 162–4
trade unions 199, 204, 208, 211, 217, 253–4, 282, 297, 322
transition to democracy 43–5, 47, 51, 54–6, 62, 64
   see also democratization
transnational
   corporations (TNCs) 7, 37
   forces 47
transnationalization 58
Trinidad 134–6

underemployment 24, 251–2, 259, 275
underdevelopment 5–7, 37–8, 41
unemployment 6, 19, 22–3, 25, 88, 90–1, 193, 251–2, 275, 313
United Kingdom (UK) 102–3, 197
United Nations Conference on Population and Development, Cairo (1994) 234
United Nations Economic Commission for Latin America (UNECLA) *see* ECLA, ECLAC
United Nations Environment Program (UNEP) 168, 179
United States of America (USA) 9, 10, 12, 38, 44, 47, 59, 63, 129, 185, 187, 189, 197–8, 202, 249–50, 266, 304, 308
   and the Caribbean 100–4
   see also North America
unequal exchange theory 7–8
urban 23
   environments 166–77

informal sector 23, 306–16
   poor 230–1, 317–21
urbanization 2, 209, 211, 244, 289, 305–24
   see also migration
Uruguay 12, 37, 41, 43, 49, 52, 54, 69, 74, 79, 81–2, 84, 129, 135–6, 189, 228, 232, 252, 262, 275, 277, 311
Uruguay Round 81, 140

vasectomy 237
Vatican 236
   see also Roman Catholicism
Venezuela 15, 37, 43, 52, 61, 69, 73–4, 79, 81–2, 84, 129, 134–6, 138–9, 141, 143–6, 154, 178, 189, 194, 200, 213, 232, 248–9, 252–3, 255, 259, 262, 277, 292, 310, 312, 319–20, 322

wages 23, 242, 244, 252–4, 292
   see also incomes
Washington consensus 10, 15, 18, 27, 83–4
waste disposal 171–3
water supply 167, 170–1, 174–5
welfare state 191
women 197, 229–30, 234–40, 247, 257, 261–7
   rural 280, 286–9, 292, 301, 302, 304
   urban 313–16, 323
   see also gender
women's organizations 205–7, 212–13, 220, 221, 289
workers 193
   agricultural 272–3, 277, 280–3, 287–90
World Bank 6, 9, 15, 29, 52, 59, 60, 77–9, 109, 112–13, 121–2, 159, 168, 179, 251, 260, 301, 312, 318
world economy 5, 6, 10, 27, 36–42, 49
   see also international economy
World Trade Organization (WTO) 9, 102, 107–8

Zapatistas 216, 219, 222, 301, 304